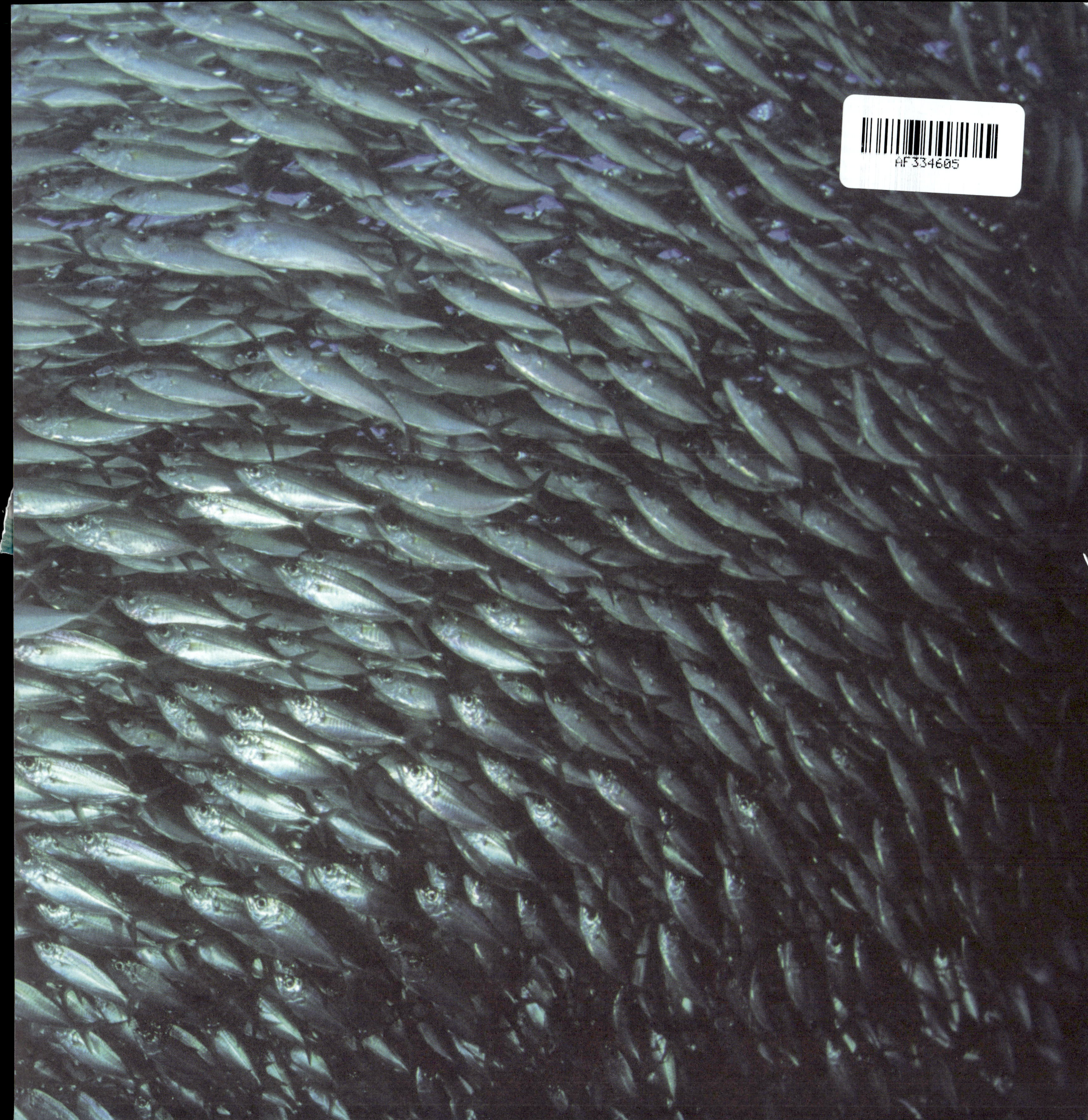

AF334605

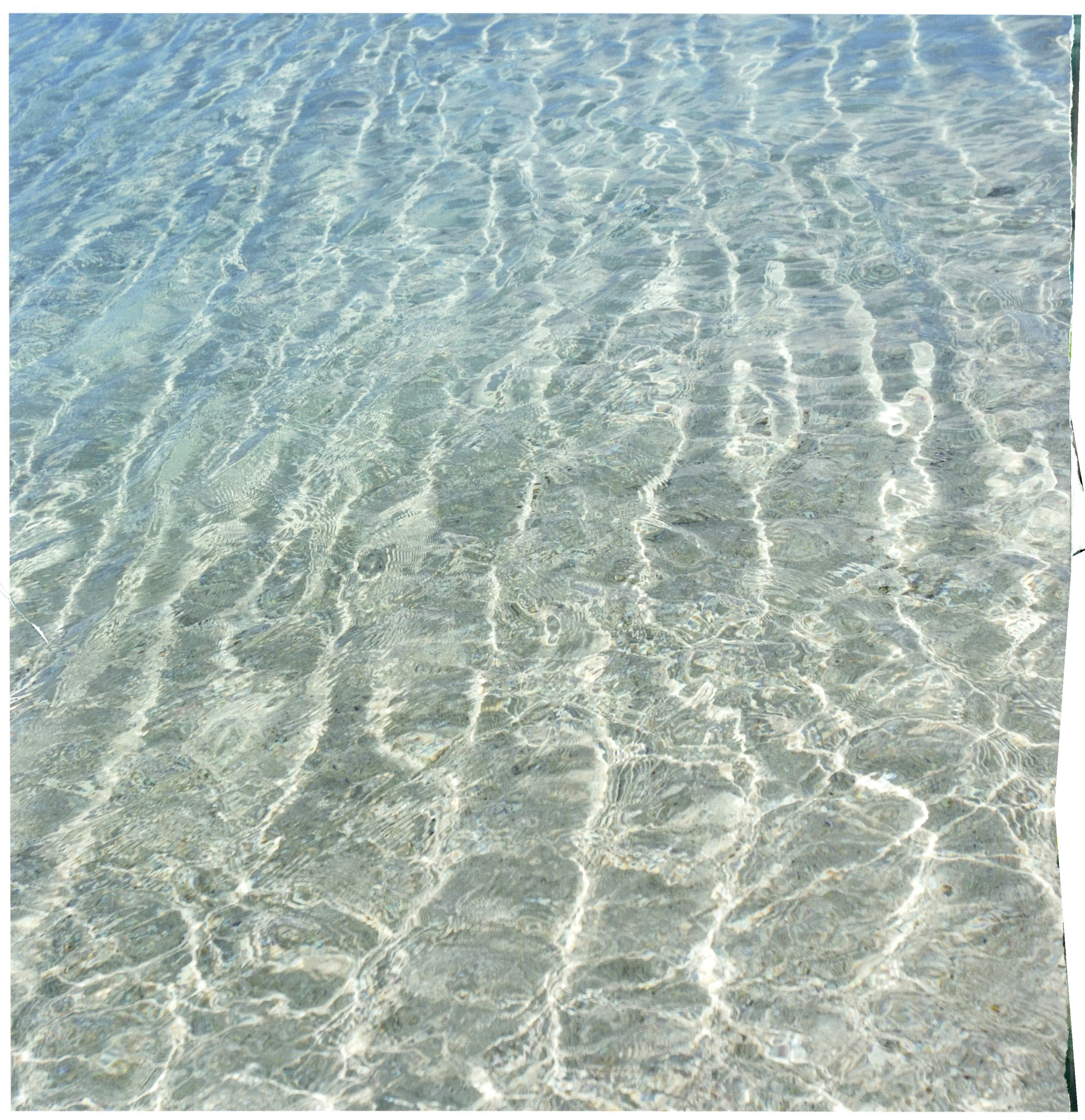

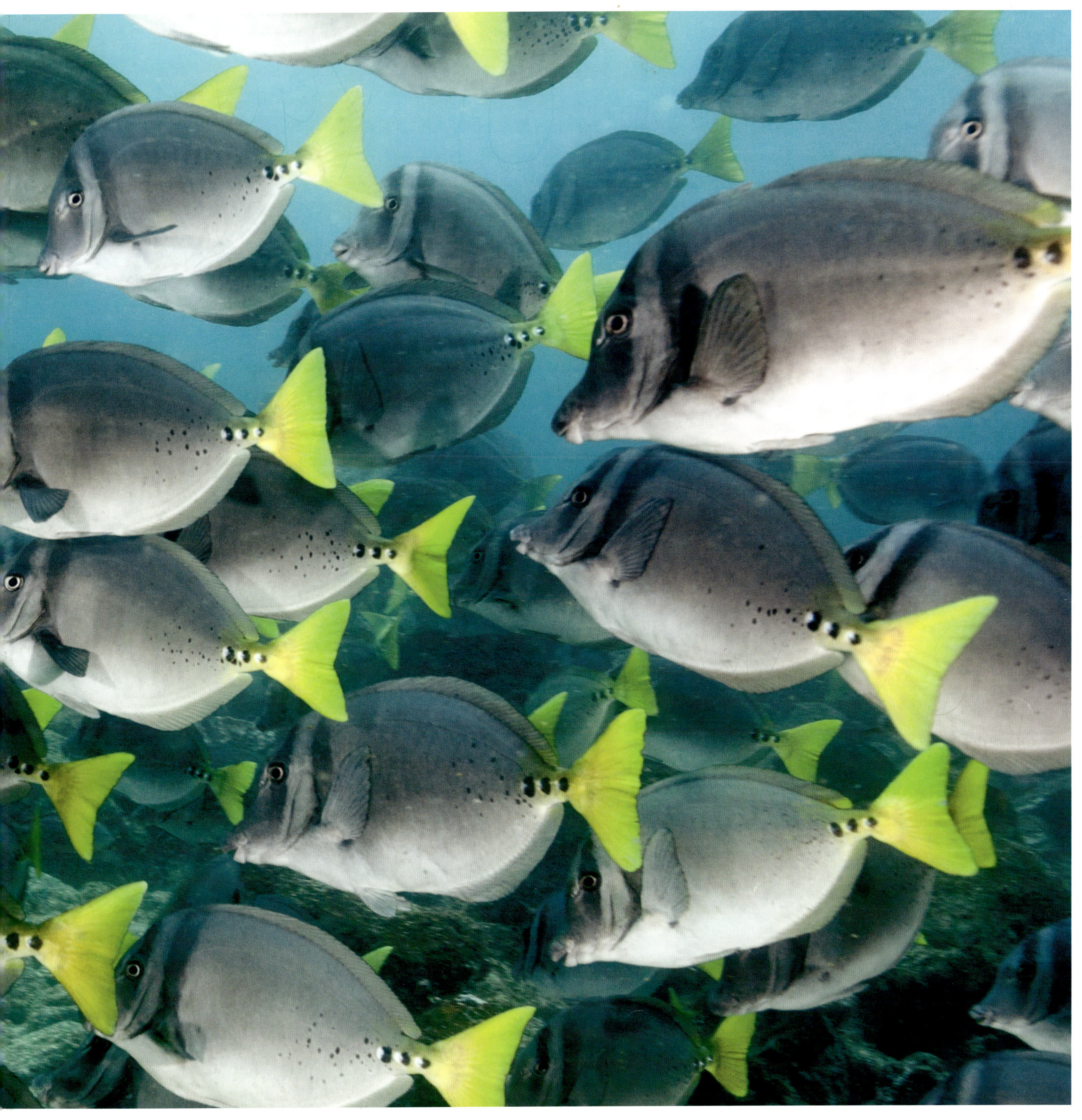

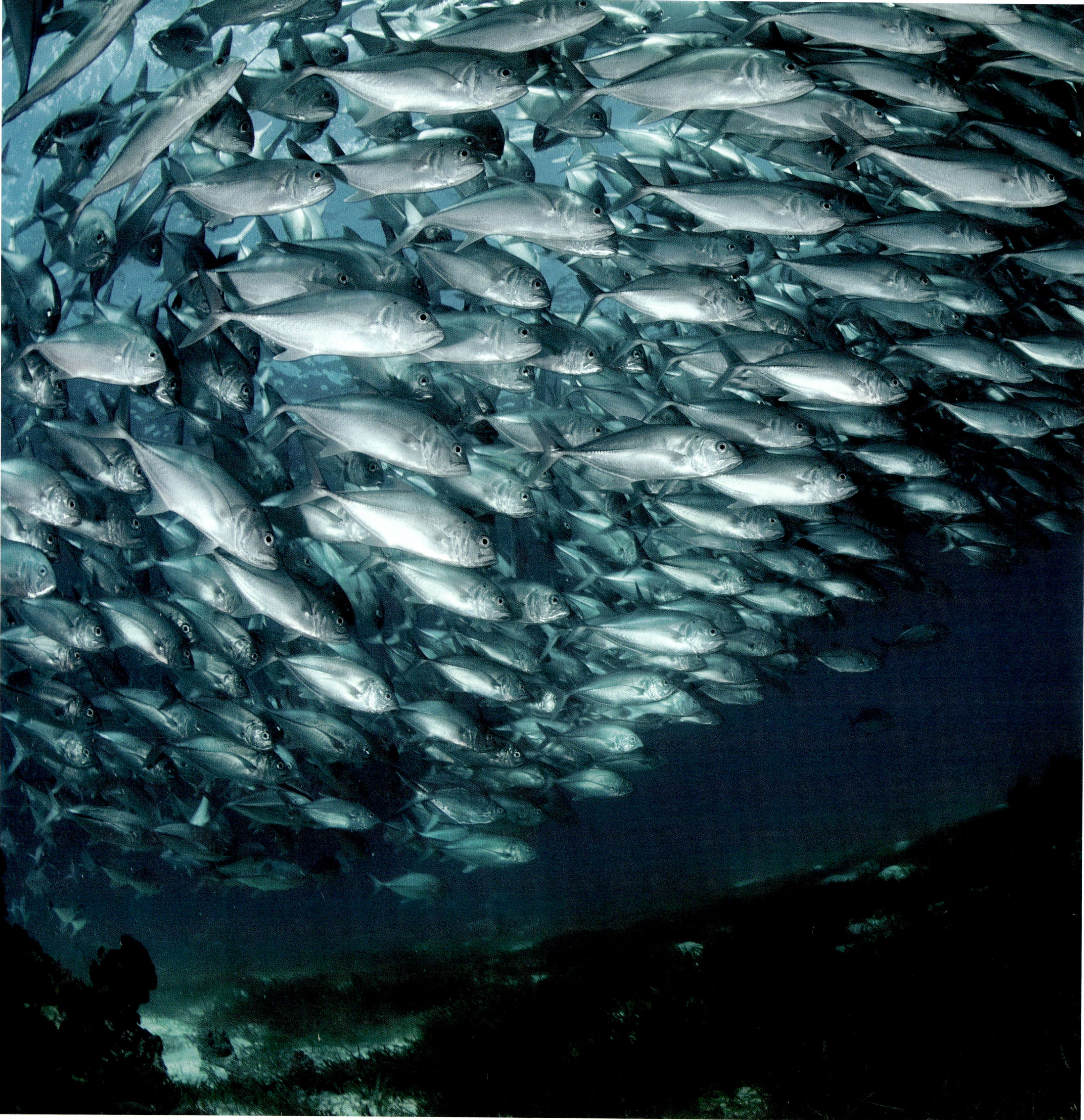

The Life & Love of

The Sea

Lewis Blackwell

Abrams, New York

in association with

Blackwell&Ruth.

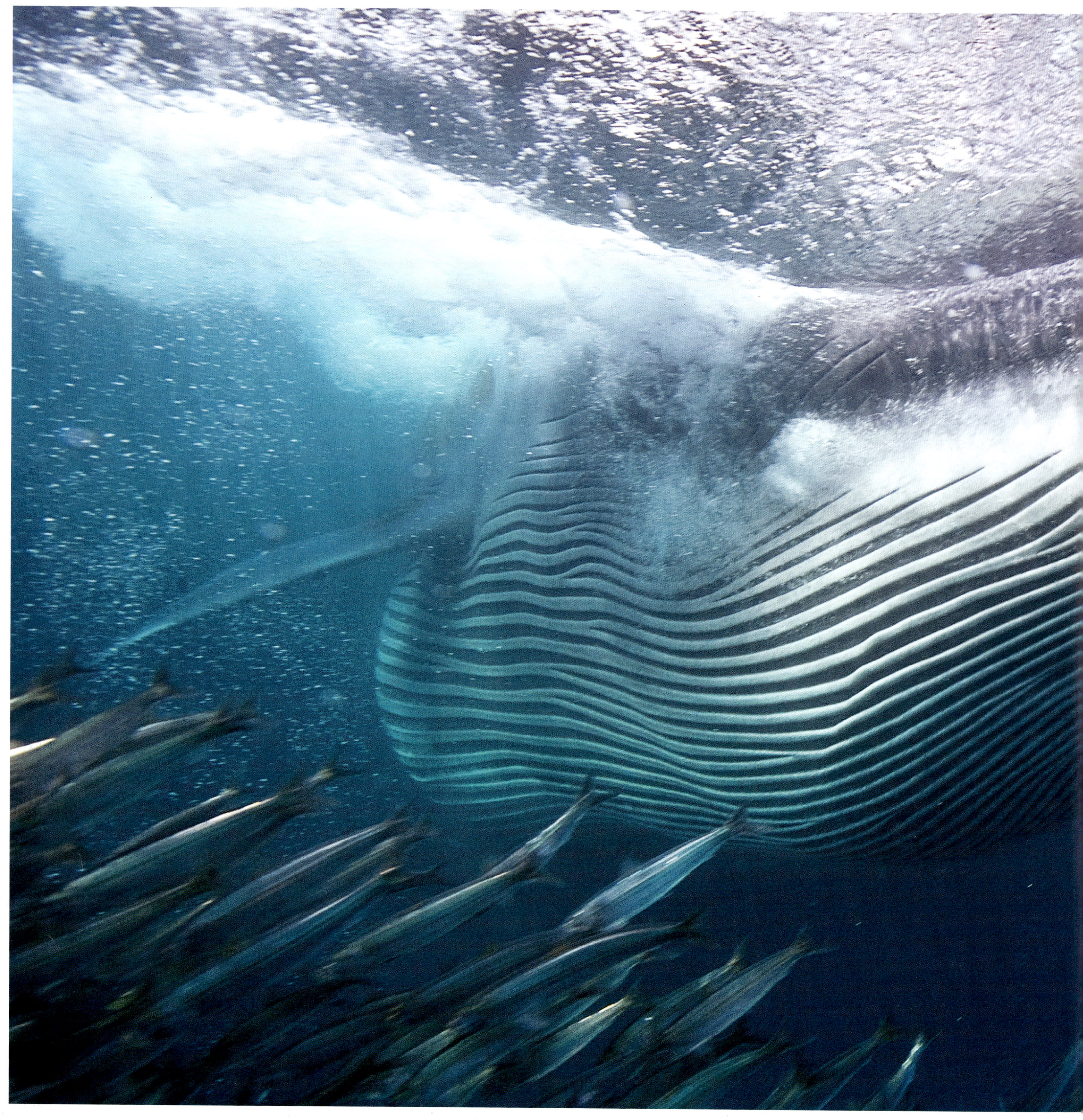

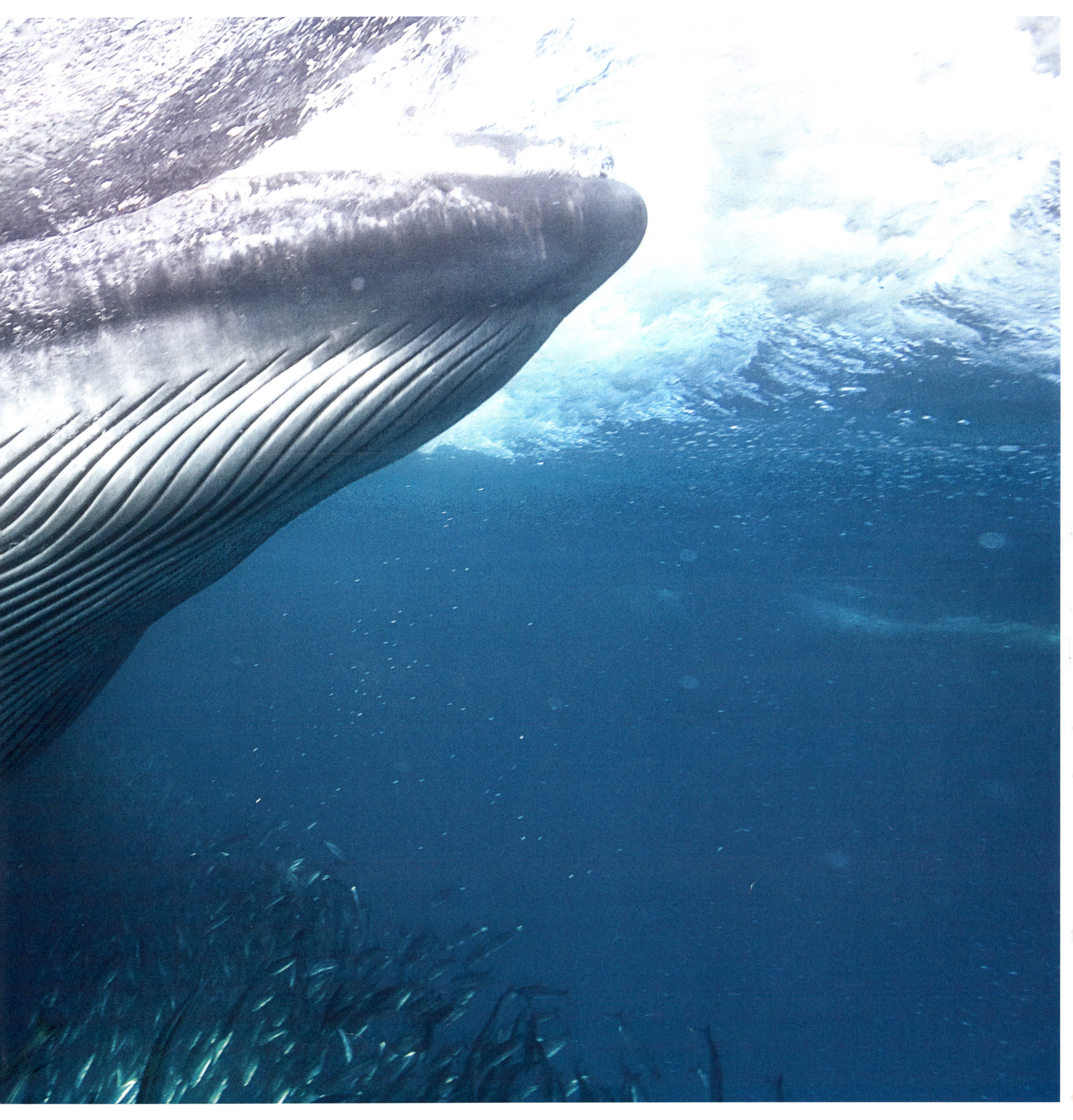

CONTENTS

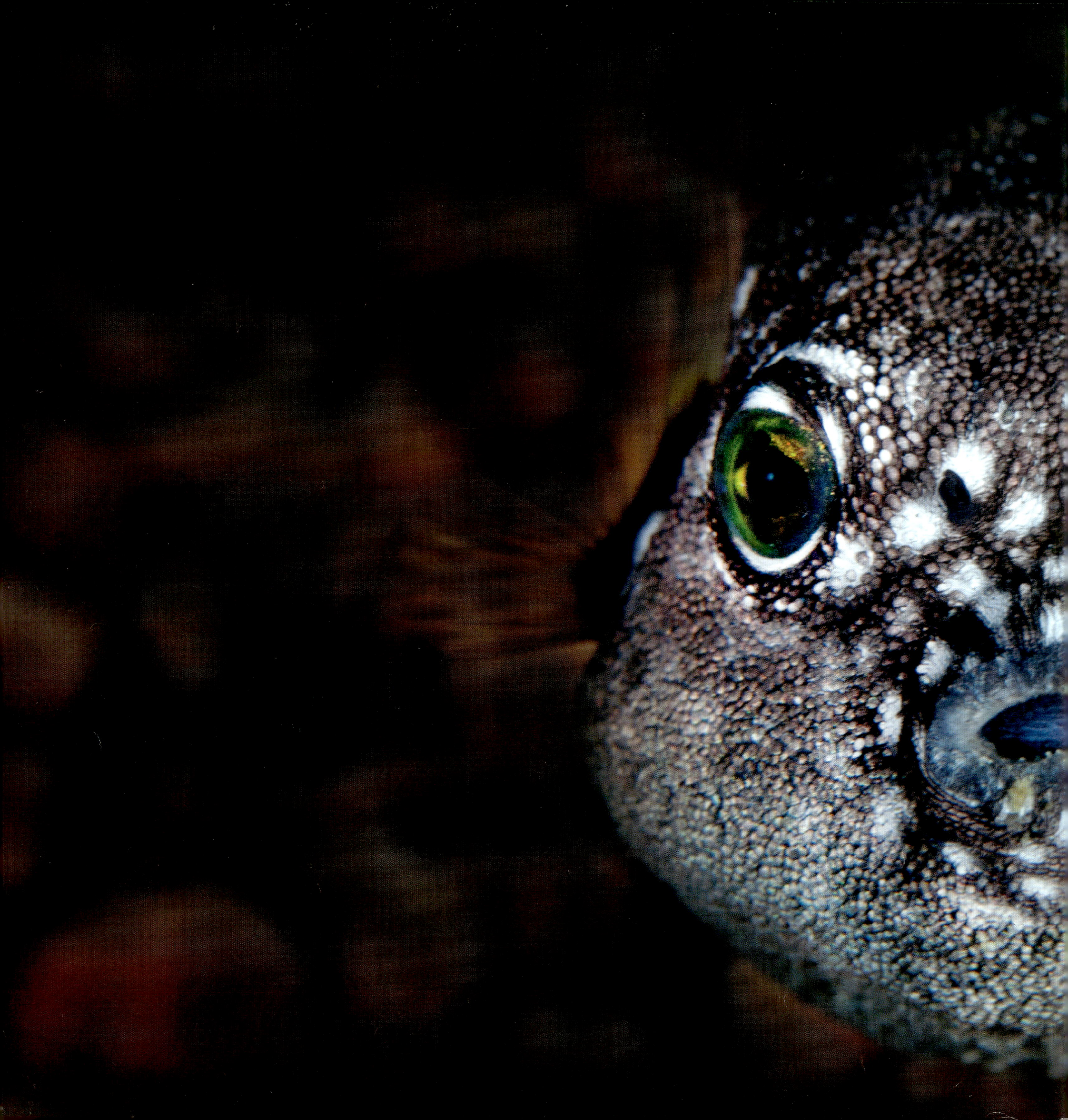

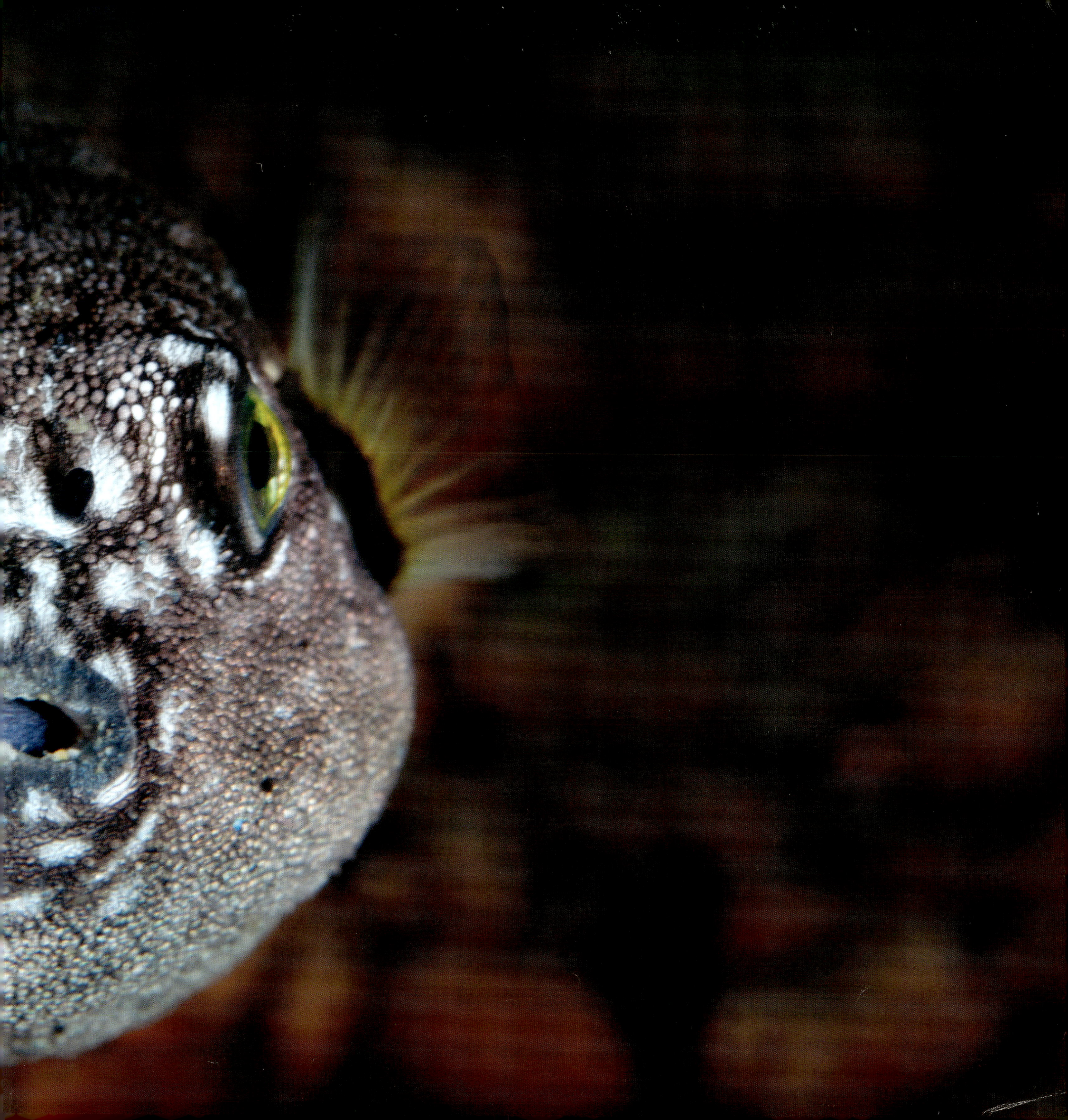

How inappropriate to call this planet Earth

when it is quite clearly Ocean.

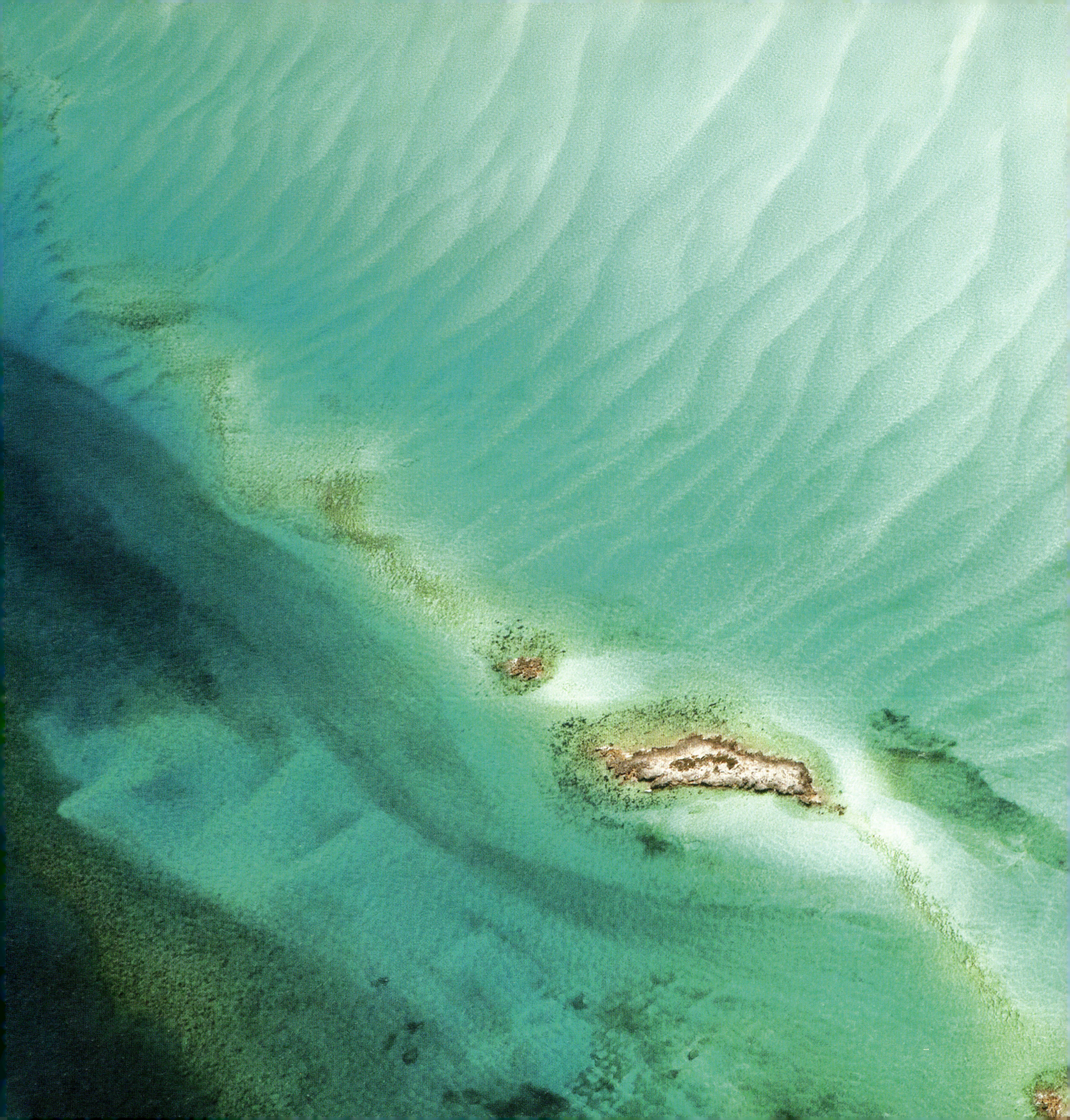

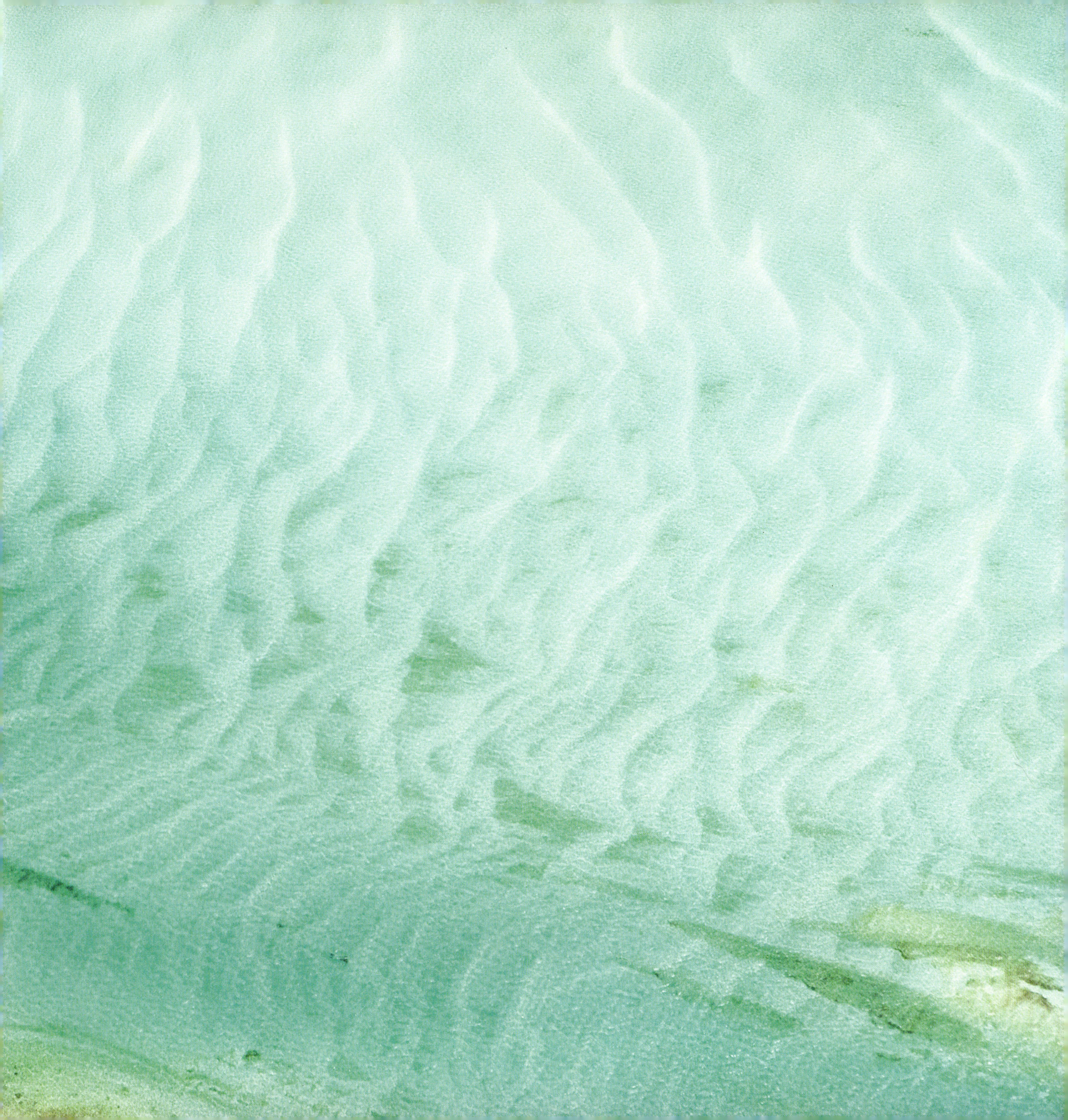

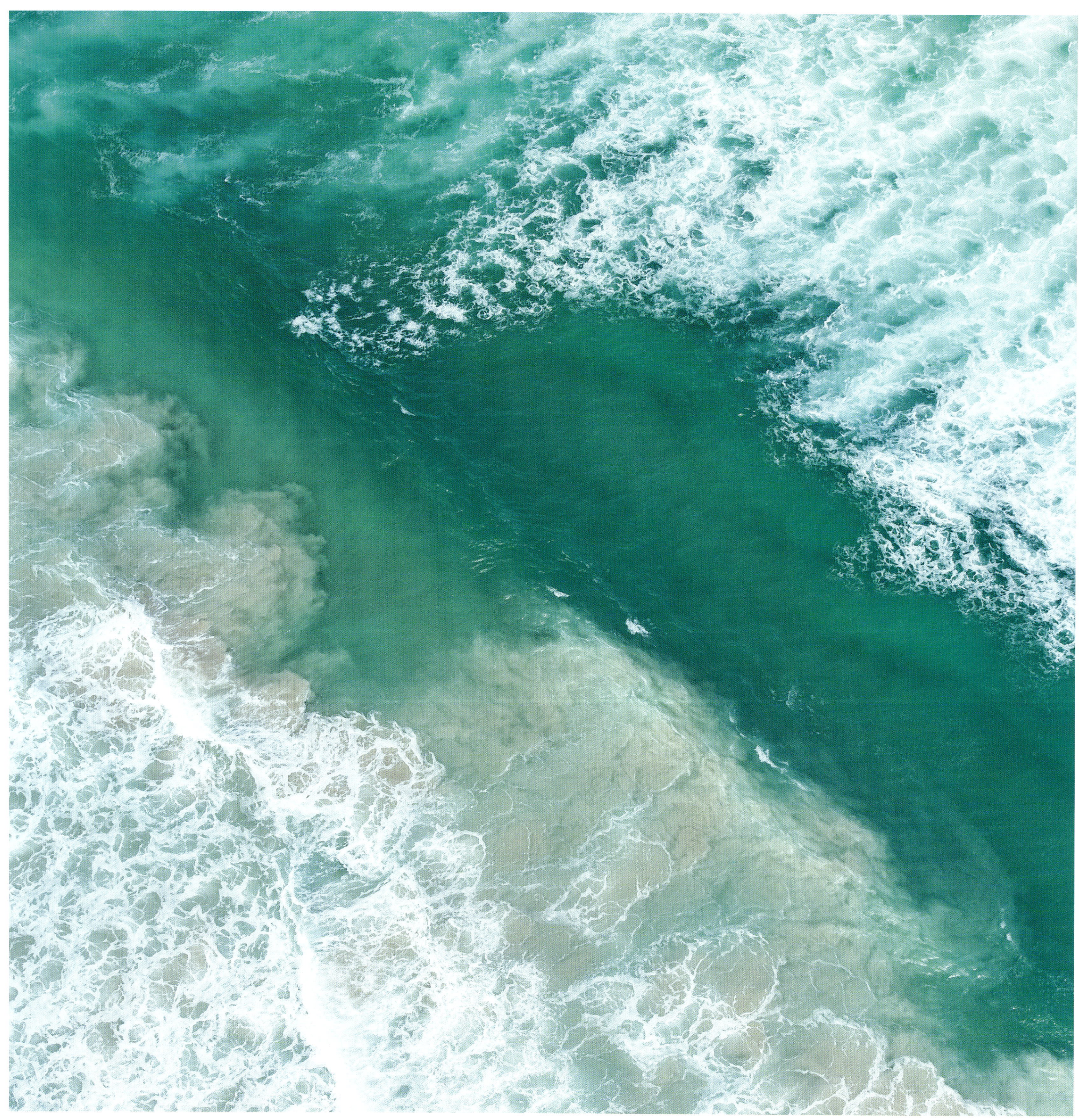

We may think that the sea needs no introduction. It is in our history; we feel it in our body. We are somehow creatures of the sea, and myths have had it that we were literally connected. The chemistry was seductive, if illusory—various wise men cited the similarities between the ingredients of our blood and that of seawater (sodium and chlorine are notable features, but they differ markedly in their quantities, along with other trace minerals that overlap). We have largely stopped endorsing such an explanation as proof of our bond, but we know now that the sea is larger than us and is a controller of our planet that we have to respect. However, the reason the sea needs no introduction is not due to biology so much, but to psychology. We love and desire and sometimes fear the sea, and it is writ large in our memories. Our minds quickly conjure up a sensation—the sound of the waves, the smell of the brine, the sensual touch of moving water—and then a story or two when the words "sea" or "ocean" are mentioned. Now I have cast them on the surface of imagination and memory, certain thoughts flicker past in no particular order: a mudflat on the east coast of Britain, where I am standing, five years old, looking out to the mystery of a distant line of retreating tide. Relocate to the Red Sea and my first visit to a coral reef. Focus on the wonder of a passing turtle disappearing with a leisurely wave of its flippers into the blue-green below. Next, scuba diving off the Mediterranean island of Giglio to a small statue of the Virgin Mary at 65 feet (20 meters) down on the now notorious Le Scole reef (the cruise ship *Costa Concordia* ran into the reef in 2012, leading to the loss of 32 lives). Oh, and now comes that ever-so-slightly crazy boat ride last year in thick fog on Clayoquot Sound, up the west coast of Vancouver Island. I feel mixed emotions as I recall the fears of others amid the uncomfortable tales from our skipper—just for a moment, a sobering reminder that not all the boats come back, that the sea is a harsh mistress. It is like flicking through the images on my phone, only for them to pop into a cinema-screen film that then wraps around me with smell and touch. You don't need a phone, just the magic word. Sea. Close your eyes for a moment and try it.

And now I think some of you will ask an awkward question. What is the difference between sea and ocean? The answer can be confusing. Depending where we live, we use these terms quite distinctly, or sometimes rather similarly, even happily substituting one for the other. The world's ocean is that vast mass of water that covers 71 percent of our planet and gets subdivided into the Pacific (the largest), Atlantic, Indian, Arctic (the smallest), and (sometimes) the Southern Oceans (we can make the Southern Ocean disappear, if we want, absorbing it back into the Pacific, Atlantic, and Indian Oceans if they are allowed to extend all the way to Antarctica). When we look at any part of those oceans we may perfectly well say, "That's a lot of sea." Yes, it can all be sea as well as ocean. Indeed, that is how it is used in the Convention on the Law of the Sea, that United Nations treaty that aims to control all countries in their use of a shared resource. Sea and ocean are interchangeable at times.

Indeed, there is the phrase "the seven seas," and it applies to those oceans with the addition of the Pacific and Atlantic being divided into north and south. But those seven are our modern conclusions on the topic. For a sailor in ancient times, or during the Age of Discovery, between 1450 and 1650, when explorers radically opened up the perception of the world as a sphere wrapped with vast oceans, the seven seas were different. Then they referred to the four main oceans, plus the Mediterranean Sea, the recently discovered Caribbean Sea, and the Gulf of Mexico. For the Europeans of classical civilization, it was different again. They had yet to discover the Pacific and Arctic Oceans. For them, the limits were the Mediterranean, Adriatic, Red, Black, and Caspian Seas, the Persian Gulf, and Indian Ocean.

However, now when we do our best to distinguish seas from oceans, we may reserve, more precisely, our use of the word "sea" to just those parts of the waters that are partly or almost entirely bounded by land. So say some dictionaries. Hence we use the word for bodies of water such as the Mediterranean, Arabian, Tasman, North, Baltic, and Barents Seas, the Sea of Okhotsk, and many more. Every ocean has many seas around its edge. But this definition stumbles a little when we have to accept that the Sargasso Sea is a sea, and yet it is not defined by any land: It is a zone of the western North Atlantic Ocean and is defined by a current, the North Atlantic Gyre. And then there are so many seas that are not called sea, but called other things: the Bay of Biscay (not much of a bay at all) and the English Channel; the Denmark Strait (between Greenland and Iceland) and the Strait of Messina (in the Mediterranean, between Sicily and Italy); the Gulf of this, that, and the other; and oddities like the Great Australian Bight, also a sea of sorts, but really just hundreds of miles of coastline, an indentation in the continent meeting the ocean. All these kinds of seas are marginal seas in that they are parts of the ocean bounded by sections of land. But we also have places we call a sea that are not connected to the ocean at all, but are landlocked or interior seas: namely the Caspian, Aral, Dead, and Salton Seas. Yes, they are all salty bodies of water. And yet, just when you think you have gotten the hang of it, along comes the Sea of Galilee, which is freshwater. It's a lake, but we call it a sea. The Great Lakes and Lake Baikal are larger in various ways, but we don't call them seas.

So "sea" is in an act of conferring; it is almost an honor rather than a precise definition. In some ways, for example, Salton Sea in California is a sea in spirit, deserving of the label by dint of being very salty. It seems to work better with that designation than making it a disappointing lake where the water kills freshwater life. Meanwhile, Lake Superior is quite big enough in size and status, rubbing shoulders with two nations, without being aggrandized into a sea. But, of course, however big and watery, it is freshwater. Putting the Sea of Galilee aside as a misnomer (it's also sometimes called the Lake of Galilee or Lake Tiberias), a key filter for being a sea is being salty. Seawater is salty. Sea air is salty. Seafood is from salt water. And so on. For clarification of subject, it may be worth saying that all types of seas—marginal, landlocked, or just as another way of saying ocean—are of interest to us here. It's the nature and culture of those massive waters, the largest part of our living world, where so much remains unexplored, that we want to know more about and will discover and rediscover over the coming pages.

We begin our story in "Beginnings," asking, Where did all that water come from? How did life start? How did we get from there to here? In one chapter we swoop over time scales that are mind expanding and at times feel like they are mind exploding. We can't conceive of the scale of life in the history of the oceans, a journey back to the start of the planet, where the entire story of our species perhaps makes for only a small snippet, a paragraph or two somewhere in the middle of a book where much remains to be revealed. Let's hope we don't get written out of the plot.

In "Connections," we look at how the oceans are fundamental drivers of all life. The chain that begins with phytoplankton and on which most creatures ultimately depend in one way or another—as food or as part of the cycle that creates oxygen—is one sequence that we set out to track. Another is the journey of energy around the world in the currents of water that influence climate and natural resources everywhere, shaping everything from the comfort of our planet to the economy of nations.

We get into and under the water in "Wild," looking at the levels of life and how they interact. From the Great Barrier Reef, to the rock pools on the beach, to the near icy depths of the rich waters of Antarctica, the sea is remarkably connected and also disconnected. There are creatures living in ways that can be found all over the planet, and at the same time there are isolated pockets of remarkably divergent life in the oceans.

This is made even clearer in the following chapter on the still sketchily explored vastness of what lies below 660 feet (200 meters), which is where almost all the ocean is located. In "Deep," we discover that the oceans do not work in the way they seem to do at the surface. There is so much else down there, increasingly challenging our ideas about the very foundations of what constitutes life. It is a very dark world, where the usual rules don't apply. As it has been said, we know more about the moon's back side than about the ocean's bottom, and "Deep" presents some of the landmarks of discovery in what is a fantastic field of ongoing exploration.

In contrast to the wild world, there is the ocean world as we have endeavored to shape it. This we set about investigating in "Tame." From earliest man, we have fed our bodies and our minds with the sea. We have become ever more sophisticated in drawing resources from the waters as scavengers, hunters, farmers, and scientists. At the same time we have become ever imaginative in speculating on the sea: from how it was formed, to what lives in it, to how we might live in and on it. But the time for helping ourselves without restraint seems to be coming to an end: How will we manage, nurture, and make the oceans more productive and sustainable? Remarkable creatures, people, and science are suggesting we can enter into a new way of living with and off the seas.

That speculation about possibilities is not all about the material world. It also spills over into the world of the arts, where the dream of the sea and what it means to us has fueled all manner of writing, music, painting, and other creative production. The sea is love and hate, fear and desire. It can offer a place for reflection or one that crushes the individual. When Thoreau said, "We need the tonic of wildness," he wasn't thinking of the sea, but most probably of his pond and woods. However, today it is the oceans that persist as the great wild frontier of our world, while so much else has become known. The sea has always been that, though, always been the wildest place. It was once the edge of the known universe and still has millions of secrets to give (in unknown species alone). This continues to make it an enchanted place for artists to reflect on, and in "Beauty and the Beast" we look to see what they discover in its reflections.

Finally, we will take a few moments to look far out to sea. With "Beyond the Horizon," it is time to speculate on what may be the future of our oceans. These dominant resources are at the core of our future and that of all living things. But their fragility gives great cause for concern. Can we move beyond exploitation to a new kind of relationship? There are reasons to be hopeful, and we do not leave you with any reason to despair. Far from it.

Finally . . . there is no finally. We aim to take you on a journey that travels in a continuous and endless path, rather like an ocean current. At no point does it really start or finish; you can join it anywhere and go full circle, but along the way there may be circuitous side currents that distract, feed your thoughts, and stimulate you differently. By which we mean that this book is designed for the reader who does not necessarily start at the front and end at the back, but likes to read and respond in whatever way takes their fancy. I know I am that kind of reader, and with a book of astounding images there is always the temptation to take a peek. You can snorkel around in these pages or dive in deep. Do it in the order that you prefer. When you have swum around a bit, climb aboard again and take a cruise to a different sea, and then drop anchor and rest awhile and appreciate the view. There is no set course, because we all know the broad story and we can work on the ending together. My hope is that I have built and crewed and equipped the ship that you can command, stocked with plenty of maps to guide your expedition. So please set sail across the pages and see where your thoughts take you. Move with the winds or ride the fast current of ideas, out and back, and may you be revitalized by the trip.

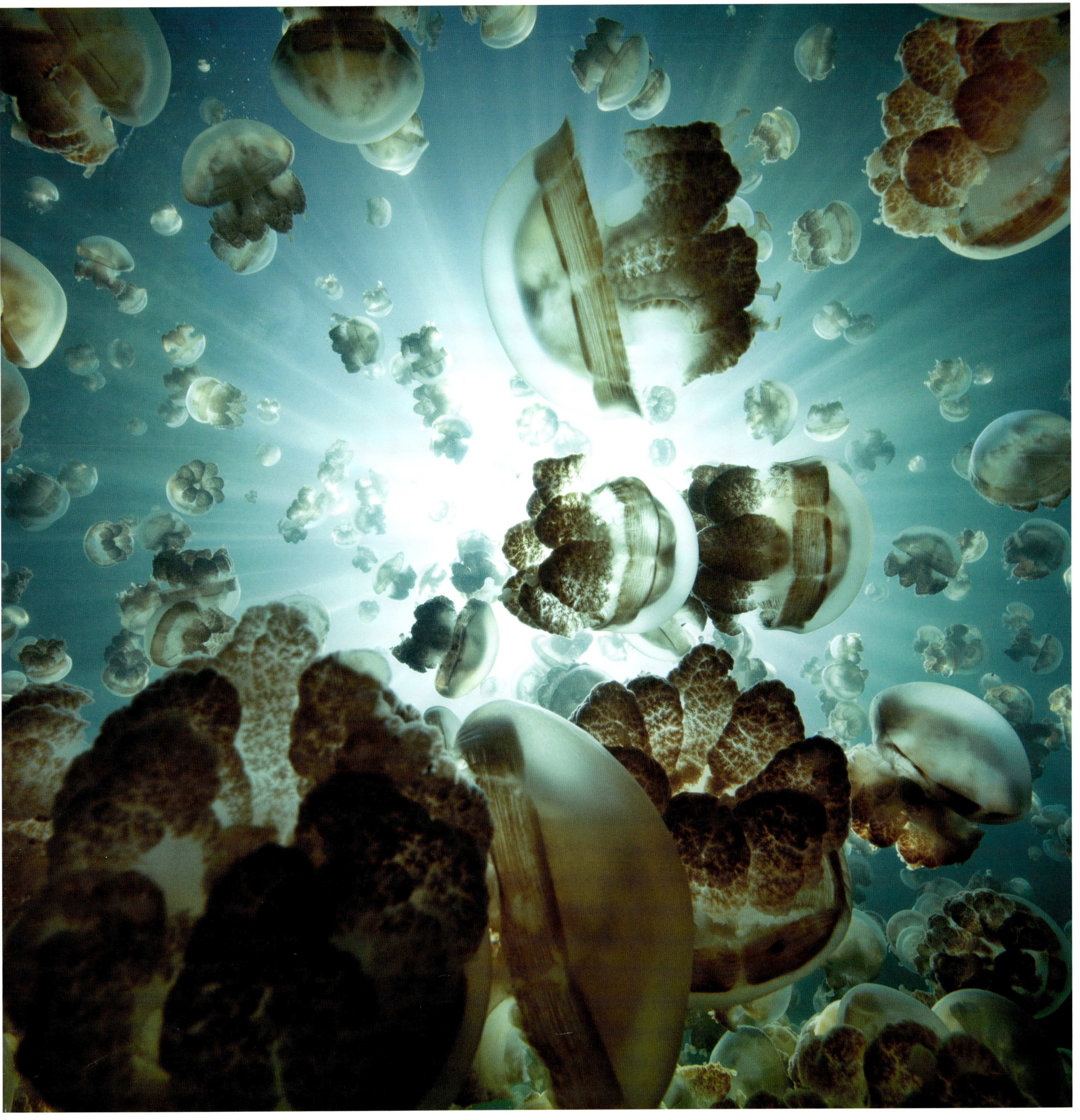

There's nothing wrong with enjoying looking at the surface of the ocean itself,
except that when you finally see what goes on underwater, you realize that

you've been missing the whole point of the ocean. Staying on the surface all
the time is like going to the circus and staring at the outside of the tent.

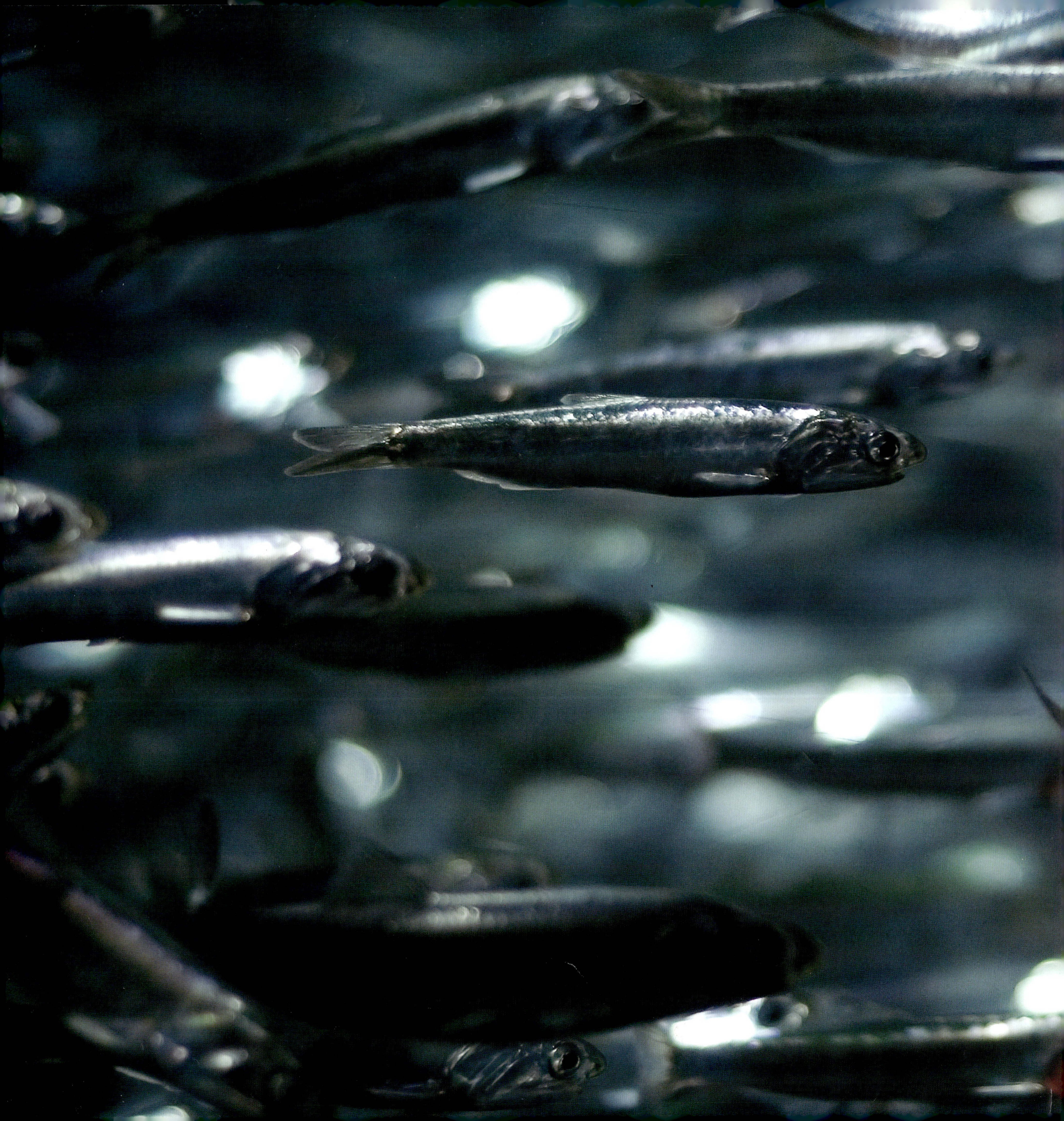

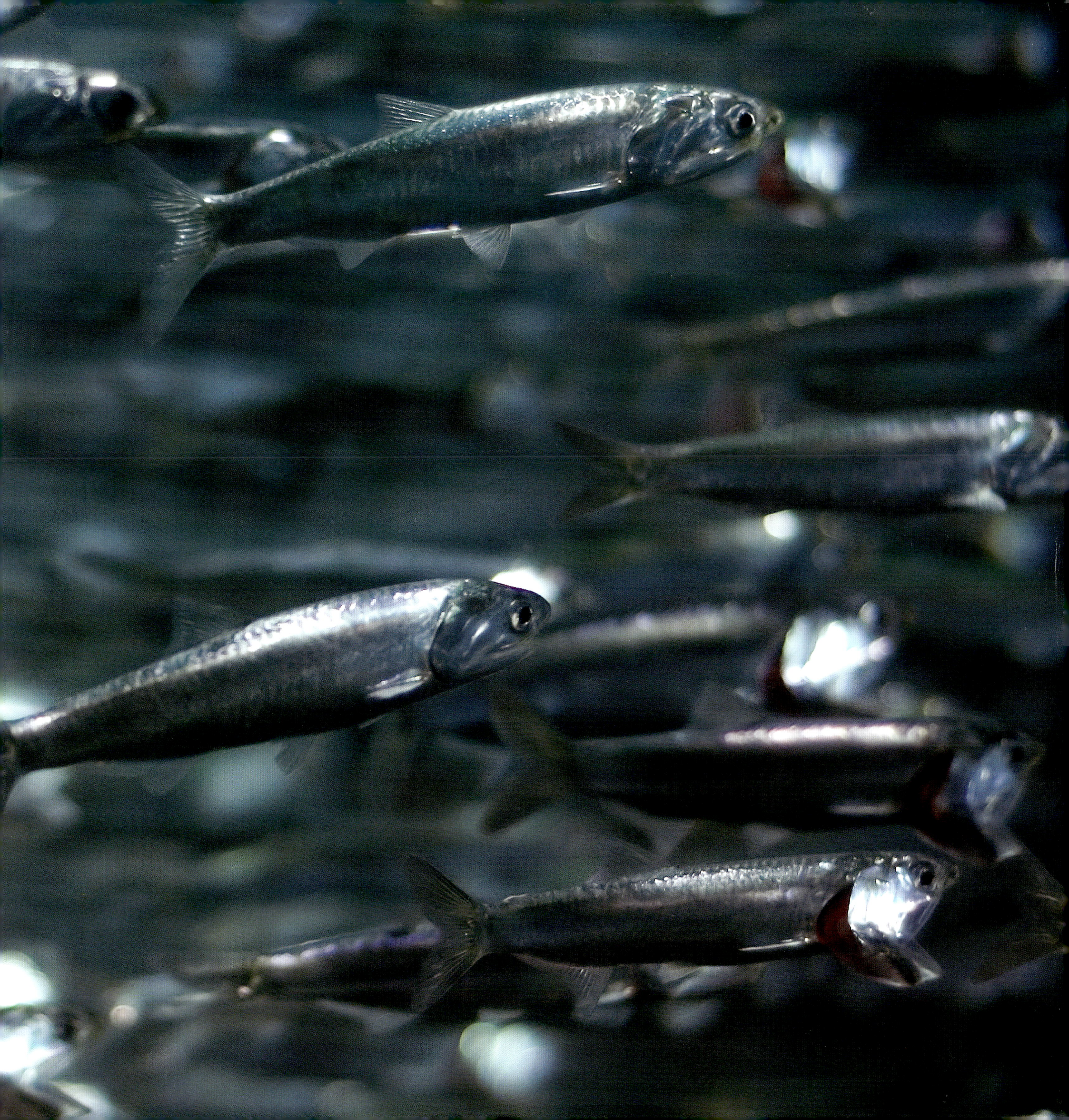

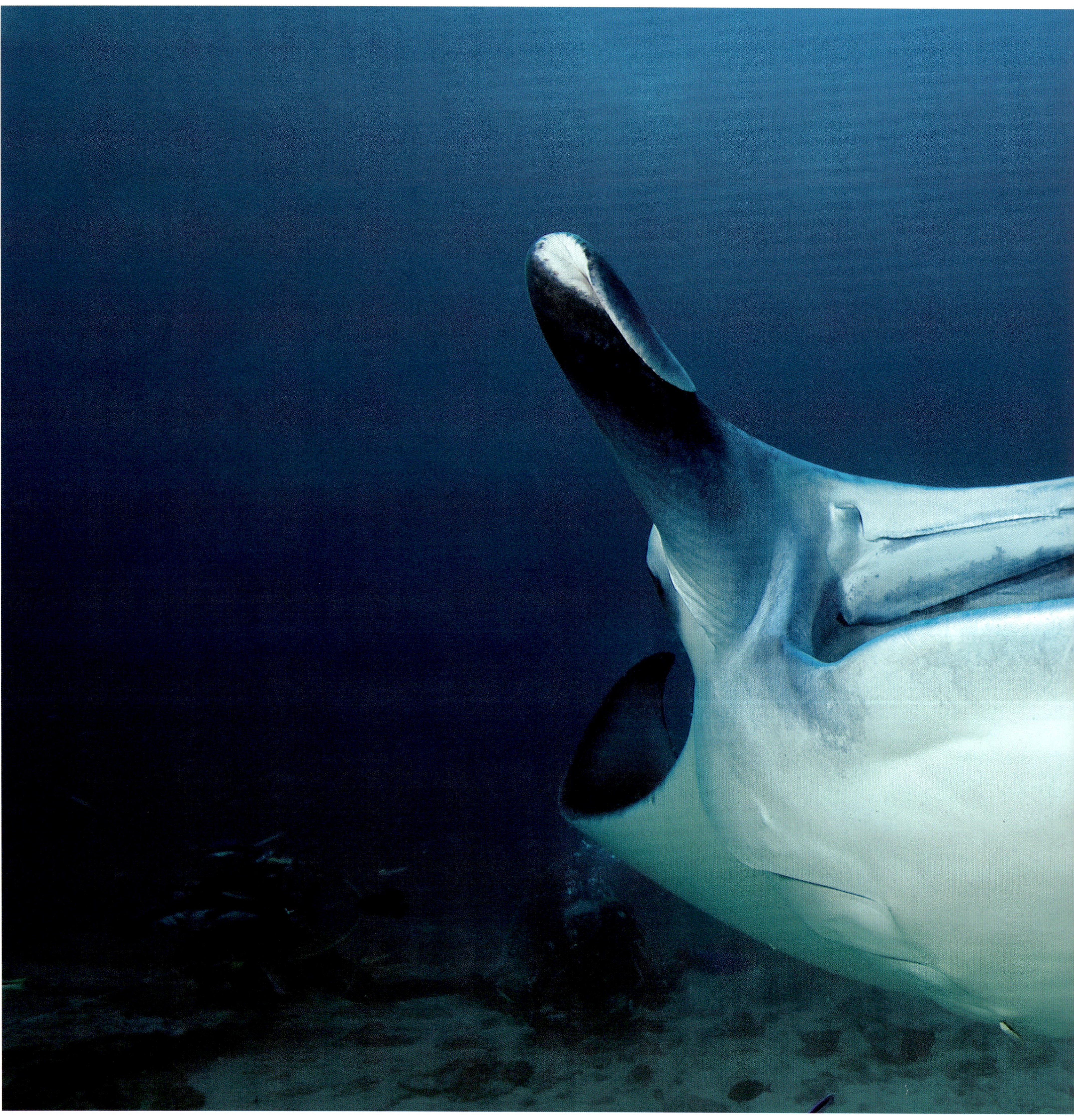

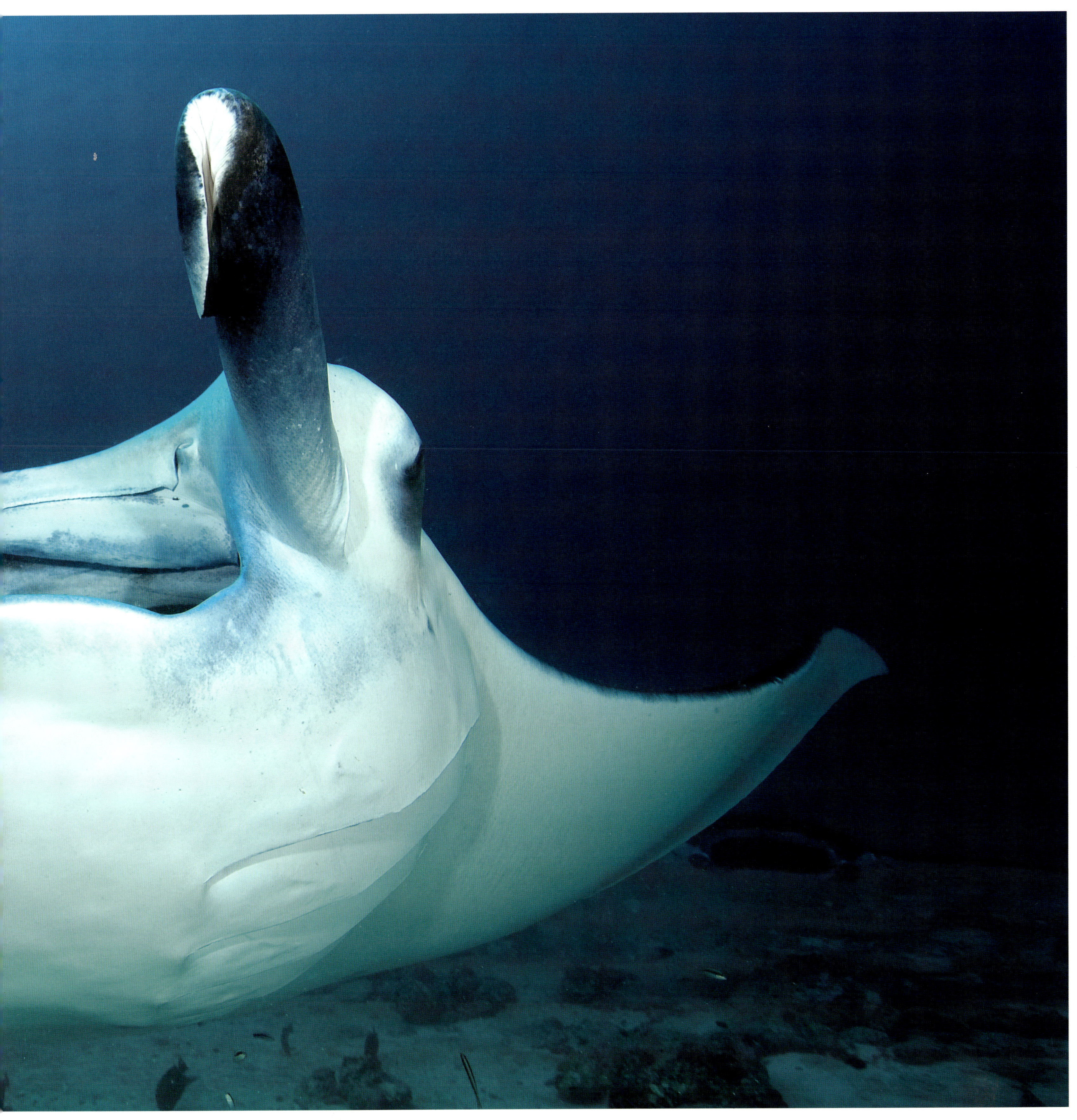

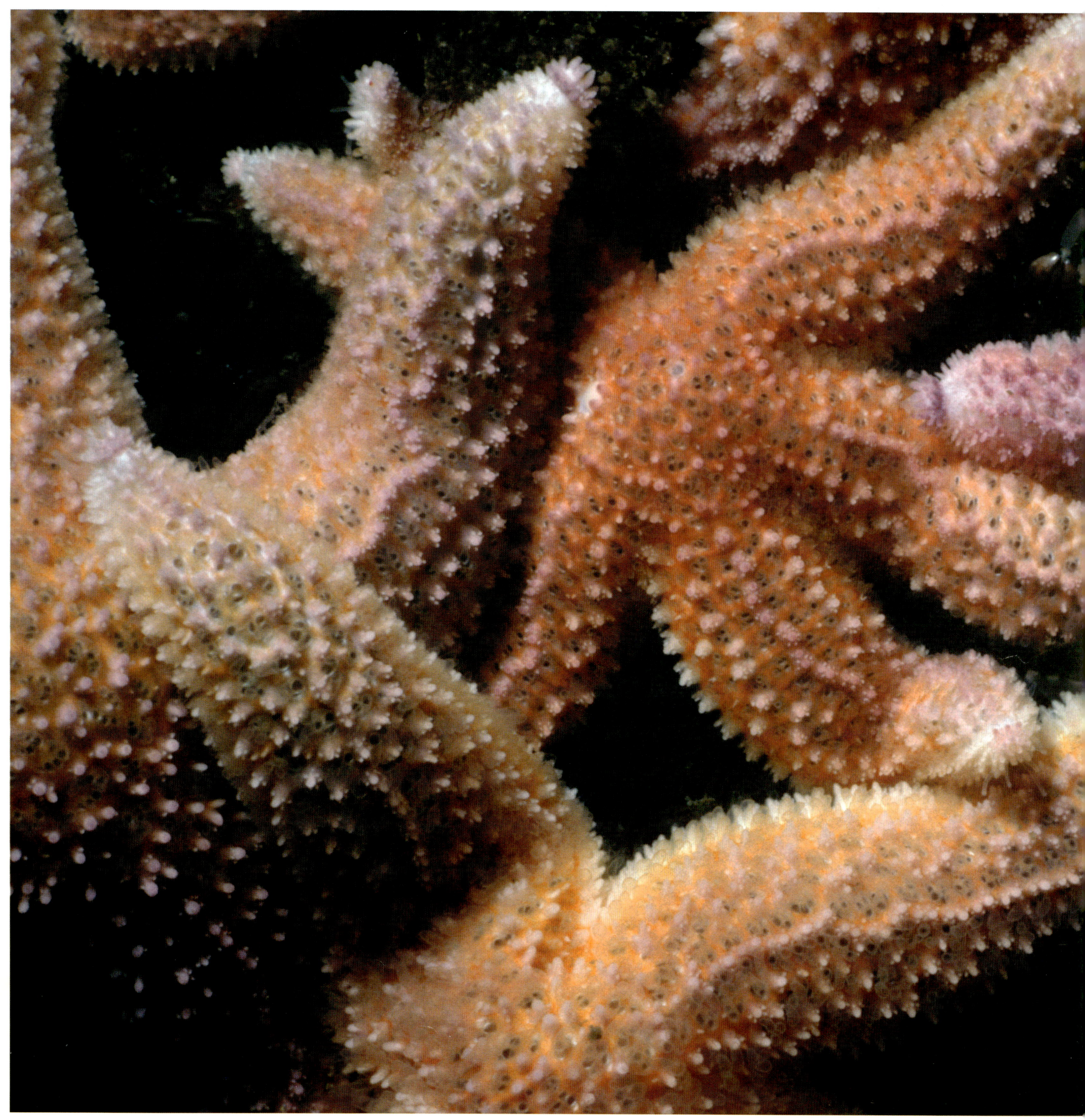

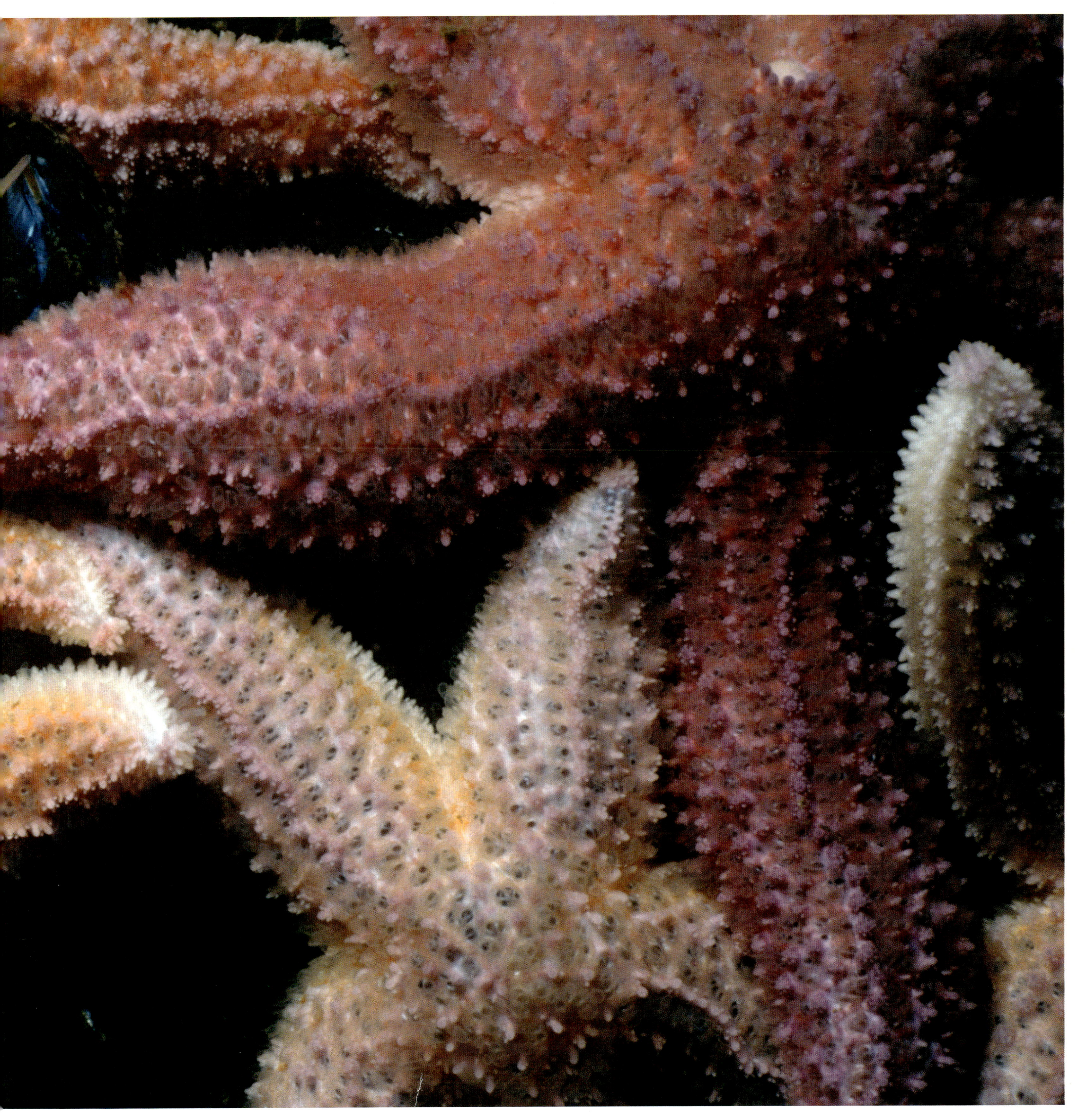

Beginnings

How we came to have water on Earth is something of a mystery. But it is a mystery we seem to have been getting a little closer to understanding in recent decades. While we don't often realize just how special we are to have water, and speculation varies widely on how many solar systems are out there that might have something similar, we do have a strengthening hunch that most planets in our solar system don't have any or at most little; nothing like the quantity that fills the earth's seas and rivers and lakes, and creates the conditions for life to be formed and thrive. It seems to have been here from the beginning of the planet. For this knowledge, for this appreciation of what was around more than four billion years ago, we are beholden to radioactive dating and one mineral in particular.

Zircon is a popular gemstone. It has a long history in jewelry, but has of late acquired a somewhat second-best reputation as a cheap substitute for diamonds. This does the mineral a considerable disservice, for zircon goes far beyond the decorative in its value—it is a powerful tool for revealing our deepest history, serving as a messenger from the distant past. This is a stone that is helping reveal the most ancient stories of our planet. In doing so, it is giving us strong indications of when and how the oceans first existed.

The variously colored stone can be found everywhere in the crust of our planet. Due to its chemical stability and its low-level radioactivity, zircon can act as a witness across time and space. With small traces of uranium and thorium, the decay of these radioactive elements can be used to date any rocks containing zircon. Despite whatever may have happened over millions of years—erosion into sand, metamorphosis through heat into a different kind of rock—the zircon crystals are highly durable, persisting in a stable form, but with the radioactive elements running down like a clock. Even better, more than one clock is on hand, as different forms of uranium (235 and 238) in zircon decay at different rates, thereby enabling a cross-check process on the time that has passed. It is in part thanks to zircon analysis that scientists have come to estimate the age of the planet at around 4.5 billion years. Furthermore, but more controversially, the oxygen isotopes found in some zircon samples are taken as evidence that there was possibly already water on the surface of the earth close to its beginnings.

Those beginnings are hard for us to conceive. We don't have a way of visualizing such a gap in time, or even the length of various stages of planetary evolution. Zircon analysis and the decoding of some other radioactive dating aside, we have no other ways of finding evidence, given the many changes that have happened to the surface of the planet in the intervening billions of years. There are no surviving rock forms older than four billion years—there is a gap of perhaps 500 million years from when the planet is thought to have begun, and in which the origin of the oceans also falls. But the clues from zircon analysis are pointing back to the existence of water being close to the birth of the planet.

There are contending theses on what those first few hundred million years, or even billion years, brought to Earth. The radioactive dating gives us increasingly accurate indicators, but we don't easily agree on how to interpret these shards of information. The experts are far from consolidated around a single version of events. There are chiefly two theories on how the oceans came about. The most widely supported theory is that the water emerged from the cooling mass of the planet. It is believed to have been pumped out in the gases and materials that erupted during the vast volcanic explosions that would have covered the planet's surface in this early phase, during which the construction of the planet emerged through differentiation of various strata. (Our planet is thought to be made up of a core of heavier material, probably mostly iron, then a mantle, and ultimately a firm crust on the outside that is principally granite and basalt.) This "degassing" process, rather like a body burping (or worse, given the sulfurous nature of these emissions), would have wrapped the globe in a noxious cloud of hot gas. It might have been similar to the state we now believe Venus is in, a surface scoured by storms of impenetrably thick clouds of extremely hot gas, never condensing. But on Earth, over a few hundred million years, cooling took place and the gas would have condensed, and via eons of rain, the surface of the planet would have started to become covered in water.

The leading alternative theory has it that water may have come from elsewhere, from collisions with other astronomical bodies, such as comets, arriving rather like massive ice balls smacking into Earth and melting on the surface. But while that would have added to the water content of the planet, it is thought that this probably would not account for more than 20 percent, at most, of our water. This leads us back to the "degassing" theory of water separating out from volcanic emissions during the cooling of the original molten mass as Earth began. (We see how this happens in current volcanic eruptions, where large white clouds of water vapor emerge, as well as lots of other hot matter.)

For all the theories of the early Earth being a molten mass, some of the most recent zircon analysis has also led to the case for a "cool early Earth" theory. This doesn't rebut the early, explosive, energetic years of Earth, but suggests that it did not stay in that state for long. Zircon-sourced evidence of oxygen isotopes indicates an earlier point than previously thought for liquid water on the earth's surface. This theory suggests that the planet had cooled down enough for oceans to begin forming within perhaps 200 million years of its origin, more than four billion years ago.

This first era of geological time on Earth has been called Hadean, after Hades . . . in other words, hell, inspired by that sense of it being a hot, volcanic, and unlivable environment. However, in a groundbreaking paper published in the journal *Geology* in April 2002, the authors, Valley, Peck, and King, dismissed the term as being inappropriate: "We suggest that Earth was not 'hell-like' then and may have been hospitable to the earliest life." They believe that the first 500 million years of the planet may have seen long spells of quite temperate conditions in which water could have been liquid on the planet's surface. This would have led to the formation of the first oceans and, potentially, the breeding ground for the very first moves toward life.

Our planet had somehow lucked out to be an unusual place, one that could be a crucible for life. It was the right distance from the sun to be warm enough not to be eternally frozen, as with planets farther from the sun, but not so hot as to be just a scorched lump

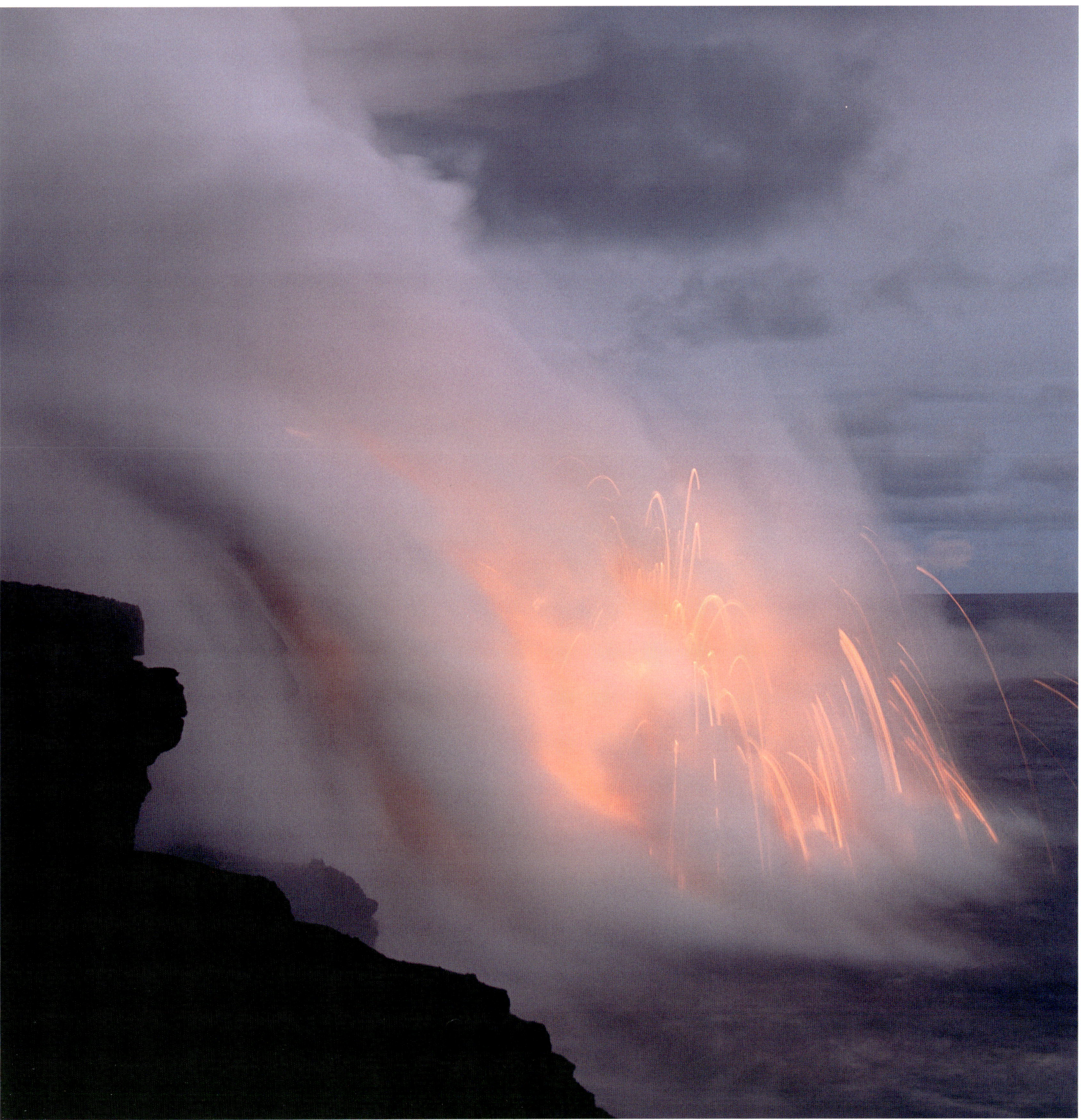

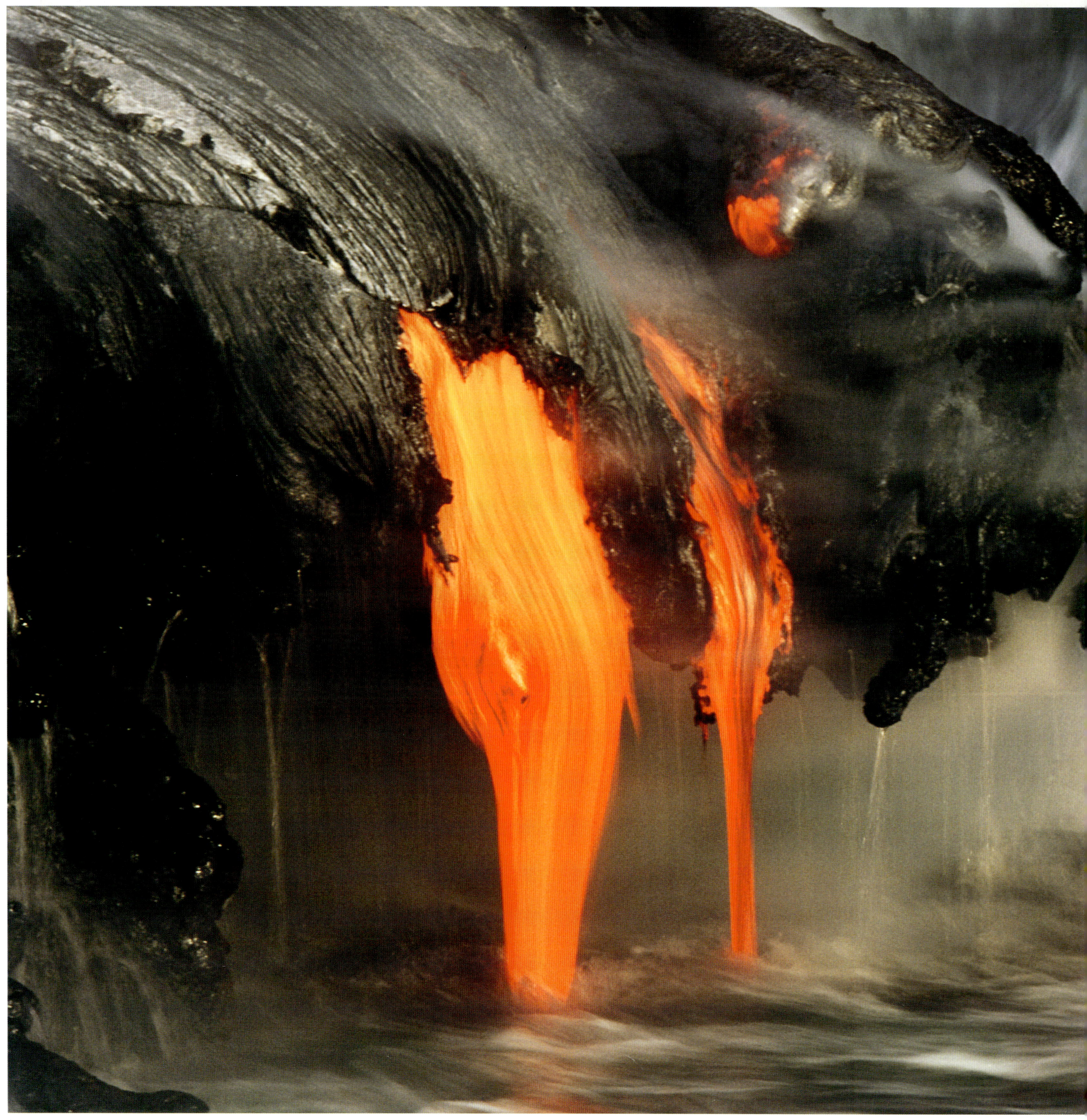

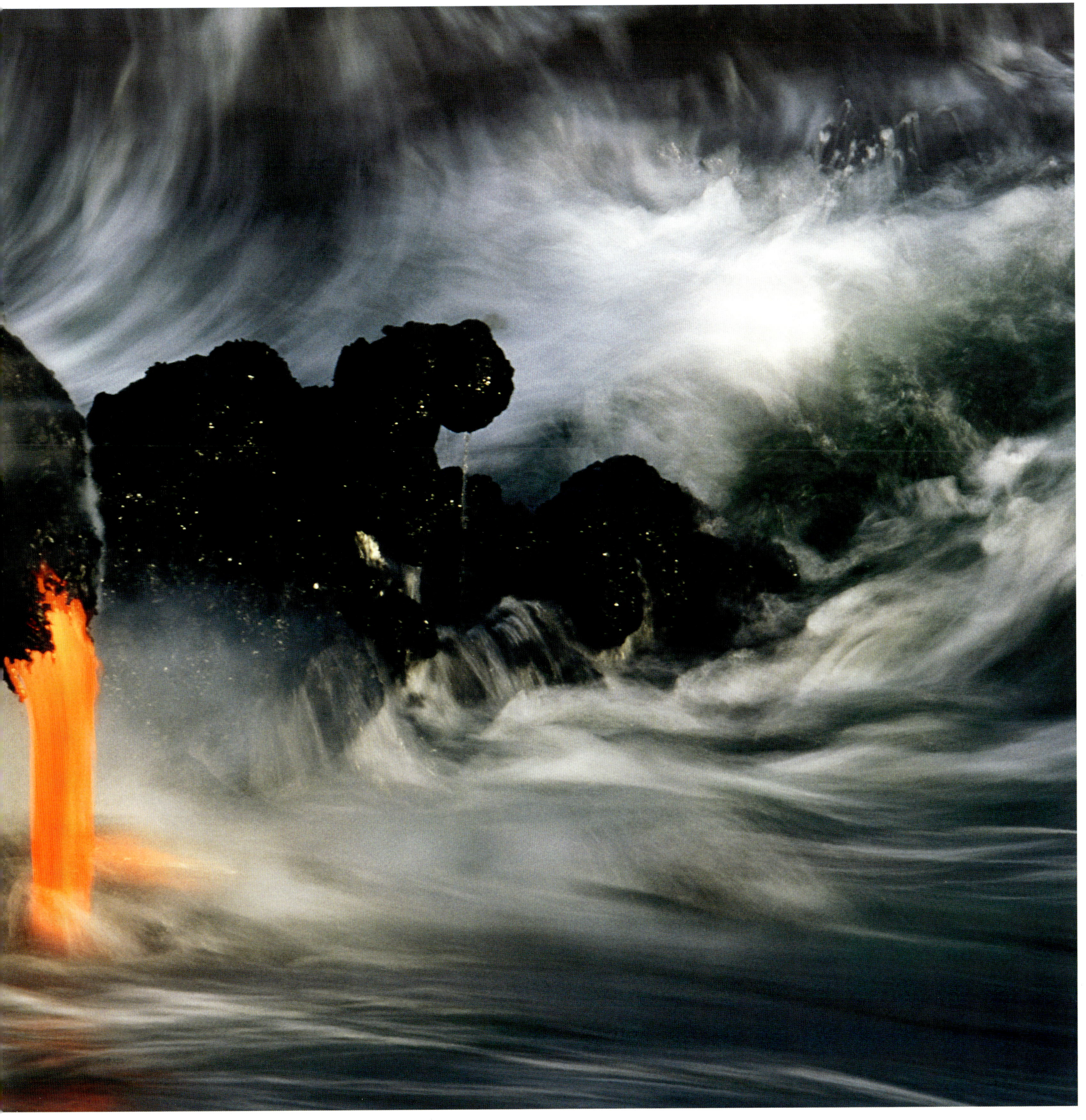

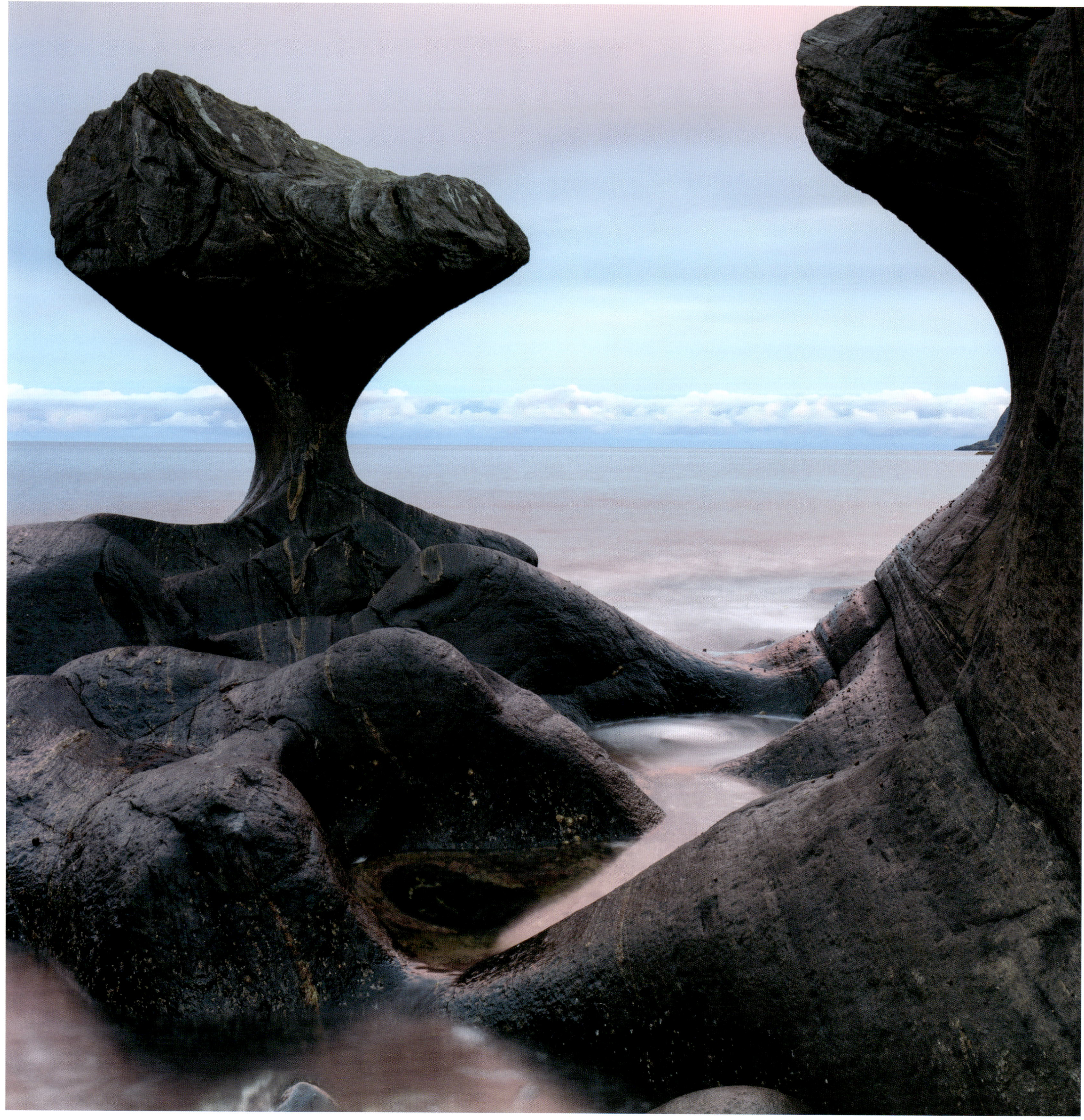

of rock like Mercury, or a hot, gaseous soup bowl like Venus. We had water, it wasn't boiled off, and the planet was large enough to have a gravity that held the atmosphere (and hence the water) in place on the surface. Furthermore, the substantial atmosphere and the magnetic field generated by the earth's core helped protect the surface from ultraviolet rays and solar radiation. There was the basis for life to emerge, but those early years of the oceans would not have been fit for life as we know it. While water vapor condensed and began to form the oceans, other gases stayed, providing a thick layer of unbreathable air. Hydrogen may have been the dominant gas, but carbon dioxide, carbon monoxide, hydrogen sulfide, ammonia, methane, and more—some of the gases typically emitted by lower-level volcanic activity today—would have made up much of the atmosphere, and it was a hot, toxic brew. The sun would have been unseen behind vast layers of poisonous cloud. Occasional collisions from meteorites, up to 300 miles (500 kilometers) across, would have caused traumatic changes, rapid heating-ups before cooling down. Rainstorms may have lasted thousands of years.

Through all this, the world was remarkable for establishing the building blocks for life. One giant ocean of water was probably in place by four billion years ago, relatively shallow compared to the depths of much of today's oceans, as the tectonic plates at that time were comparatively level and had not crunched themselves into the undersea mountain ranges of today. As for land, there was almost none. The early planet had a surface that was more than 90 percent water.

The earliest confirmed landmass that geologists agree on is the supercontinent of Ur. This was a supercontinent that stood out only in that it had no rivals. Evidence for it exists today in rocks found in Madagascar, India, and Australia, but it was less than half the size of Australia. It was a landmass that existed for around three billion years and, in the history of continents, is probably going to retain its title of being the longest lasting of all. It was succeeded by other super-continents, generally getting larger and often emerging with periods of cataclysmic change. Our planet repeatedly became "Snowball Earth" as periods of severe cooling led to extensive ice cover.

From the origins of Earth to now, most of the time the surface of our world has looked very different from how we see it today—indeed, unrecognizable. From Ur onward, we have seen the emergence of various continents that have come together and then split apart again, wandered around the surface of the planet and then collided into each other again to form new supercontinents, only to split up again a few hundred million years later, then drift back together in fresh combinations a few hundred million years on from that. We may project that something similar will happen to our continents of today in the distant future, because the continents remain driven by the convection currents of magmatic rock in the hot mantle underneath the earth's crust. The tectonic plates that make up the crust are being pulled around by forces underneath, and this will continue to happen, causing volcanic eruptions and earthquakes along the way. Indeed, much of this is happening right now, as I write and later you read, without us even realizing a thing, as it occurs miles beneath the surface of our oceans. From time to time we get disturbed by the shiver of an earthquake.

Our weather systems today remain affected by similar forces as drove the first emergence of water and then life on the planet. There's no reason to think this will stop—at least not before the supernova of the sun vaporizes our planet in five billion years' time. So let's look on the bright side: We need not worry any time soon, and our species will have evolved greatly or died out long before then.

But with all that land emerging and shifting around and gradually building up to nearly 30 percent of the earth's surface as we see it today, what was happening to the sea? For one thing, how did it get to be salty? This is thought to be simply a by-product of water eroding the minerals around it, which leads to the salts being suspended in the water. Freshwater falling on land washes away minerals, causing them to swirl around in the ocean water, while the scouring effect of the ocean on the seabed also adds to these impurities. However, the ocean does not get ever saltier. It appears to have been largely stable in its mineral content for a long time. The current theory is that ocean salinity has been the same for billions of years, as a result of a cycle of new salts being added while others are removed through sinking to the bottom or reacting out of the water. This balanced-cycle theory scuppered earlier notions that the age of the planet might be revealed by assuming a constant measured supply of additional salt and thereby being able to calculate back to a pre-salty origin (a theory first proposed by the scientist Sir Edmond Halley in 1715 and occasionally revived, notably by Creationists today, in attempts to argue that Earth is much younger than science suggests).

All that shifting around of plates and emergence of landmasses also saw the oceans start to get deeper. As plates and landmasses crashed into each other, underwater levels began to vary, gradually leading to the vast undersea mountain ranges that now exist, typically along plate boundaries. What we see on land, with mountain ranges such as the Himalayas being derived from intercontinental collisions (which are often ongoing—Mount Everest continues to rise), has been taken to much greater heights and depths under the surface of our oceans. This has happened many times over before arriving at the current patchwork of plates with their resulting bumpy seams. For example, the Mariana Trench in the western Pacific, the deepest point on our planet at almost 36,000 feet (11,000 meters) down below sea level at its greatest depth, is probably no more than 170 million years old. It is formed by the Pacific Plate being thrust below the Mariana Plate. We also see plates pulling apart and creating the conditions where volcanoes occur. It's apparent on land in a few locations—for example, the Kamchatka Peninsula in Siberia, and Iceland. But it is much more common on the deep seafloors of oceans, where the process of "seafloor spreading" occurs. This means plates are gradually pulling apart in deep ocean trenches and near the mid-ocean ridges, allowing magma to leak up out of the crust in almost ceaseless, hidden volcanic eruptions.

The secrets of deep marine environments, such as the Mariana Trench, have influenced our theories on how life may have begun on the planet. The bottom of the deep ocean was once thought to be lifeless, as being absolutely dark and cold and with little oxygen. But in recent years, explorations probing the depths have found numerous strange life-forms, with many probably still to be

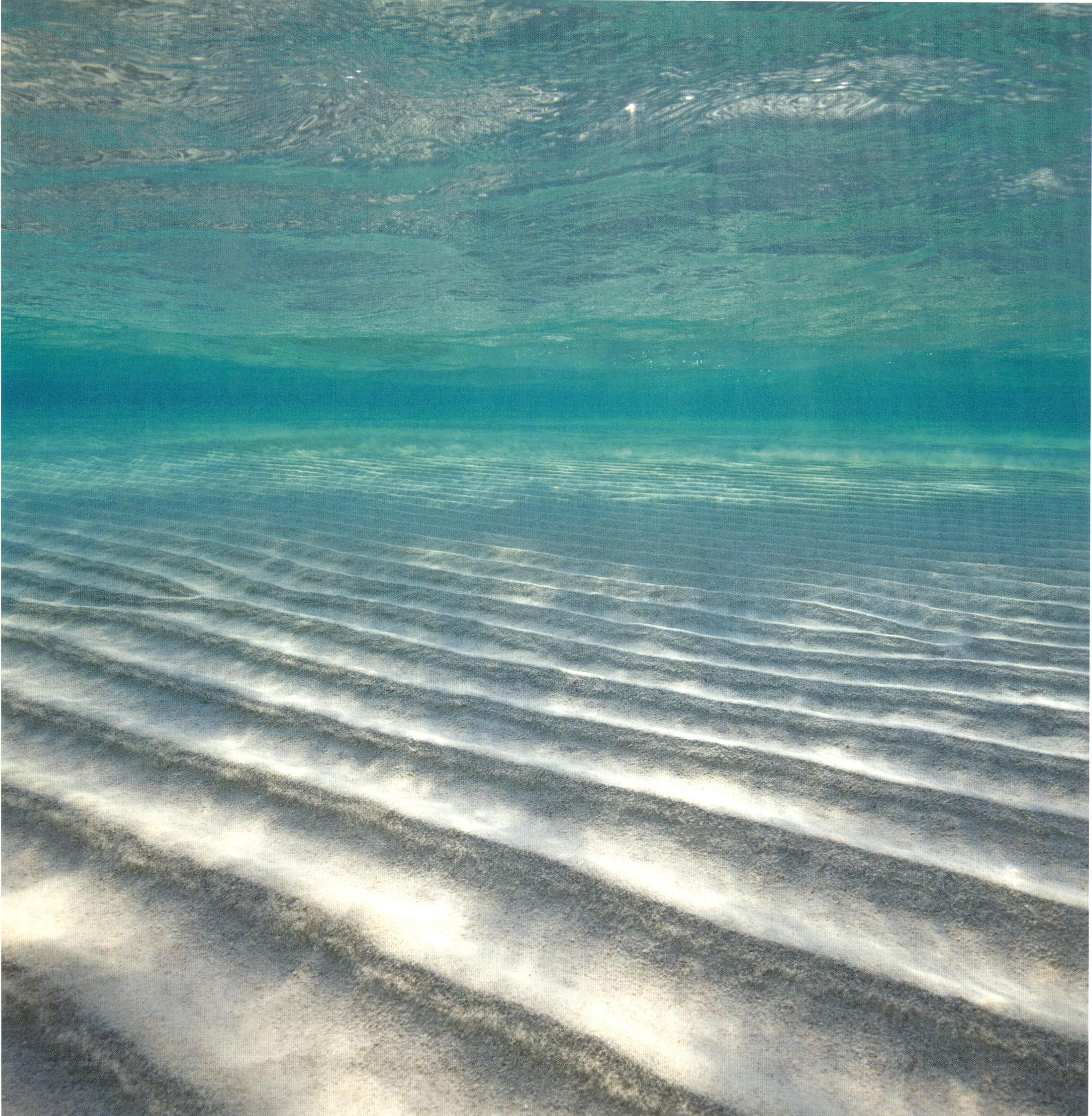

Limitless and immortal,
the waters are the beginning and end
of all things on Earth.

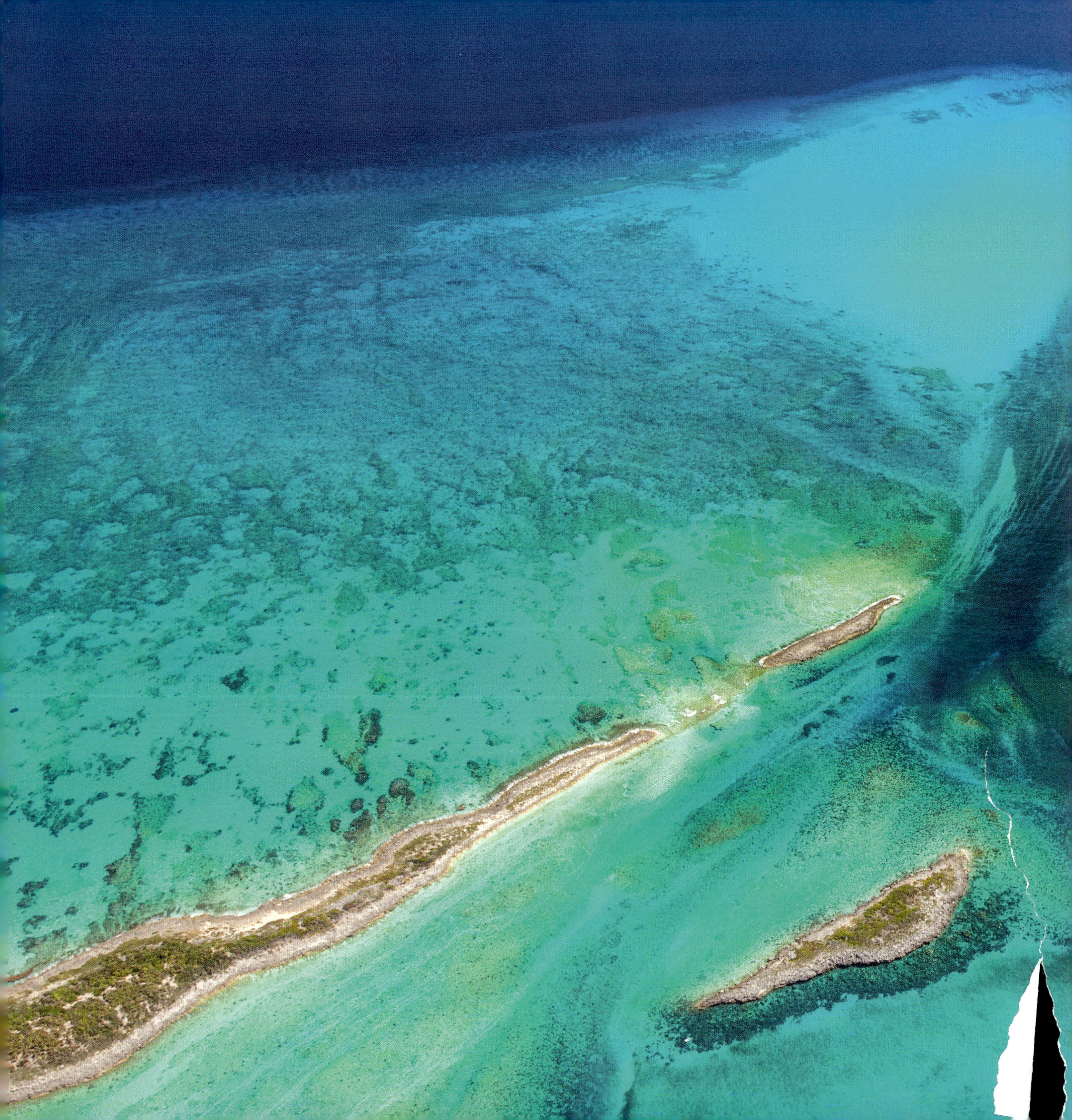

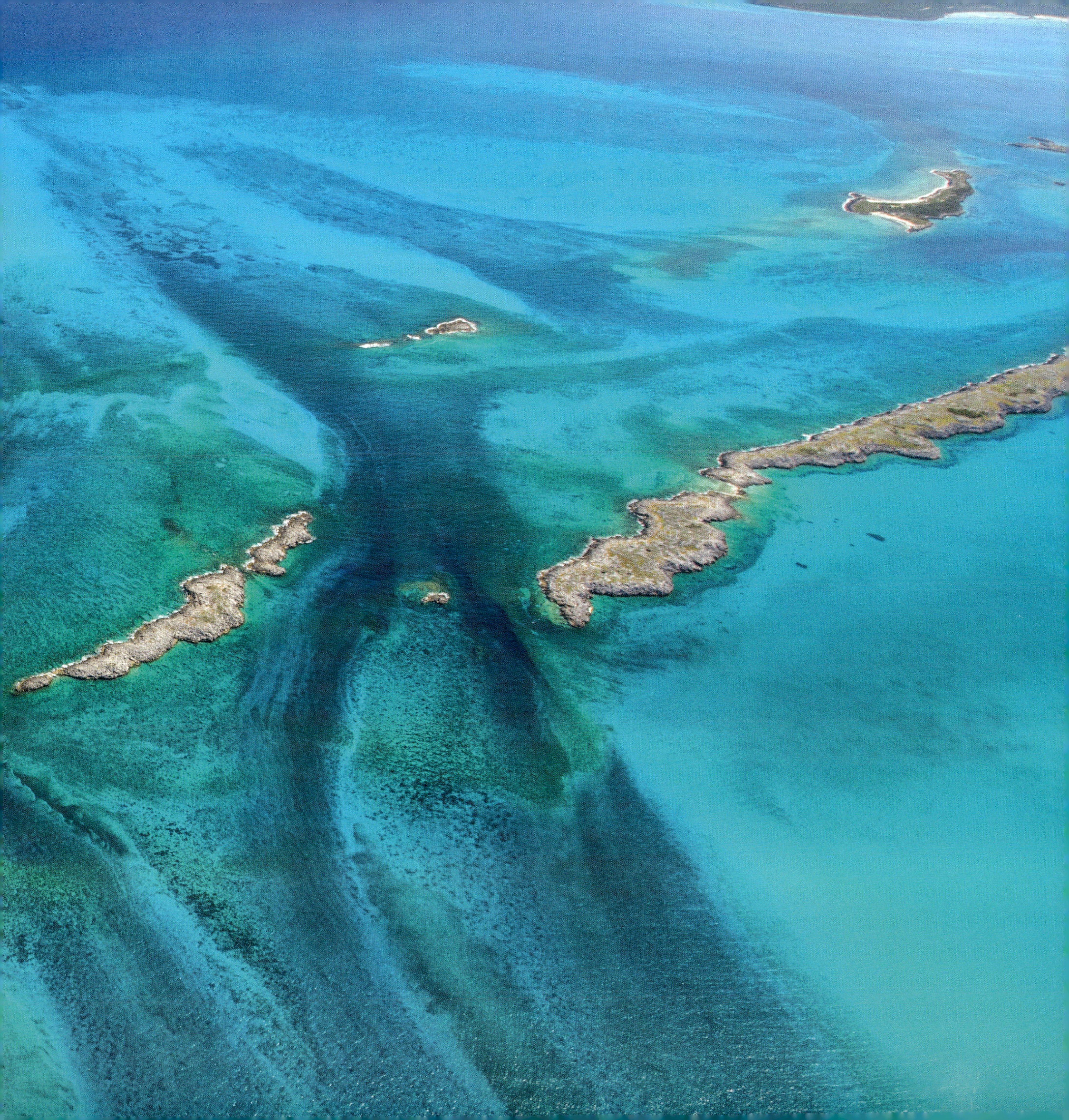

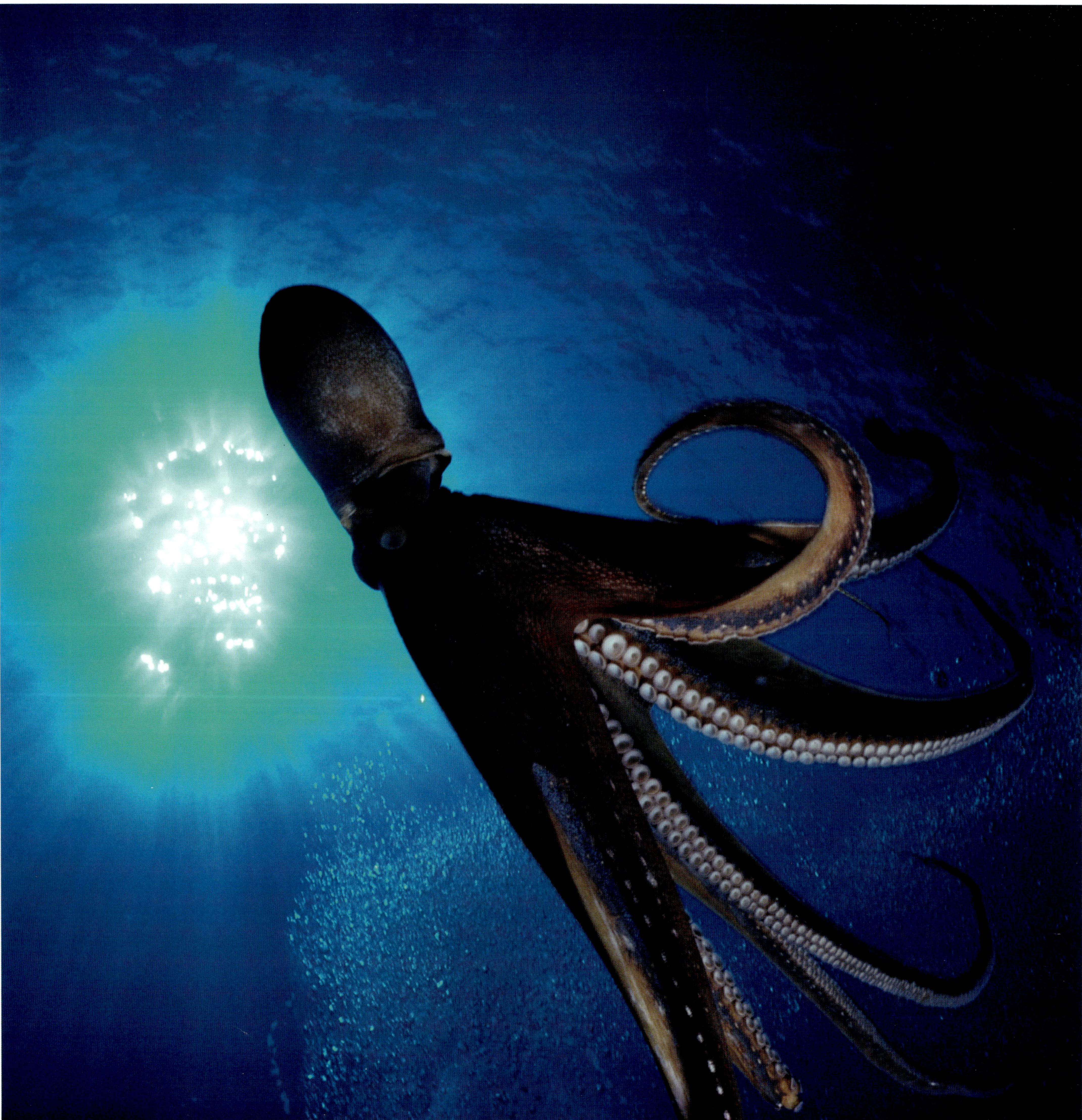

discovered. They seem to thrive where most life could never go. While some of the larger creatures are weird and wonderful, and not a little bit scary to our eyes, it is the seemingly less exciting territory of bacteria found down there that gets scientists particularly excited in considering the origins of life. If basic bacterial life-forms can make a home around deep marine hydrothermal vents, which are driven by the massive geothermal activity, then it could be that early life also emerged in environments where it was not dependent on oxygen, but somehow made sense of this early primordial soup. Biogenic materials—materials that could only be derived from life-forms—have been found dating back to 3.5 billion years ago or earlier, in rocks and fossils, and it may be that they existed in circumstances very inhospitable to current life, in waters that were unlike today's oceans. It may have been a kind of mud, rather than water, that early life emerged from. For example, Gustaf Arrhenius of the Scripps Institution of Oceanography has found evidence of life in sedimentary rock with biomarkers going back more than 3.87 billion years, and his team continue to look to push it back earlier. Other researchers have shown that the conditions for life to form may be more likely not in sodium-rich oceans but in the potassium-rich environments near volcanic mud flows (which happen underwater as well as on land). In 2012 a team led by Armen Y. Mulkidjanian reported: "Our analysis argues against the widespread belief that the first cells evolved in marine habitats." But while the first cellular organisms would have more likely formed in these steamy mud flows, they added that it was their evolution into a more robust structure that presaged lift-off for life-forms: "Only the invasion of the ocean by membrane-encased organisms transformed life into a planetary phenomenon."

These early life-forms may well have emerged multiple times: Earth appears to have gone through repeated cataclysmic changes from other space bodies colliding with the planet, and these events could have wiped out early life several times over. But at some point life hung on and started to evolve from its single-cell origins.

At this point the deep story of evolution passes right under my nose. Throughout the summer, and with the addition of a bracing Christmas Day race, I love to go swimming in ponds on Hampstead Heath, London. It's a wonderful experience of cool, deep water and open air, with ducks, swans, and the odd heron for company. In the warm months, though, there is a growing problem with blue-green algae growing over the surface, which can prohibit swimming for health reasons. Blue-green algae is toxic to humans and so a nuisance to pond users, but perhaps we should be more welcoming of this distant relative, for cyanobacteria, as it is more properly called, is the earliest multicellular life-form we have identified. It can be found everywhere: from the oceans to the soil and rocks of all continents, including Antarctica. Its emergence is seen as a key step toward life as we know it because it was the earliest form of photosynthetic life. Photosynthesis is the way in which our plants work today, converting energy from the sun into fuel to live from. Cyanobacteria and its photo-synthetic followers started to generate an oxygen-rich atmosphere and thereby created conditions suitable for new life-forms.

This shift in the atmosphere may also have killed some earlier kinds of bacterial life—evolution always has winners and losers. When this started to happen is a matter of ongoing debate. There is evidence that cyanobacteria may have been around as long as 3.5 billion years ago, based on its presence in the fossils of rock-like sedimentary structures called stromatolites. The oldest undisputed evidence of cyanobacterial presence dates to 2.1 billion years ago, but it may have significantly predated that point, as it seems to be a key precursor for the "Great Oxygenation Event" that occurred around 2.3 billion years ago. Previous to that, the oxygen level of our atmosphere was less than 1 percent—today it is around 21 percent. Photosynthetic life reduced methane and increased oxygen and created conditions for the diversification of life and the arrival of life-forms that we wouldn't need a good microscope to spot. But life was still mostly microscopic and had little multicellular diversity until about 540 million years ago, when there was a rapid change. In what has come to be called the "Cambrian explosion," multicellular life emerged rapidly, with numerous new species. While this may seem a very long time ago, it is actually about 80 percent along in the history of life on our planet.

The Cambrian explosion, or "Cambrian radiation" as it is also known, led to a great expansion in the range of life-forms, setting down the base for the major groups of species that we see today. In a period of around 20 million years it would seem that the earlier, simple single-cell and early multicell life diversified dramatically into most of the major variety of life-forms that we see today. While those early species are long gone, biologists can trace back to fossils of that period the origins of the different creatures and plants we have today. Many of these earliest emerging marine life-forms can be found in one particularly remarkable fossil source—the Burgess Shale. This treasure trove for paleontologists was only discovered in 1909 in the Canadian Rockies. It is an exposure of mudstone more than 500 million years old, probably laid down in a movement of reefs and eroded cliffs battered by stormy oceans. Over time, this material disappeared under other formations, but then later reemerged at certain points along the Rockies. The riches of the Burgess Shale continue to be excavated, and over the years have prompted various theories. Stephen Jay Gould suggested that the fossils indicated the diversity of life was even greater in the Cambrian era than now, but this has been rebutted with arguments that all the fossils can be seen as linked through to the groups of life we have today. The connections are often visually immediate: fossils of sponges, worms, various crustaceans, starfish, and jellyfish-like creatures all show the ancient forerunners to the familiar sea life of today. The more recently discovered, remarkable, rich fossil beds of the Maotianshan Shales, at Chengjiang in China, which date back to around 525 million years ago, also corroborate this picture of emerging diversity. Some of the most important fossils from there show early vertebrates—our (very distant) ancestors.

But why and how did life suddenly blossom into such rich variety at this time? There is no simple answer—indeed, there are many theories, and as some get dismissed, so new ones emerge. It is worth noting that it wasn't exactly overnight—the period of the rapid growth is currently dated from 541 million years ago to 515 million years ago—and major groundwork for the emergence of this life may have happened earlier, in the pre-Cambrian era. A group of long-lost marine life-forms called Ediacara biota could hold the secret to the shift from few to many living things, but they have long since disappeared, with little trace. From the fossils we have, dating back

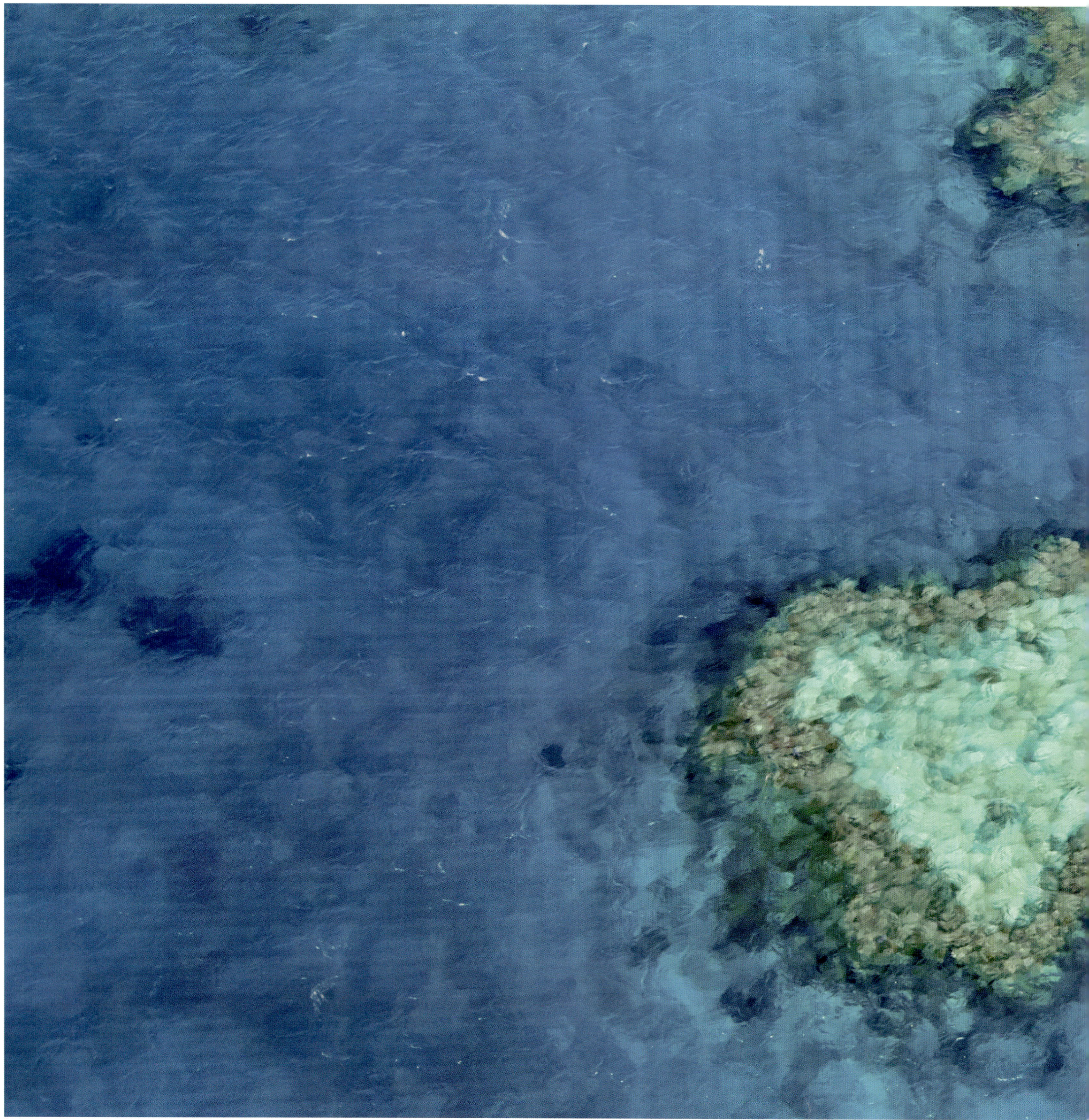

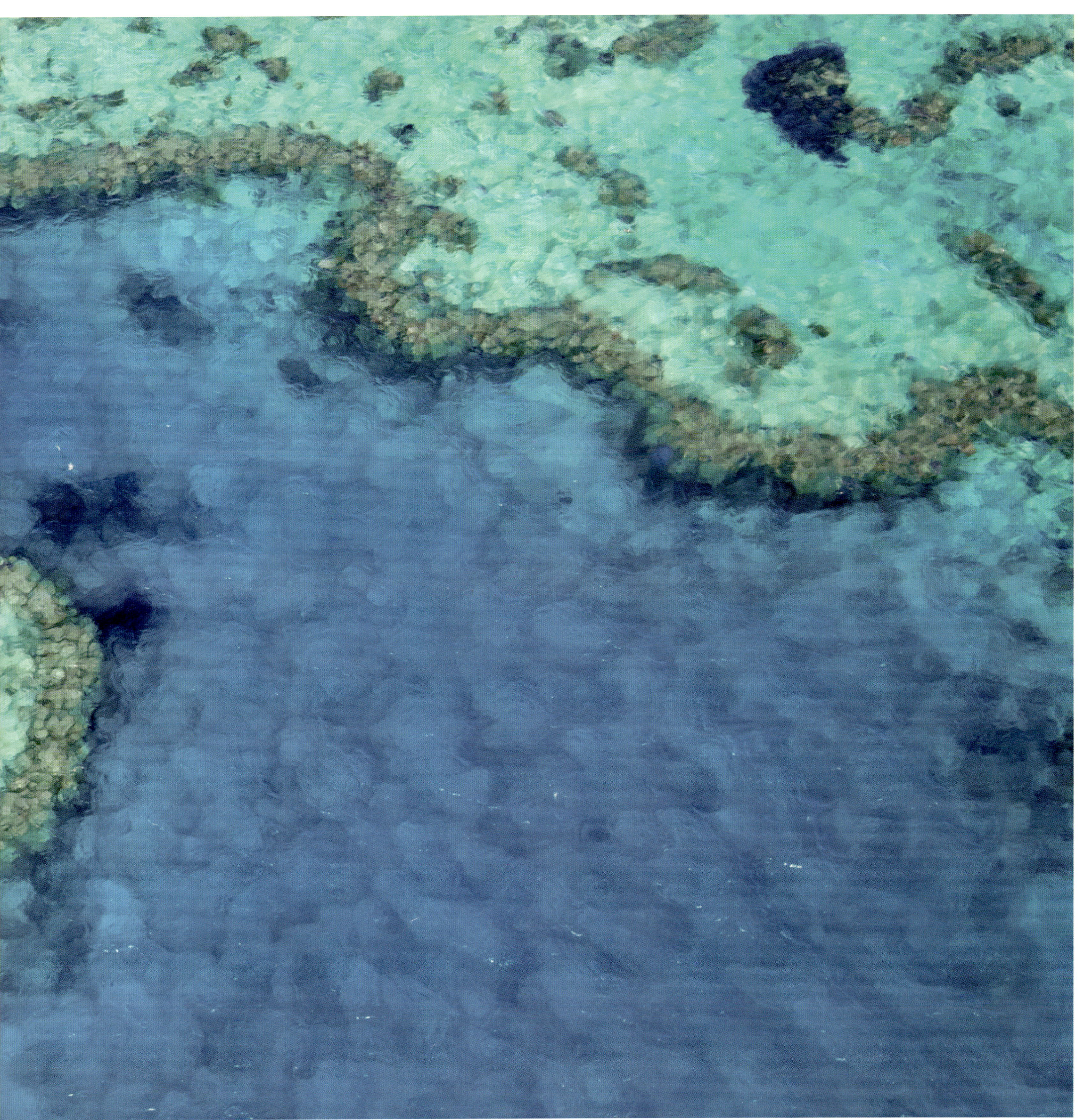

more than 600 million years, we can get clues of a lost richness of emerging life, but we do not even know for sure if they were plants or animals. Other unknown life-forms that have gone unpreserved (and many weren't, being in the wrong conditions for fossilization) may have been the predators or the prey of the Ediacara biota and may have influenced what we do see happening in the Cambrian period. The Ediacara biota are probably the earliest complex multicellular creatures. They occurred in times of major changes on the planet. Huge swings in temperature, changes in atmosphere, movements of continents, and rising ocean levels could all have been involved in giving birth to this period and then repressing most of its life before there was a renaissance in the Cambrian period.

These early multicellular creatures were, until recently, thought to be made only of soft tissue. But the vision of these distant times is shifting. On a trip to the Namibian desert in 2013, a team of scientists examined a rocky formation that turned out to have been a reef on the seabed in the late Ediacaran era, just before the Cambrian explosion, and which had in part been built by some of the earliest sea creatures with a skeleton. Called *Cloudina*, these pencil-shaped animals, up to 6 inches (15 centimeters) long, were wormlike in form and yet made of a bony substance. The lead geologist on the team, Rachel Wood of the University of Edinburgh, described the creature as being like "a series of hollow ice-cream cones all stacked up," in which only the last cone would have been living matter—rather like how coral exists today. *Cloudina* has characteristics that suggest it is an ancestor of coral, anemones, and jellyfish.

So, the indications are that somewhere around 650 million years ago our seas and oceans started to fill up with more and more life, which evolved relatively rapidly over the next 150 million years. The oceans were the birthplace of life, as it is thought that those early continents had a thin and largely lifeless crust of soil at best, with some microbial forms spreading, but no life of any great size or diversity.

Instead of seeking one dominant reason behind the explosion of life, scientists now are more likely to argue for a combination of the various hypotheses working together. There are exotic biological reasons for the rapid change (such as the so-called "evolutionary arms race," whereby species develop new features and diversify in order to survive and thrive) and also other imaginative suggestions, such as climactic shifts, or effects from extraplanetary influences (such as radiation bursts, rather than the arrival of aliens). More subtle, evidence-based theories of causation are starting to gain weight, though—it is accepted that rising sea levels definitely eroded more minerals, notably calcium, from which bones are made, and as these mixed into the waters it created conditions for new life-forms. The oxygenation of the air, and its mixing into the water over time, may also have reached a tipping point in providing the conditions for more and different life. These and other factors, which are still to be understood and placed in their correct position in the Cambrian timeline, are thought to have combined into what became an evolutionary cascade.

Now, in my quick catch-up on the life of our planet, we are a long way along in the story of the living oceans—about three billion years in and only 500 million or so to go—but at the same time we seem to be only at the beginning of life as we know it. Not that there are any surviving species of the Cambrian period, but at least a fossil jellyfish of 500 million years ago does look like something that could be in the sea today. We can start to see how we got here.

The group of creatures that emerged from this period and have since dominated all known species in terms of sheer diversity are the arthropods. These are exoskeletal animals with segmented bodies. Yes, we are talking about the relatives of the humble prawn, or its fairly close ancestors, which date back more than 500 million years. Around 80 percent of all species are thought to be arthropods (beetles being the most multifarious on land). It is likely that some kinds of arthropods were among the first creatures to move out of the seas and onto the land, with their hard, jointed bodies able to adapt and survive relatively easily. We can perhaps see the descendants of these early transitional creatures in the form of a crab, which is equipped to operate successfully at the bottom of the sea or on land in the intertidal zone.

One early marine arthropod that appears to have been among the most numerous of creatures emerging from the Cambrian period is the trilobite, whose long-distant descendant might well be the horseshoe crab. Varieties of trilobites can be found in great quantity in fossil deposits. Trilobites operated in various ways, some probably as predators while others were more into scavenging. Some swam and most likely depended on early plankton for food. They thrived for a long time, but then became extinct around 250 million years ago in what is called the Permian-Triassic extinction, or more poetically, the Great Dying. A name given for good reason—it has been estimated that 96 percent of all marine species that had developed by then were suddenly made extinct. This event wiped out most life-forms at the time. What happened? We don't know for sure, but it may have been that a massive volcanic eruption caused an immense greenhouse gas effect on the planet, leading to a major rise in temperatures that caused oxygen to be released from the water, killing most ocean life. As the water vapor mixed with volcanic gases, acid rain may have developed, destroying the forests, washing nutrients into the seas, and further damaging the marine ecosystem.

One type of humble marine creature that has been very useful to scientists in understanding these abrupt changes in our planet is the form of plankton known as radiolaria. It has mutated through numerous species, but as a type dates back to that Cambrian explosion. Microscopic fossil evidence of the existence of radiolarian, in greater or lesser degrees and varieties, is a useful measuring aid for understanding just how severe the changes were at various times. Researchers find that radiolarian persist, but at dramatically different volumes, influenced by the level of oxygen available in the ocean waters at different times. An indication of just how extensive and all-pervasive such microscopic creatures are on our planet can be sensed by knowing their decayed corpses cover a large amount of the ocean floor as a thick layer of slime that ultimately is compressed into rocks.

The Great Dying is the most severe of numerous extinction events that have beset our planet and caused great ruptures in the evolution of all life. And yet life has bounced back. Today it is thought that 99 percent of all species that have lived are extinct, including whole families. The last major extinction event was around 65 million years

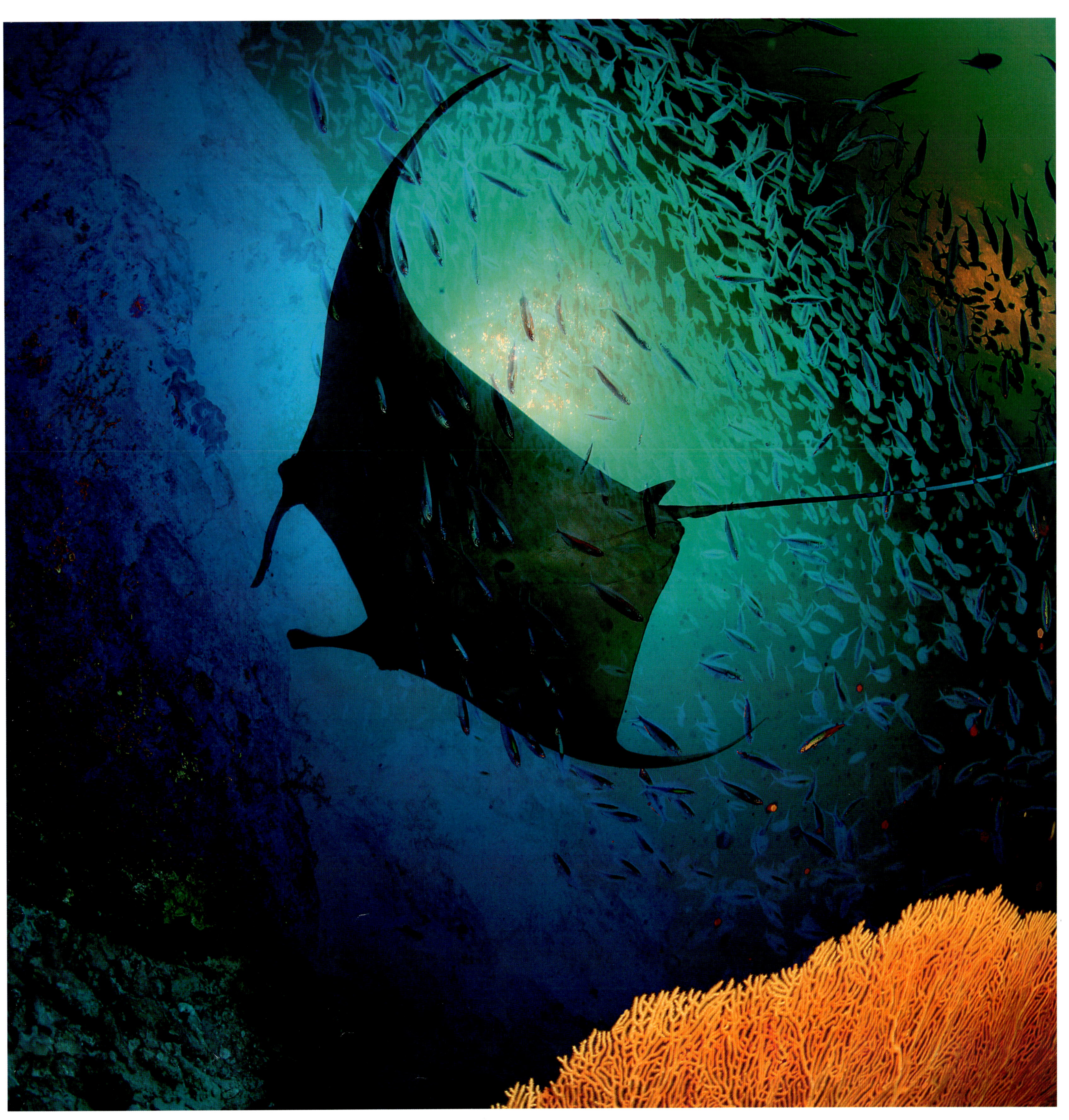

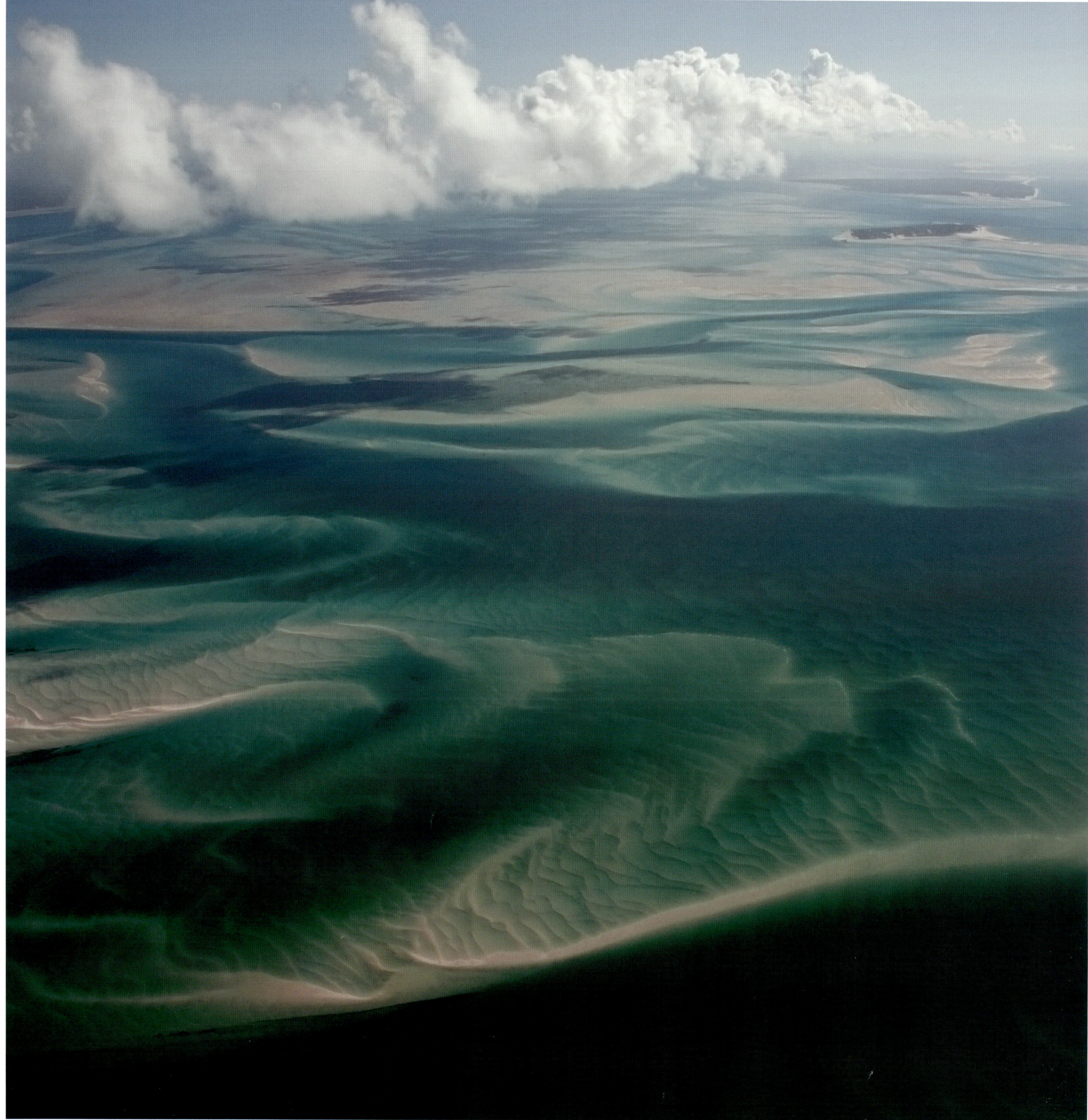

When one has been long at sea,
the smell of land reaches far out to greet one.
And the same is true when one
has been long inland.

ago, and finished off the dinosaurs, but there may have been as many as twenty such events of various scale that have greatly shaped the evolution of ocean life since the Cambrian explosion. And there's one extinction event you get to have a ringside seat for. It is called the Holocene (or sometimes now the Anthropocene) and is the one we are living through. The cause of the current wave of extinction? Humans. But we can return to that topic later. Meanwhile, we might fast-forward through these expansions, contractions, and revisions of life to the last few frames of the film and see where current marine life started appearing.

The sponge is perhaps the oldest life-form we know, with fossil evidence as old as 760 million years. While this is not the same species, it is a clear relation to the sponges of today. Its descendants have lasted through all the periodic catastrophic collapses since. The jellyfish is also a wily veteran, perhaps more than 500 million years old, and the horseshoe crab is not much more youthful. The species of these types have of course changed, but the basic form continues. One rather magnificent, and now endangered, ancient creature is the nautilus. It is often called a "living fossil" because it stands apart as the only survivor of a much larger group. This dates back 500 million years, a late Cambrian arrival that is the living ancestor to the class of creatures known as Cephalopoda, which contains squid, octopus, and cuttlefish. Nautiluses are the only cephalopod where the bony structure is externalized as a shell into which it can retract. The shell is so striking, with a geometric coil and pearl-gray marking, that the nautilus is increasingly endangered as it is hunted for the value of the shell as an ornament. What the trinket collector doesn't note is that the real magic is what goes on inside the living nautilus. It is found around coral reefs in the tropical waters of the Indo-Pacific and has been spotted at depths around 2,000 feet (600 meters) deep. But unlike most creatures of the deep, it can also survive nearer the surface in areas where cooler water can be found in the tropics, in as little as 16 feet (5 meters) of water. When other creatures of the deep are brought up to the surface very few can survive because the massive change in pressure causes internal rupture. However, the nautilus seems to be designed to cope and shows no ill effects. Exactly how and why this happens is not understood. The nautilus has the power to propel itself through the water and adjust its buoyancy. But perhaps most remarkable for such an ancient creature is that it possesses an early form of the intelligence that is more highly developed in its cephalopod relations, such as the octopus. Tests show that it has limited powers of memory, both short- and long-term, when tested with food stimuli. While an octopus shows the ability to remember for weeks, the nautilus doesn't get beyond twelve hours. The nautilus is a remarkable window back in time that may not have changed much over 500 million years. Everything around it has changed immensely, including the very nature of the sea and its breadth and depth; the creatures and plants that it once lived among have long gone.

As most species have become extinct, there are other marine creatures that stand out as markers back. Although not as old as the sponge or nautilus, there are several which stand out as remarkable survivors. The coelacanth was long thought extinct, a member of the order of fish named Coelacanthiformes, which dates back 360 million years, and was believed to have disappeared with the dinosaurs. But one was fished up in 1938 off Madagascar, causing a worldwide stir as something seemingly coming back from the dead. Occasionally more are now found in the Indian Ocean. This is a result of deep-sea trawling becoming more invasive. Two species of coelacanths have been identified and both are on the endangered list. The coelacanth is particularly interesting for being a suggested link between fish and walking creatures—it has numerous fins that allow it to have remarkable mobility in the water. This structure is thought to have been the base for evolution into the first four-legged creatures.

Another ancient fish family that we are familiar with is the Acipenseridae, commonly known as sturgeon, which goes back at least 200 million years. The twenty-five species in the family vary between those that live in rivers and near coastlines and those that stay in freshwater. They are a relic of a much wider group of fish in the past. Our fondness for eating their eggs—caviar—used to put them on the at-risk danger list as a species, but increasingly the farming of sturgeon may work to ensure its survival, at least for some of the species, in captive environments. There are even attempts to introduce sturgeon into parts of the world where they have never been native— all species come from the Northern Hemisphere, but they are now being introduced into rivers in Uruguay and South Africa.

Another group of creatures that is ancient and endangered due to us is the sea turtle. These reptiles date back 150 million years, with links to dinosaurs. Their journey as a family also indicates that evolution is not a straight line—ancient as they are, they are also descended from land-based creatures that evolved back into the sea from whence their ancestors had once emerged. If we consider the massively changing conditions imposed by the near total wipe-out of life from catastrophic extinction events, we can understand that evolution at times involved a restart or resetting of direction. Most sea turtles are omnivores of the sea, eating plants, jellyfish, sponges, and other creatures, which makes them at home in a vast range of oceans. Relatively little is known about their lives because they are such nomads—the females only come ashore to lay eggs and the males spend their whole life in the ocean. They may travel thousands of miles, seemingly knowing where they are going through similar means as migrating birds, reading slight differences in the earth's magnetic field.

Sharks are another group that have some deep history. In particular, the rather frightening-looking (even by shark standards) frilled shark, a denizen of the deep that is almost snakelike, can claim a lineage back 150 million years and more, while the slightly less scary goblin shark has been dated back 118 million years. The broad group of sharks, living and extinct, goes back at least 420 million years and is immensely diverse, including more than 500 species, ranging from the tiny dwarf lanternshark, which grows to a maximum of about 8 inches (20 centimeters) long, to the largest fish in the world, the whale shark, which can exceed 40 feet (12 meters) in length. The latter might sound terrifying, but it is a filter feeder, sifting plankton and other small life for food, and is no threat to us. The largest shark ever was megalodon, a truly terrifying beast that probably snacked on large whales and reached 59 feet (18 meters) in length. It was a forerunner to the great white shark, but it makes that terror of the seas today look almost friendly. Megalodon was one of the greatest

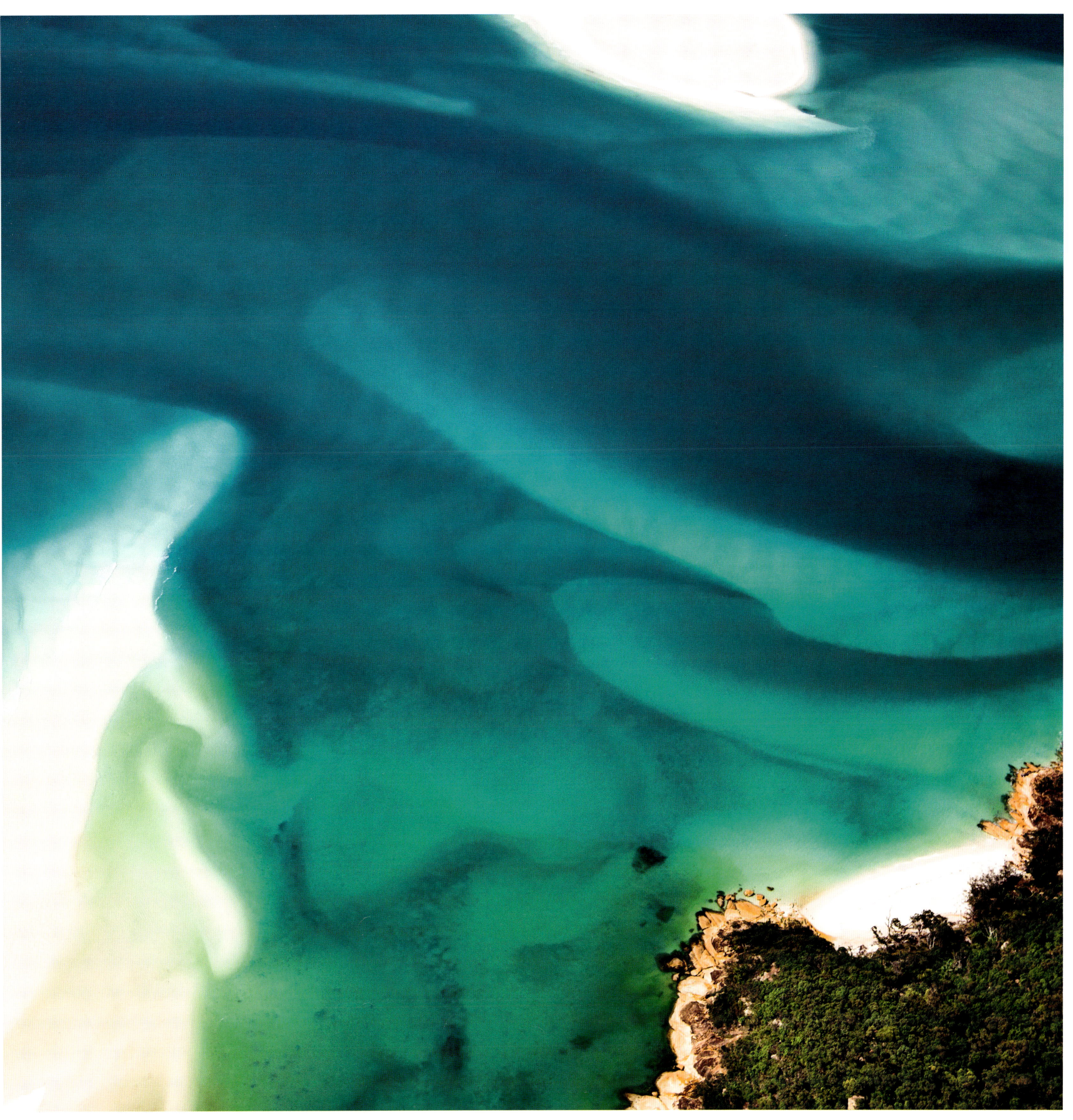

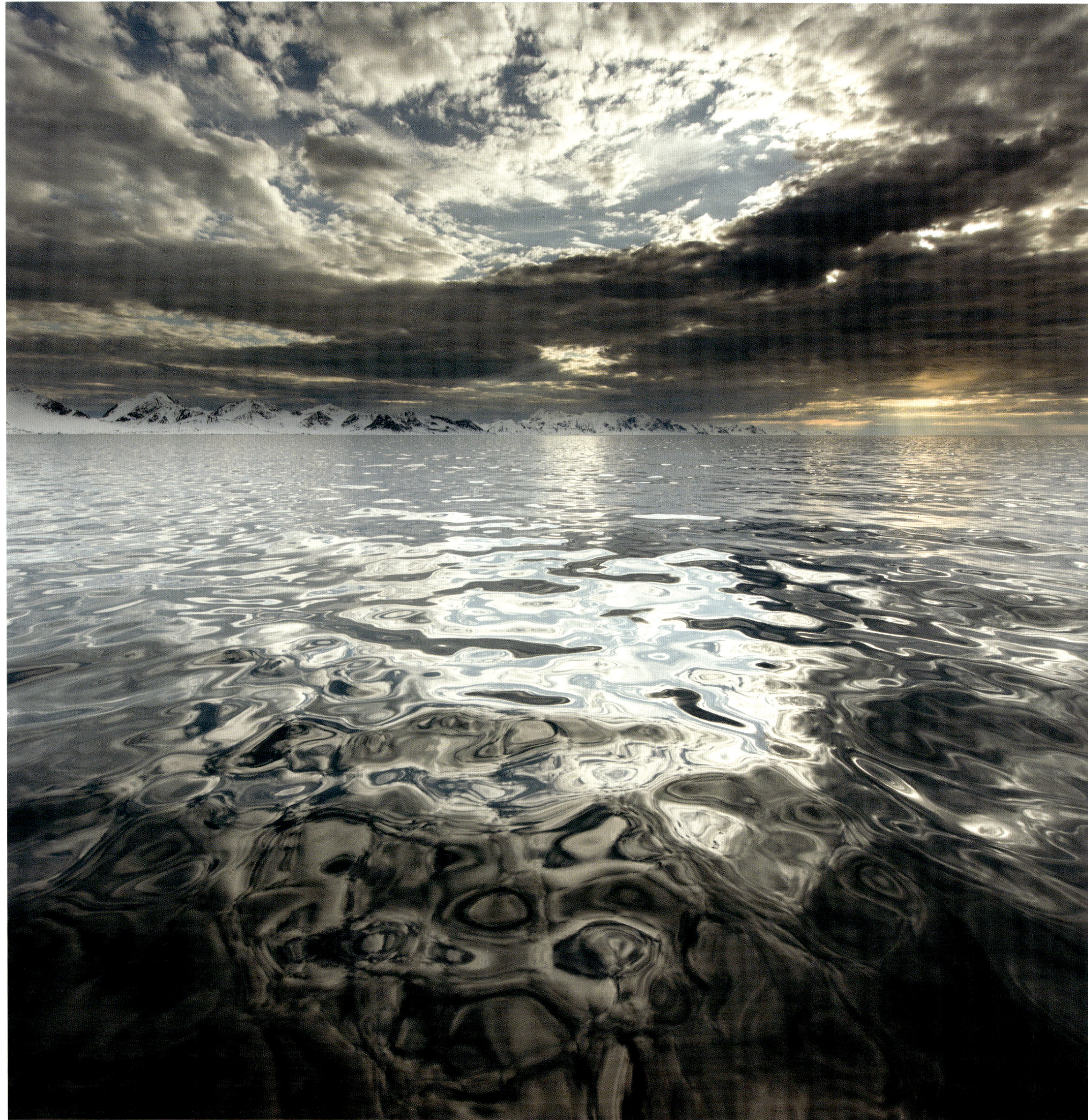

predators ever, a machine for killing, as it were, but it couldn't handle change and became extinct around 2.6 million years ago, before our ancestors emerged from the trees. Reasons for extinction may have been cooling temperatures that affected its reproduction and also the species running out of food when a less productive environment reduced supply.

Despite the largest shark being not quite as large as it once was, the evolution of marine life conforms to what is known as Cope's rule. This states that species have tended to get larger and larger since the Cambrian explosion. The theory that evolution has this directional quality is named after the nineteenth-century American paleontologist Edward Drinker Cope, who did not specifically argue such an idea, but did believe in directional trends in evolution and had noticed the tendency in size increase across a clade (the name for a complete group of species on one branch of the evolutionary tree that have all descended from one common ancestor). It was only in early 2015 that a team at Stanford University completed an exhaustive project that involved comparing fossils records from more than 17,000 groups of species—covering more than 60 percent of all species that are known to have existed. Over five years—and with the help of numerous researchers, including high school interns—scientists were able to confirm that there is a clear drift to gigantism among marine species. For some reason, or reasons, big is beautiful when it comes to evolution, at least definitely in the oceans. This might seem counter to our thoughts when we think how dinosaurs once ruled the earth but then came to a sudden end. However, sifting through all the evidence supports Cope's comparative studies of the nineteenth century and confirms and extends his theory. Since the beginning of the Cambrian era, the average size of sea creatures has grown by a factor of 150. So that means the small crustaceans then were much smaller than now, on average, while at the top end of the scale we have whales, sharks, and giant squid that are much larger than the apex creatures of 540 million years ago. It is not that all species became larger, but that the larger species divided into other species and so spread their gigantism farther and faster than small creatures. As to how this path prospered, there are various theories: that the larger creatures saw the benefits of eating bigger prey, or of being stronger and faster. It may also have been a by-product of increasing oxygen levels that supported the development of larger bodies. As the Stanford analysis was only recently published, much more discussion and research is likely to stem from this insight over the next few years.

We have dwelled here on the emergence of the first complex life-forms and ecosystems. In doing so in these pages, we can only dip into moments in a vast span of time, and sample some of the species that came and went as a way of indicating the forces at work. In these beginnings we connect with something fundamental about our ocean life. The vast waters of the earth are the origin of all our lives and all species. In reflecting on them we can sense, through the scraps of knowledge and theories on the evolution of early creatures, how the planet works in its four-dimensional richness.

What started in the massive ocean that covered most of the world and evolved to oceans, continents, islands, seas, and rivers, and which went through numerous catastrophic extinction events, has brought us to today and the infinitely complex, ever-changing ecosystem of our oceans and seas. As some of the references in this chapter show, we are still very much both piecing together a sense of how we got here while also discovering insights into current and extinct species. One thing we can be sure of is that there is much about the oceans that we still have to discover, from ever-shifting shorelines to those still uncharted depths.

The sea is everything. It covers seven-tenths
of the terrestrial globe. Its breath is pure and healthy.

It is an immense desert, where man is never lonely,
for he feels life stirring on all sides.

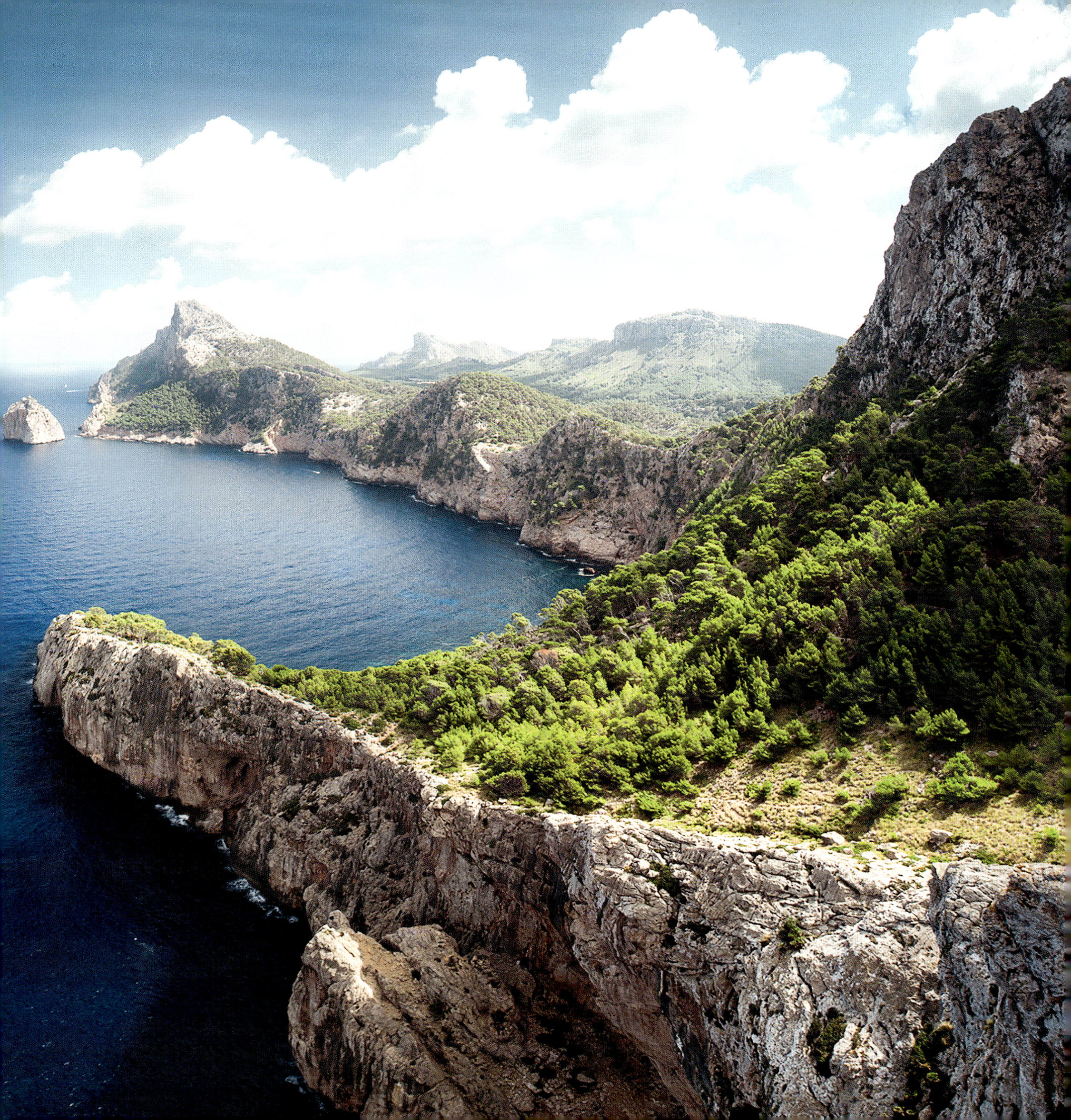

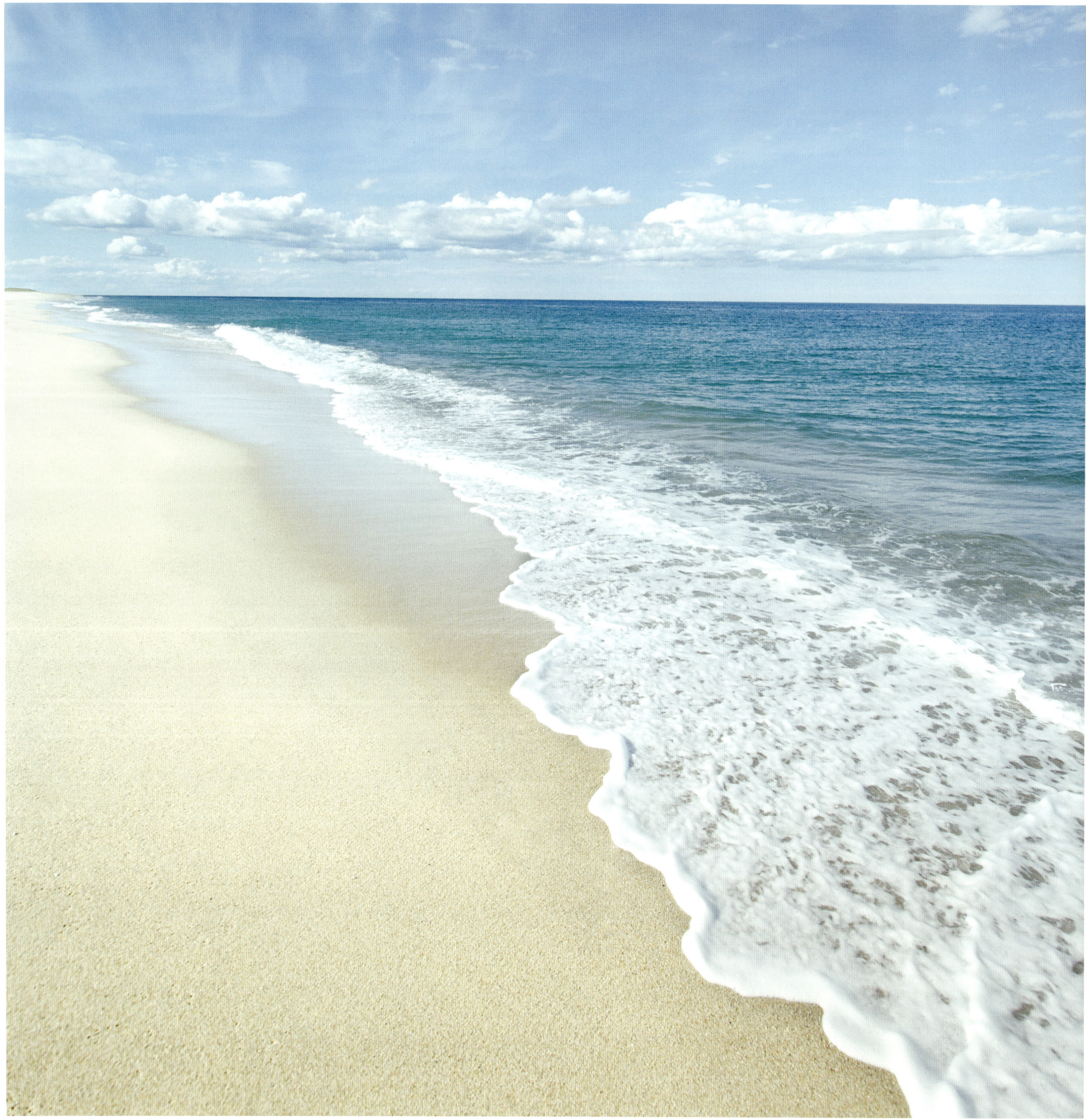

There's nothing more beautiful than the way
the ocean refuses to stop kissing the shoreline,
no matter how many times it's swept away.

Connections

Our seas and oceans are incredibly varied and yet also linked by consistent factors. There's a lot more than salt in seawater. It is also a restless, ever-changing entity. Let's look a little closer by looking from afar.

It is springtime in the Bay of Biscay. Many miles overhead, a camera on a NASA satellite takes another snap: It shows the curve of the land, puffs of cloud like sprinkled snow, a trace line of sandy beaches, and then the dark bay. But it is as if some of the white clouds and yellow sand have mixed with the blue-reflecting water, for there are swirls of light greens and shifting blue-grays like vast clouds of firework smoke drifting over the seas off western France.

What we are seeing is not the seabed, or pollution, but a seasonal bloom of phytoplankton. These tiny microorganisms react to the lengthening daylight and warming waters and suddenly explode into tremendous rapid growth, creating this multicolored patterning of the sea that can be seen from space. It is a phenomenon that can be noticed around the world as the seasons turn and bring more of the sun's energy to an ocean and feed this spurt of photosynthetic life.

This is a fundamental force that makes our world habitable. The microscopic, mostly single-celled plants that make up phytoplankton—more than 200,000 assorted species of which very few have been studied—are arguably the most important building blocks of all life on our planet. They are "primary producers," generating biomass from inorganic material, and thereby providing the starting point for the pyramid of life. Some get eaten by zooplankton, themselves microscopic creatures, and they get eaten by other tiny animals, which may also be classified as plankton (the definition of which is simply organisms that live in the water and cannot swim against the current—so include, for example, jellyfish). And the larger plankton get eaten by bigger creatures, and they in turn are consumed, all the way up to apex predators such as sharks, killer whales, and humans. But it's not just that phytoplankton begin the food chain. Even more fundamental than that, they create much of the oxygen in the air and the sea. Through photosynthesis they consume carbon dioxide, output oxygen, and store carbon. Half of all the oxygen in our atmosphere is down to the work of plankton. It is the chlorophyll in the organisms that creates the discoloration of the Bay of Biscay when seen from high above, and it is the nutrients in the water—such as nitrates, phosphates, and calcium, which might be washed out in concentration from the land at river estuaries—that also help fuel their spring efflorescence. The phytoplankton have lives that extend only a few days, while the massive blooms in the ocean typically last weeks. Finally, as conditions alter, they die back, and those not eaten will drop through the sea to the ocean floor. The bloom may range from several feet to a few hundred feet deep. That which is not eaten and retained near the surface falls away and descends to the seabed, becoming ultimately a silicate slime that locks up carbon. This is happening all around the world—you can see it on many a satellite image if you look carefully. NASA estimates up to 10 gigatons of carbon are taken out of the atmosphere and sequestered to the ocean depths in this way each year. And the final flourish of plankton's value to us is that those layers of fossilized microscopic life past become buried over millions of years, crushed and locked into rock from which we get the oil reserves under our seas. What we think of as mineral oil is actually not a mineral at all—it is 100 percent organic by origin.

It would be hard to think of anything that has made a larger contribution to life on Earth today than microscopic plankton. This invisible life feeds all life in the oceans, they feed the air with oxygen, and their ancestors indirectly created much of the energy we use. When we think of sea life we tend to think of the fish and mammals, corals and plants—the things we can see. But we must look closer: 98 percent of all life, which is the basis for feeding the other 2 percent, are the innumerable plants and creatures at this microscopic level. In every liter of seawater is a teeming mega-city of plankton and other life. There may be up to 10 million phytoplankton in that liter and a million zooplankton (the microscopic creatures that feed on the phytoplankton). There are also vast numbers of bacteria and viruses that in turn play an important role in the balance of sea life, controlling the plankton populations in ways that we are only just beginning to understand. Just a drop of surface seawater might contain 10 million viruses. These bugs are feeding off bacterial hosts, also to be counted in the millions in every drop, and as they break them down, this helps phytoplankton get nutrients from the sea. So even before we can actually see any life with the naked eye, a richly connected ecosystem has played out a battle royal, which in turn creates the base for all other living things on the planet.

Plankton not only supports all life today, but in its origins it formed a crucial evolutionary step in all life coming about. Fossil evidence goes back 3.5 billion years for photosynthetic cyanobacteria, but some kind of simpler life must have come before, something that would have led to the cyanobacteria, predating any fossil records. In 2003 a team of Danish researchers published their analysis of sediment rock from Isua, Greenland: In this they identified signatures of photosynthetic life that were even older. "What this demonstrates is that the earth had a functioning biosphere before 3.7 billion years ago," claimed Professor Minik Rosing.

While microscopic, the world of plankton is visually remarkable once you are squinting through the eyepiece. In particular, the group of phytoplankton known as "diatoms" is a world of strange geometric organisms that look, in their two-dimensional form (squashed on a microscope slide) at times like polished stones and almost like jewels. But they are sophisticated and four-dimensional, albeit living brief lives of around six days. Diatoms are particularly important for oxygen creation and carbon storing: They are the lead "fixers" of carbon dioxide, processing it more effectively than any other photosynthetic life into its component parts. It is estimated they alone are responsible for more than 20 percent of the world's oxygen. They come in a spectacular range of shapes, essentially all kinds of cylindrical forms: circular, elliptical, tubular, lobed, and more. As an organic structure they exist as if within a "glass box," a hard silica shell that has piercings, allowing water, gases, and other solids to permeate and be processed. Diatoms occur in all kinds of water, and even moist soil or on the backs of seabirds. But given the extent of the oceans, this is by far their main environment and the reason they are so ubiquitous and influential on the atmosphere and all life. Every fifth breath we take is oxygenated through the work of diatoms, and they sequester more carbon than all the rainforests.

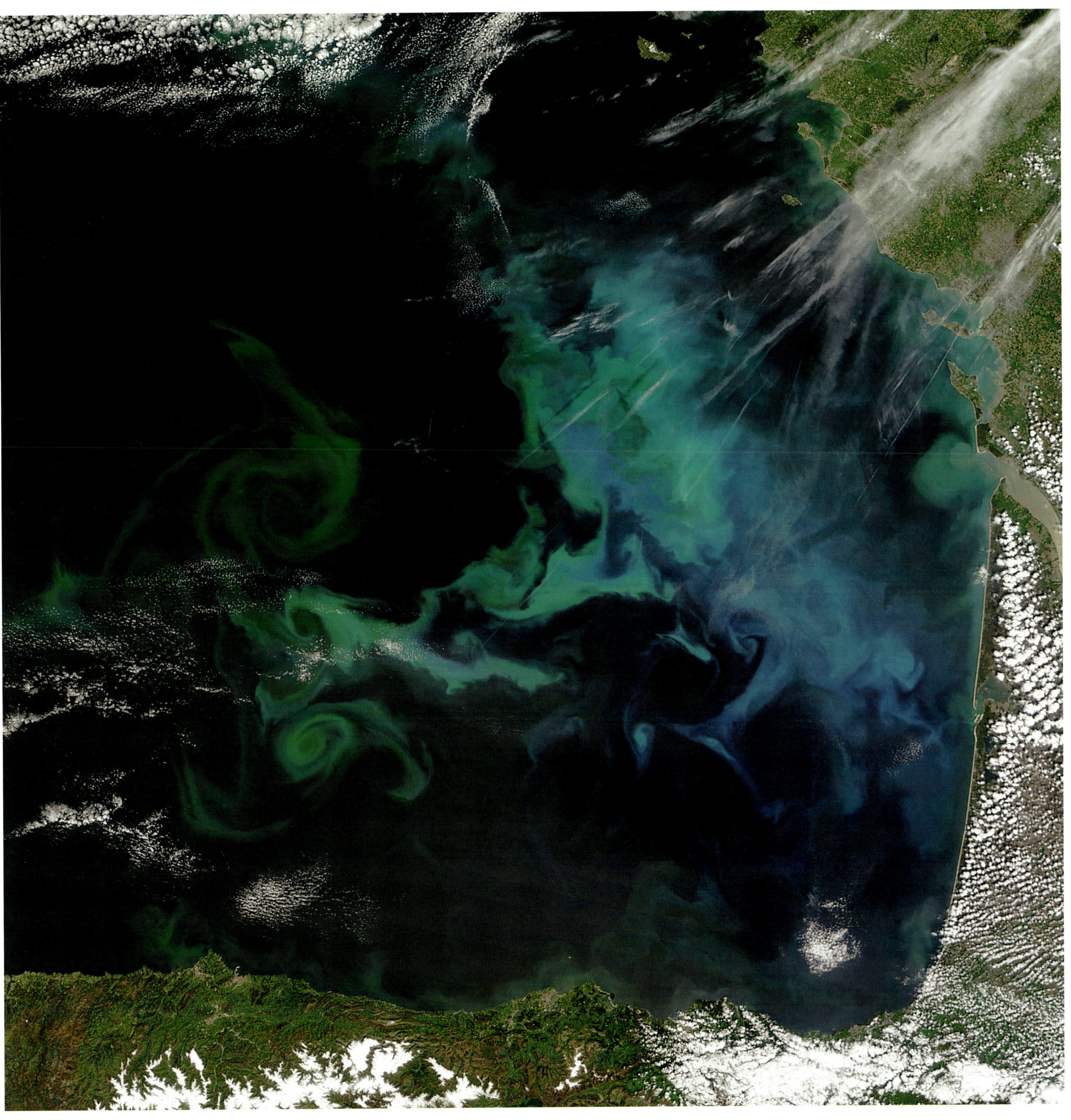

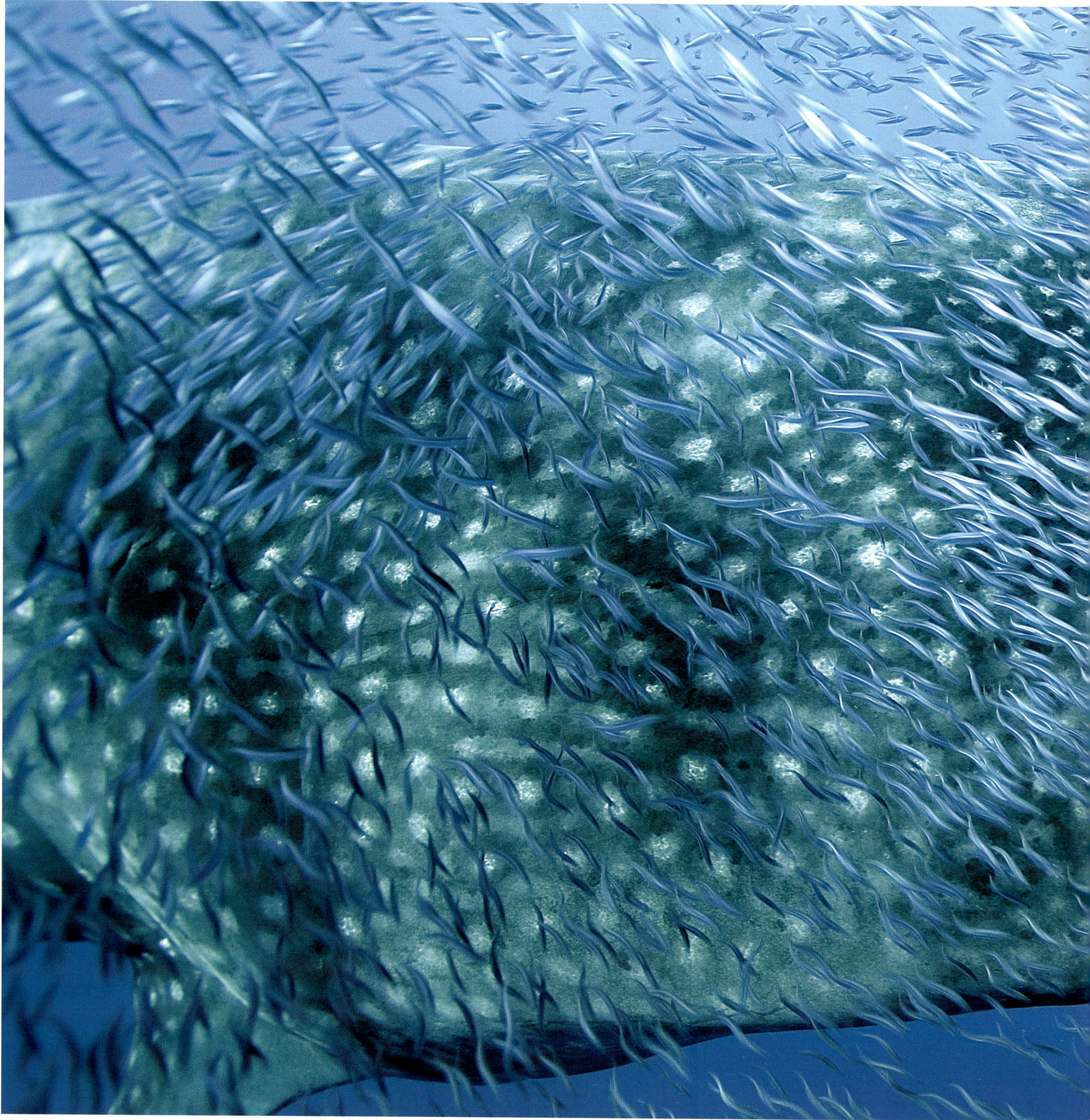

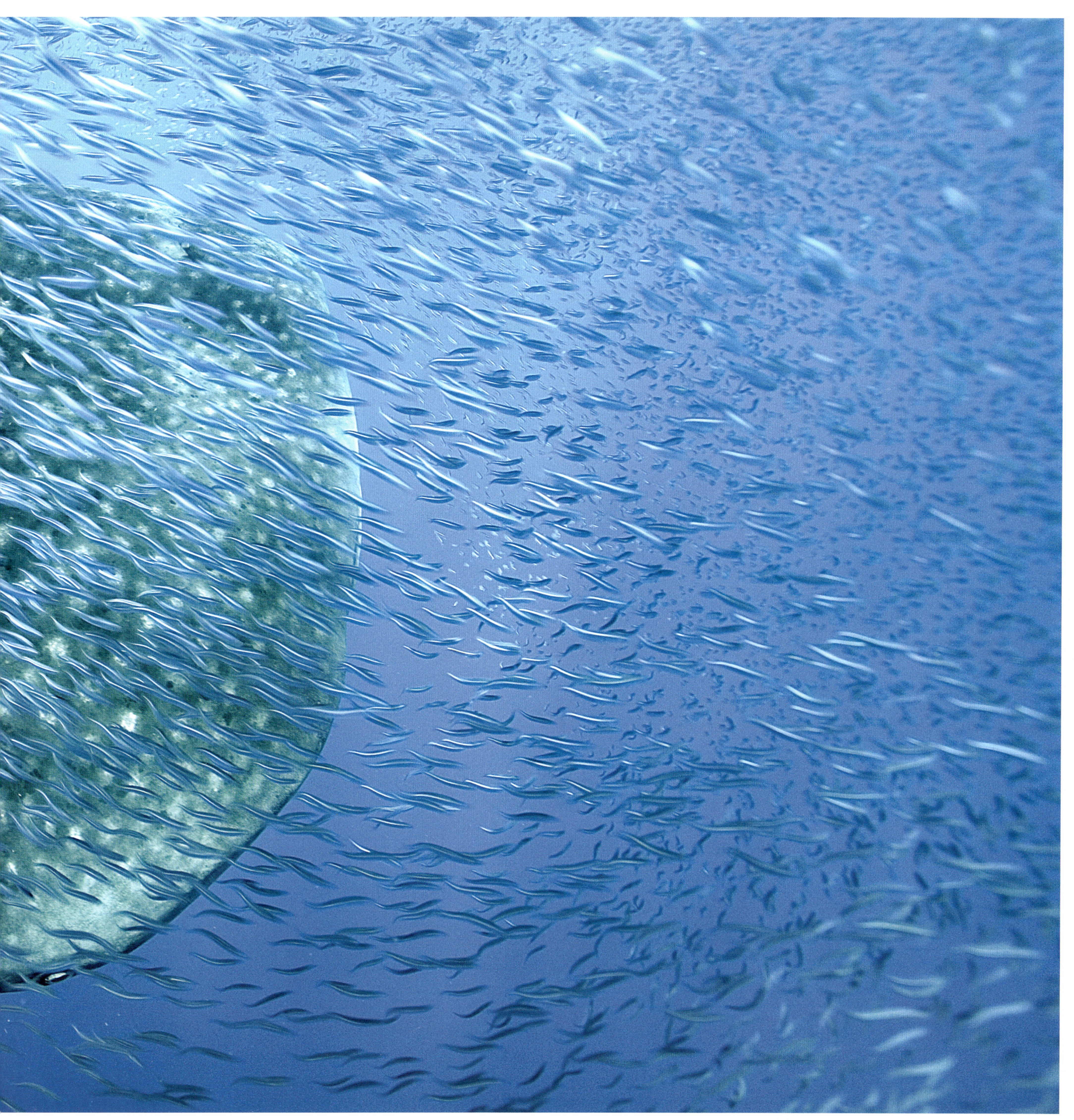

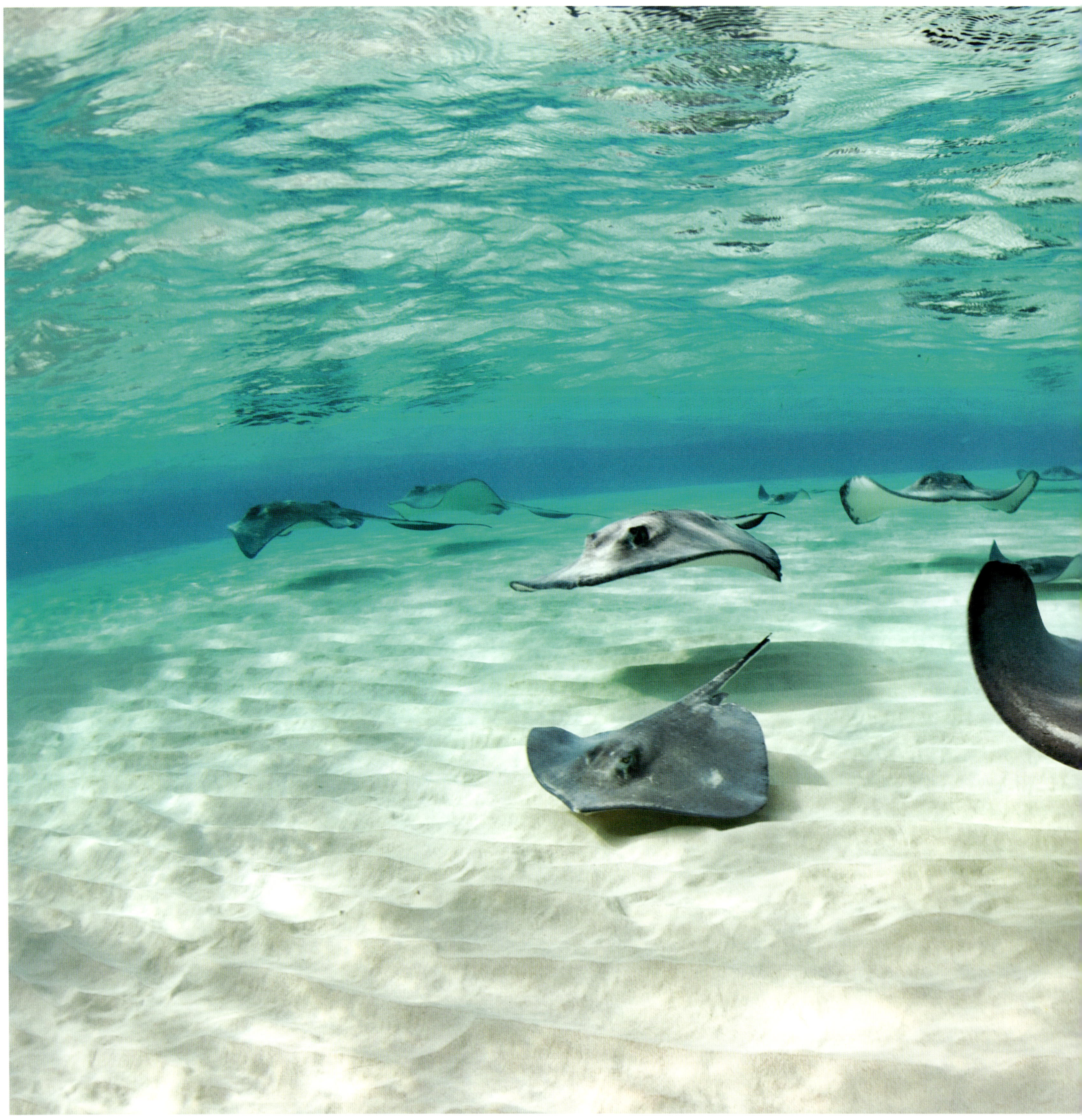

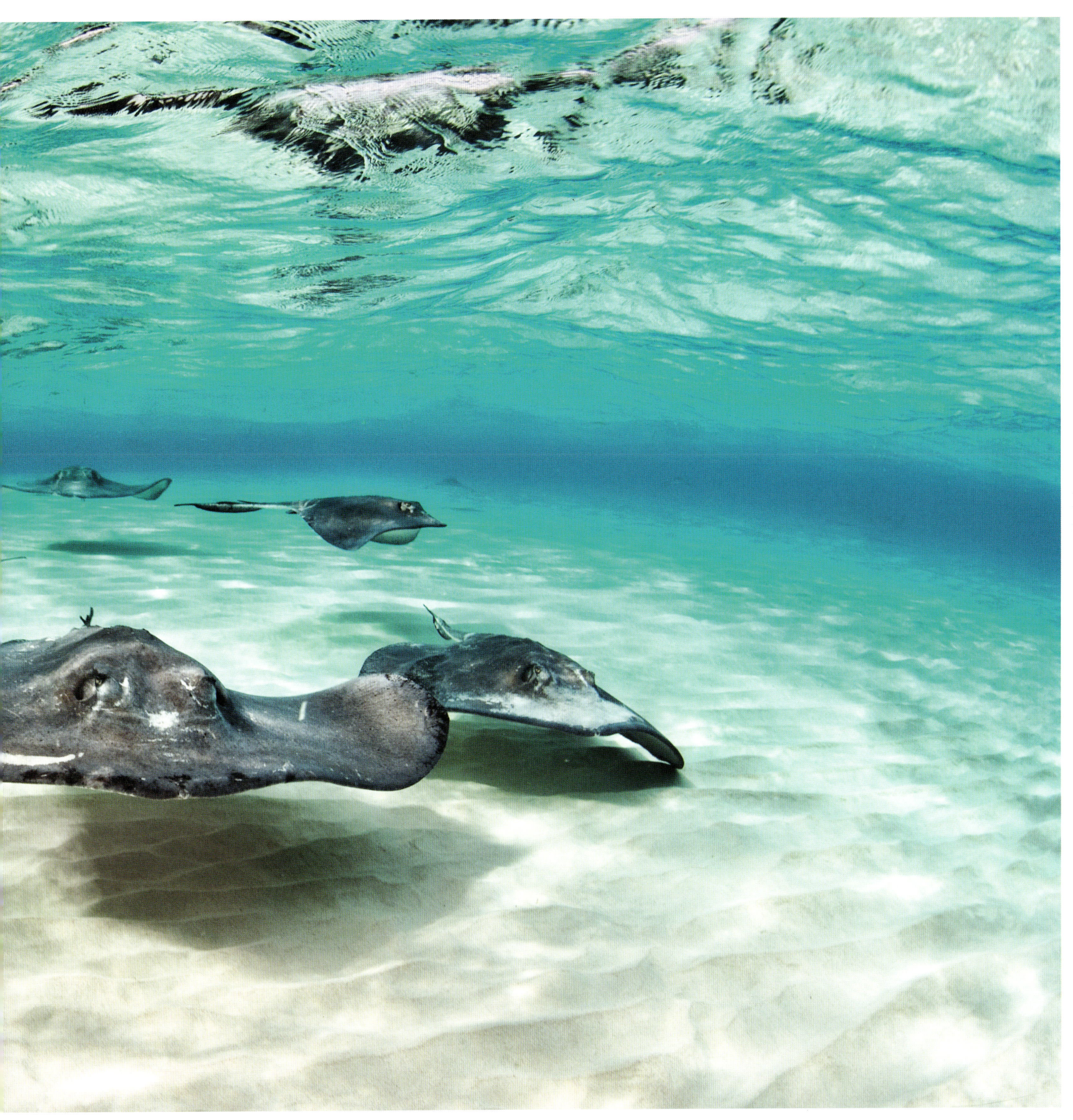

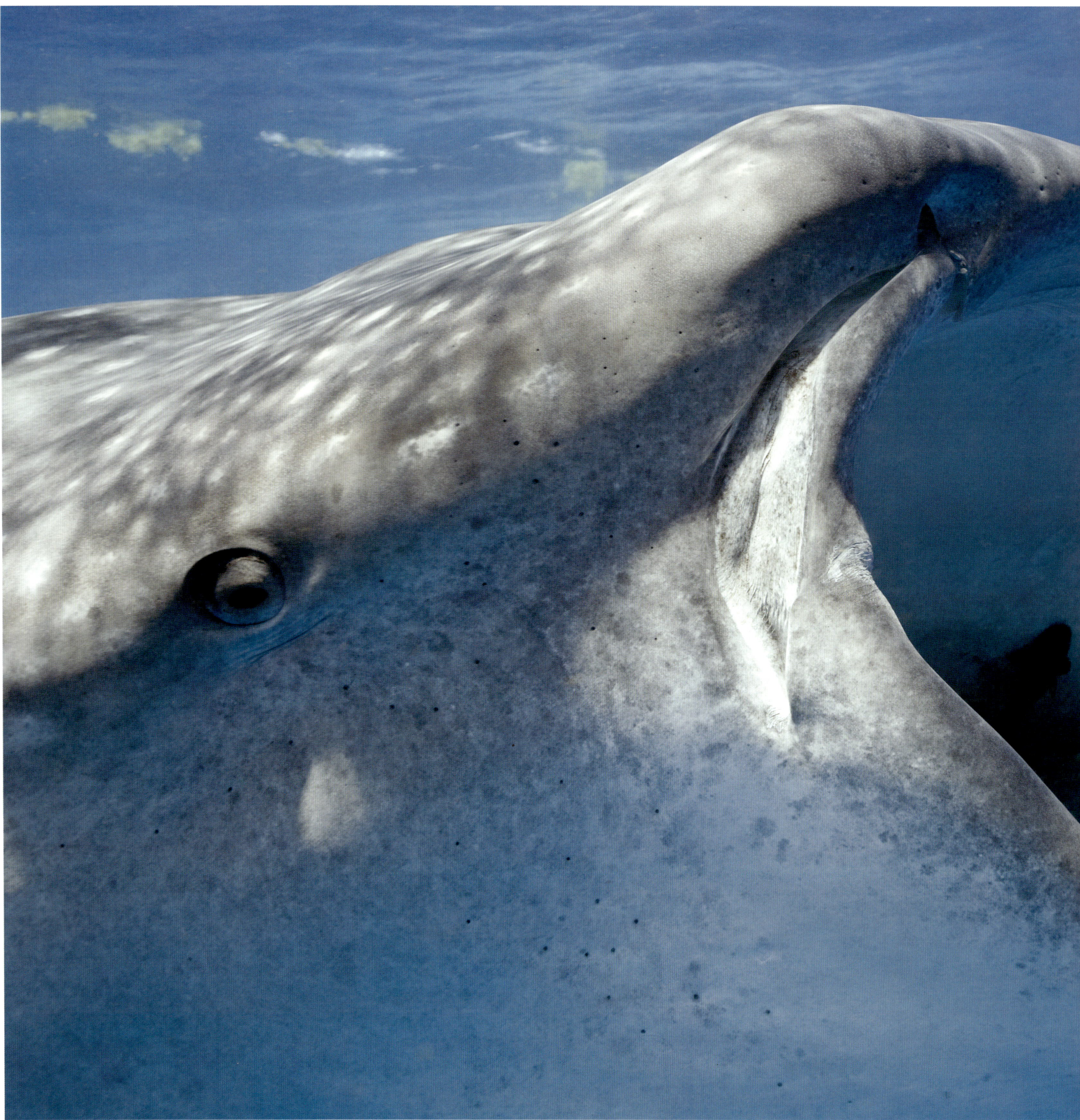

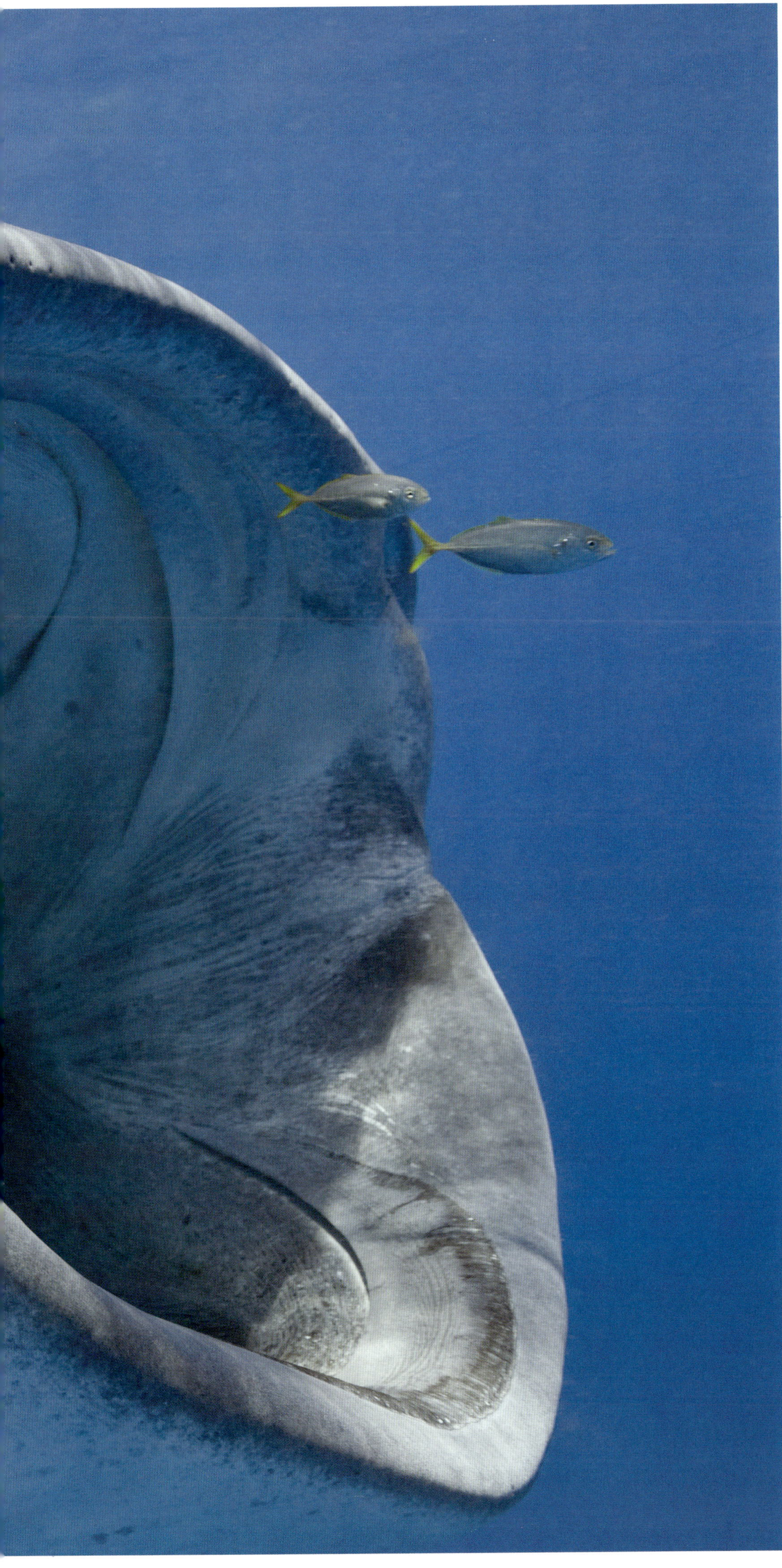

There is an even more invisible power at work in the oceans than the mysterious microscopic world that fills every drop of seawater with vital life. This is the world not of matter, however small, but of energy—all parts of the world are greatly impacted by how the ocean currents move heat around. Given that the majority of the sun's energy that reaches Earth lands on the oceans, this is no surprise. We are all impacted, whether living on the coast or a thousand miles inland, by the effect of ocean currents on climate. The ocean is like the central-heating system of the world, storing and circulating the sun's energy from one part of the globe to another, in ways that significantly influence the life that grows there. The major pattern is that around the world the ocean currents take warm water away from the equatorial region, cooling it down and moving it away to warm up the polar regions. When it gets there it cools down, sinks down in the water, and circulates slowly back toward the equator. In doing this, it is in a close relationship with the atmosphere, which also stores heat. The winds move to similar thermocline dynamics, from hot to cool areas, from high- to low-pressure systems, driving across the surface of the water and aiding the current. Water is constantly evaporating from ocean surfaces nearer the equator, causing rainy conditions and heating and cooling areas. The major ocean currents influence the climate, the winds, and precipitation, but these in turn have an influence on the oceans, speeding the surface water along. There are also other forces at work, notably the rotation of the earth and the tides caused by the gravitational pull of the moon. The ocean and wind currents tend to move clockwise in the Northern Hemisphere and counterclockwise in the Southern Hemisphere as a result of the Coriolis force. This is the effect of the earth's rotation from west to east and the fact that a point on the equator has farther to travel in a day than points nearer the poles. As a result, there is an apparent deflection of things moving over the earth's surface, bending to the right in the Northern Hemisphere (causing clockwise rotation) and to the left in the Southern Hemisphere (generating counterclockwise rotations). There is actually no force—it is simply the fact that things not rooted to the surface of the earth will move at a different speed to the earth's surface as they move away from the equator, unless they slow down to match the speed of the surface. The combination of the Coriolis effect and the pressure gradients explains the cyclonic patterns that develop in weather systems.

There are many currents at work in the world's oceans, influenced not just by the Coriolis effect, but also by the heat gradients and then further adapted by landmasses and changing depths. Vast amounts of water are constantly in motion in different ways, but there are six important currents that flow around in circular patterns, or "gyres." These are the Antarctic Circumpolar Current (or West Wind Drift), which rotates around Antarctica and is the largest ocean current, the Kuroshio Current off Japan (the second largest current, estimated as the equivalent of 6,000 large rivers), the North and South Equatorial Currents, the Peru Current in the Pacific, and the Gulf Stream, which flows from Florida to Newfoundland and then extends across the Atlantic to warm the western shores of the United Kingdom, all the way up to northern Scotland. The transformational effect of such currents can be seen by how, even as it peters out on the shores of the Scottish Highlands, the Gulf Stream is able to support the growth of the subtropical gardens of Inverewe, a quirky little

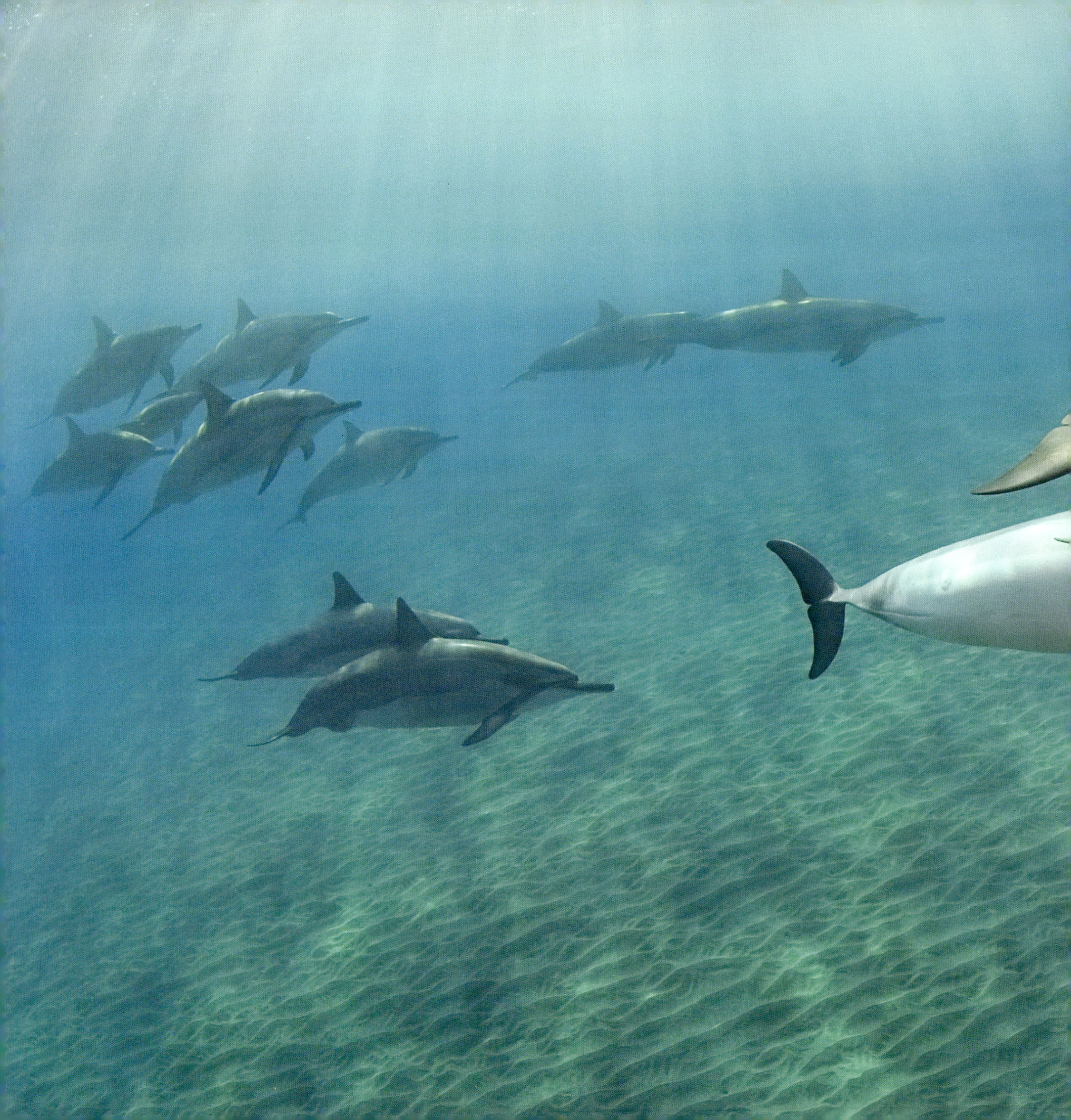

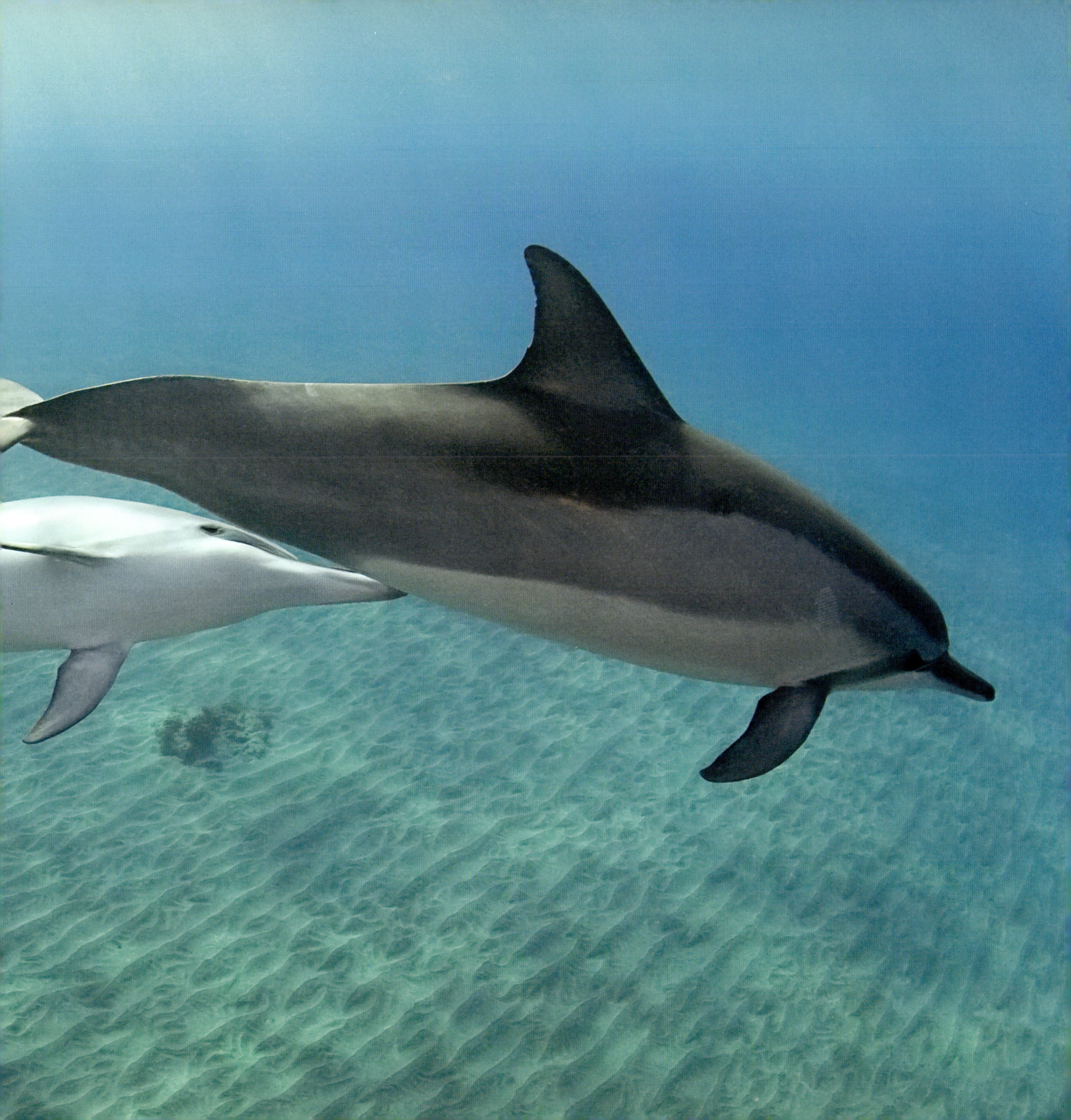

"Master, I marvel how the fishes live in the sea."

"Why, as men do a-land; the great ones eat up the little ones."

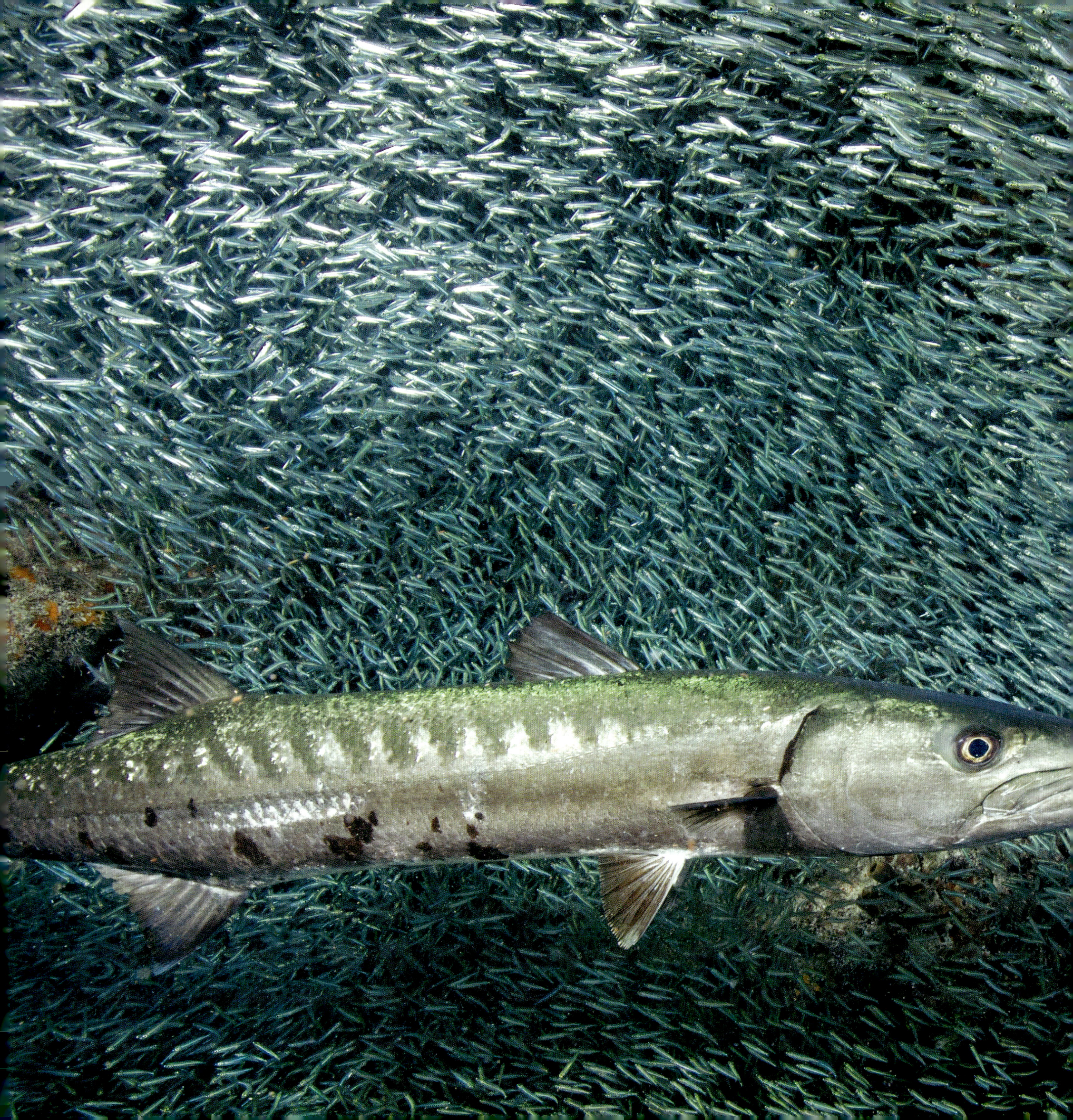

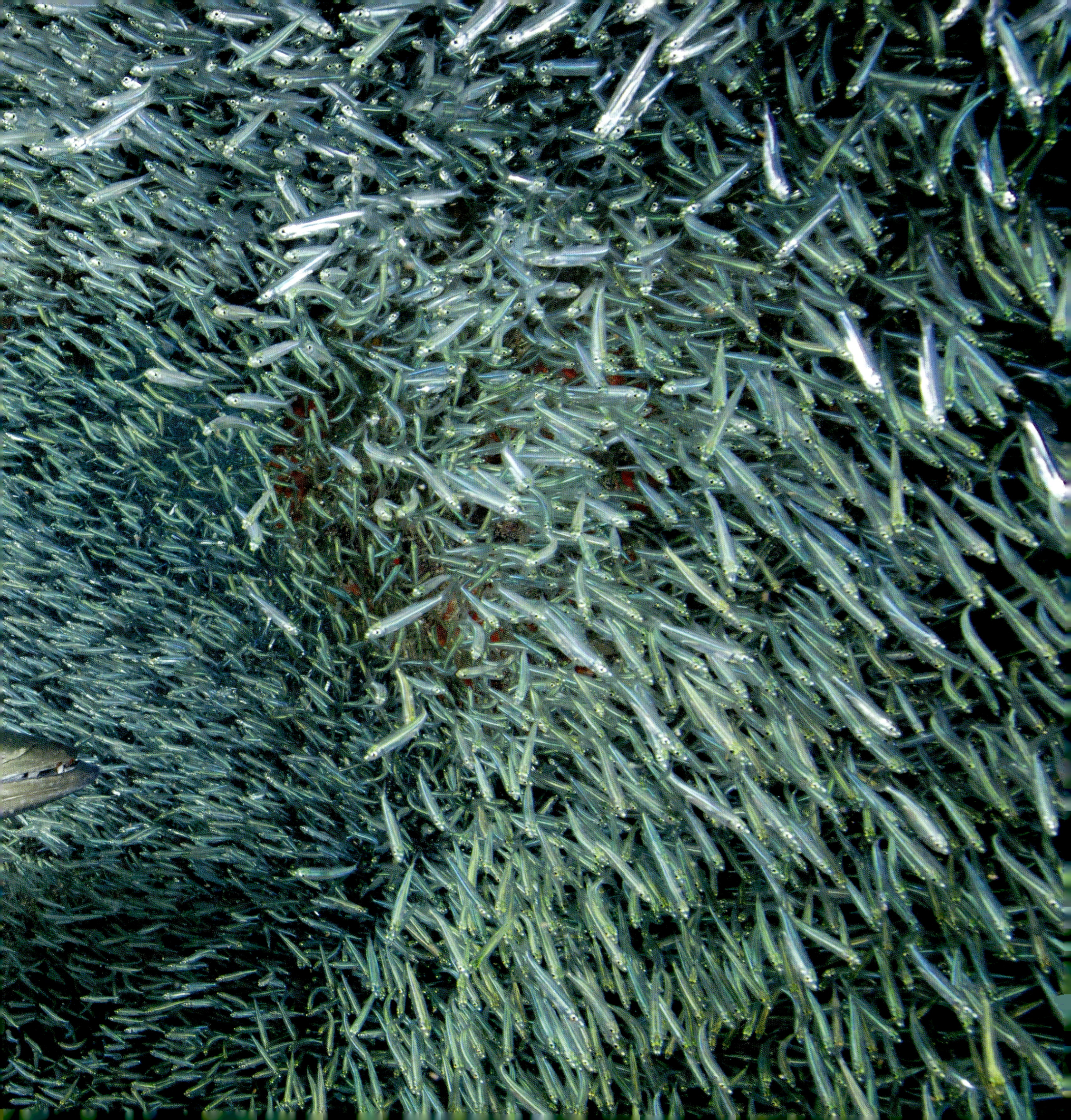

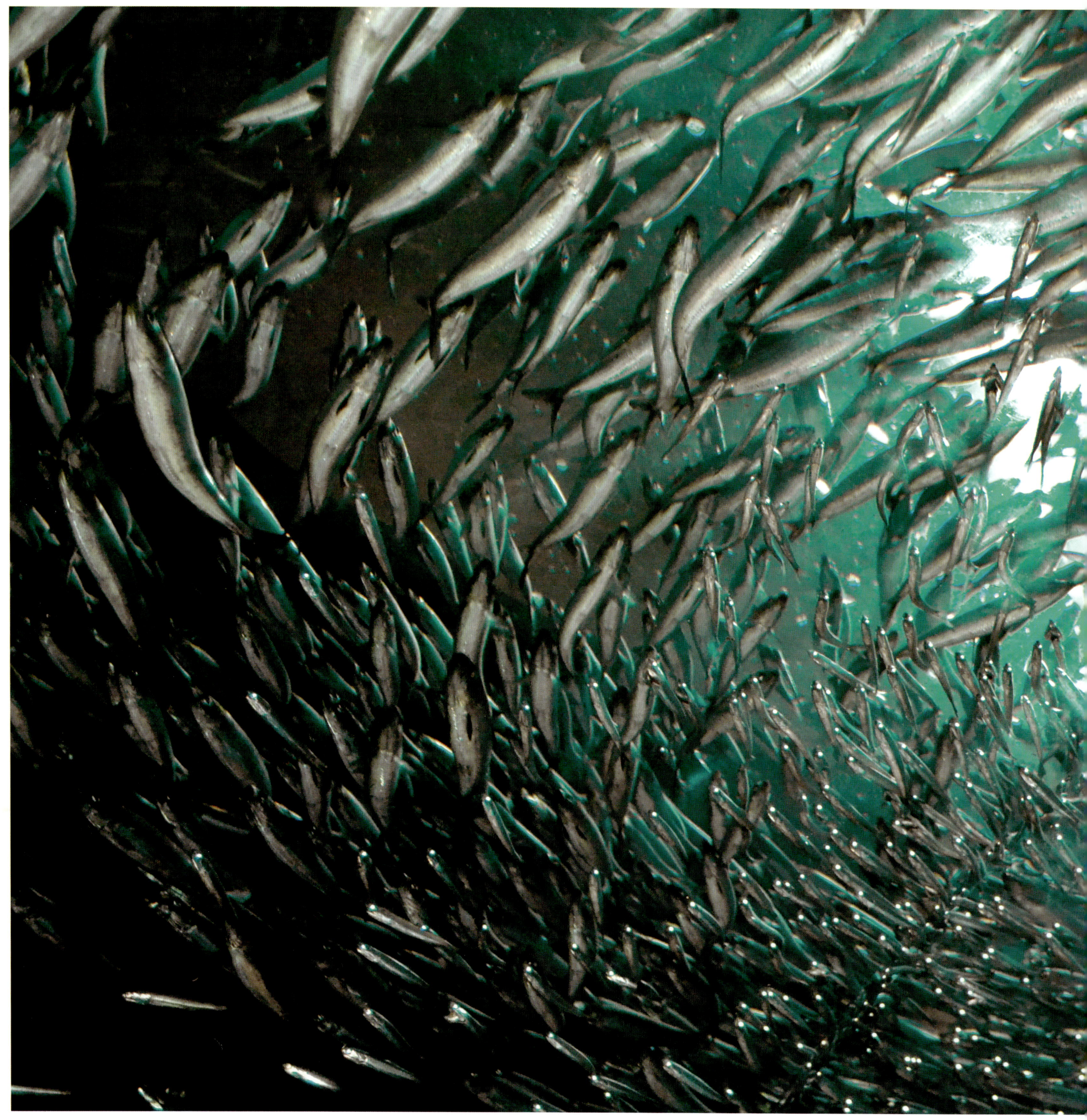

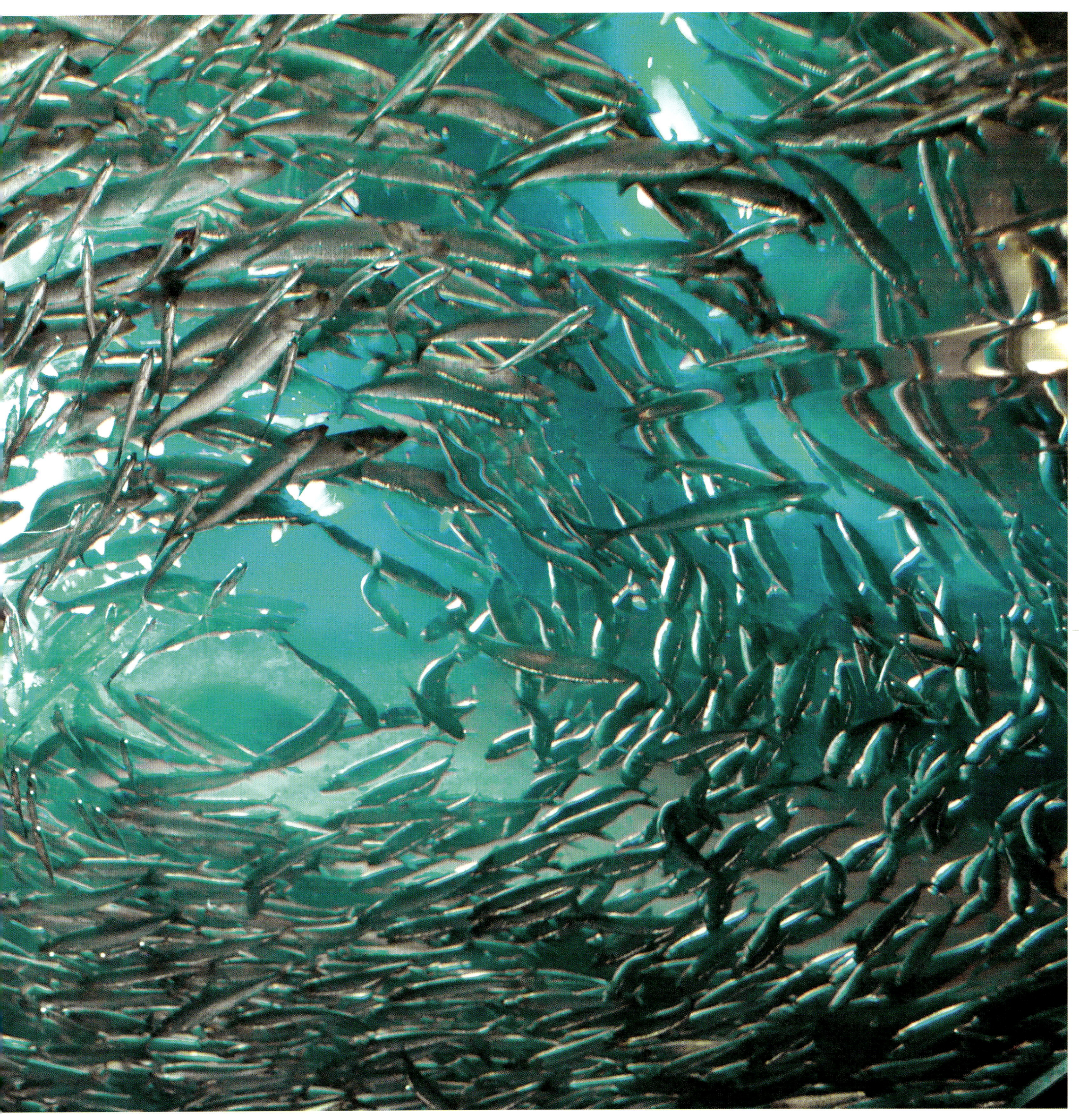

paradise of plants that would otherwise seem to have little chance of survival there. A few miles to the north or south or east and they would not live, which only goes to show just how the livability of our world is governed by the oceans. Climate-change experts, who fear that the Gulf Stream may shift or die out to a degree, project grimly cold, stormy weather for northern Europe if this happens, despite the general warming predicted.

These massive movements of water around the oceans happen at differing speeds, faster on the way out from equatorial areas when driven by heat energy, but slower on the migration back as deep, cold water. It has been estimated that for water to move through the whole cycle of ocean currents it could take almost a thousand years. This global conveyor belt has perhaps two key points of force: the heat shift from the equator, and then, at the opposite extreme, the impact at the poles of highly salty water being created as water freezes into ice, some of which is purged as freshwater ice, while the remaining, heavier water sinks. This creates a constant down-swell movement, driving deep currents slowly back toward the equator where they gradually rise to the surface, rise in temperature, and begin the cycle again.

But there are other forces at work, and one that we are perhaps most likely to notice is that of tides, where the waters of the earth are pulled by the magnetic forces of the moon and sun. Tidal movements are in effect massive waves, with the waters of the oceans being pulled into two peaks of high tides on opposite sides of the earth. The moon and sun move their relationship to the earth over the year and this affects the strength of the tides. When the moon is nearer it has the strongest influence on tides, but it is the combination or otherwise of both the moon and the sun that influences the strength of the tides over the year. Spring tides occur when the two bodies complement each other to make the strongest gravitational pull, creating the largest tides. Neap tides are what happens when the sun's and moon's pulls cancel each other out to a degree, creating the lowest tidal movement of the year. By no means are tides the same the world over: Some places have two fairly equal tides a day, which is called semidiurnal; some, only one tidal reach a day, or diurnal; and some, unequal tides, which is known as a mixed tide. And there is a big difference in the height of tidal movement. In the Mediterranean, Baltic, and Caribbean Seas there is little tidal movement to see at all in many parts, whereas the Bay of Fundy and Ungava Bay on the east coast of Canada are rivals for the title of the largest tidal height differences, in excess of 55 feet (17 meters) between the high and low points.

Tides shape a unique world at the shoreline, the intertidal ecology. This is an ecosystem that is sometimes of the sea, sometimes of the land, and can be both at the same time. It has its own unique flora and fauna while also sharing organisms with both sea and land. It is an ecosystem that is the most readily available for us to get a clue about what's going on in the oceans, and for many of us is that first real encounter. This is the world of rock pooling, of beachcombing, and much more. It's where many of us discover a fascination with things of the sea. The life that we see here has adapted to deal with some of the most stressful conditions: It may be submerged or exposed, it may be pounded by waves or fried by sun, it may be subject to varying tidal ranges over the year, it may be attacked by predators with very different attributes, depending on whether they are waterborne, land-based, or of the air. The very nature of the water may change, with salinity varying if water dries up, becomes mixed with brackish water from a river mouth or pure water from rain, or due to other land-brought pollution. With approximately 40 percent of the world's population living within 62 miles of the coast, there are great pressures on these precious environments.

These factors have stimulated the evolution of a range of unique life and also a gathering of life that varies over the period of the tide. The intertidal zone is like a city-center plaza, with changing occupants and functions during the day: for the office workers eating lunch, the afternoon tourists enjoying an ice cream, the brisk commuters, and finally as a hot spot for nightlife—instead imagine fish seeking food being replaced by crabs scavenging as the tide drops, and birds picking over the rock pools or plucking worms from the mud. The analogy of a busy city center is also appropriate because the intertidal zone is often extremely crowded—there is a real pressure on real estate, with the growth of a community highly influenced by the extent of rocks, mud, sand, and vegetation occurring between low and high tides. Intertidal space goes from being a three-dimensional environment of water over a seabed to being a more two-dimensional experience, restricting or certainly changing the living space available. This land-sea interface is also one of the most stressed areas in need of conservation, as human influence eats away at precious edge environments, such as wetlands and mangroves, while river pollution is at its most concentrated in its effect on the sea. The careful balance of life that exists in these environments can be demonstrated by looking at a mussel bed, often to be found growing extensively on rocks or a jetty. In this apparently simple clustering of a member of the mollusc family— a bivalve that filters the water to feed off microscopic life for its food —we can see the nexus of a complex range of other life. Mussels attach themselves to rocks by what are called byssal threads (cooks call this aspect of the mussel a "beard," but it is actually the base rather than the head of the body). They can move very slowly by making and releasing these tether points and creeping along a surface. But they have no need to move fast—the food comes to them as the water washes over and is dragged through their filtering system. When the tide ebbs they close up and hunker down, hoping to avoid the attention of any passing gulls. As the mussels get established, so other creatures start to live on and around them: Mussel colonies form a three-dimensional habitat for other small creatures and plants to fit around them, such as snails and worms and small crustaceans. The mussels are not entirely defenseless. They can use those byssal threads to tie down slow-moving predators, such as the dog whelk, rather like Lilliputians tying down Gulliver. This small-scale, slow-moving link in the chain of life feeds off even smaller life while a whole range of other creatures feeds off them. Delimiting the range of the mussel bed overall will be issues such as the availability of a firm surface, and also the range of predators, such as sea stars, who will not want to be left overly exposed above the tide.

This daily movement of the water between its low- and high-tide points, in turn fluctuating in cycles over the month and year, creates a variable that has shaped a whole range of unique ecosystems that have adapted to specific intertidal conditions around the world. What

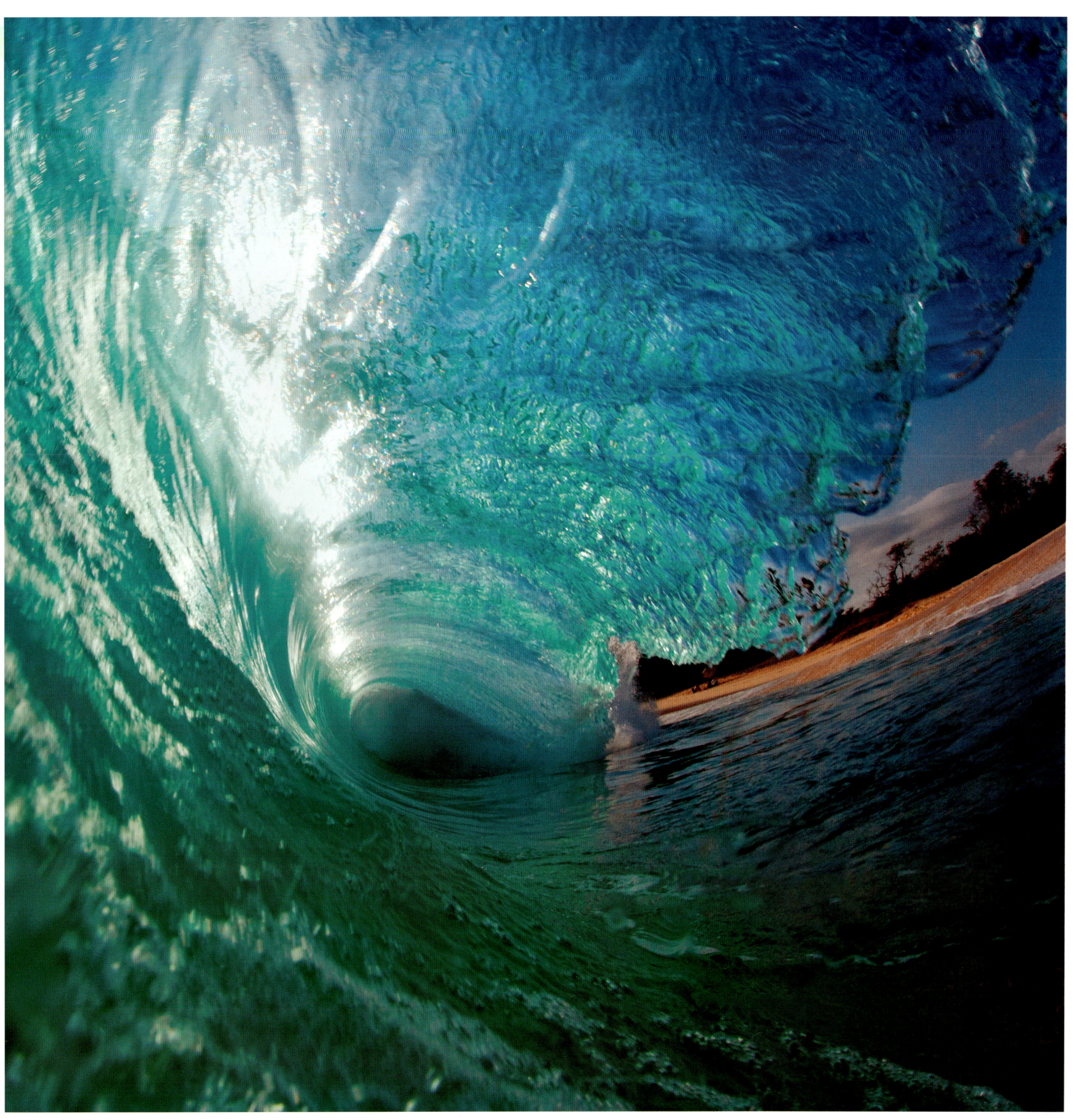

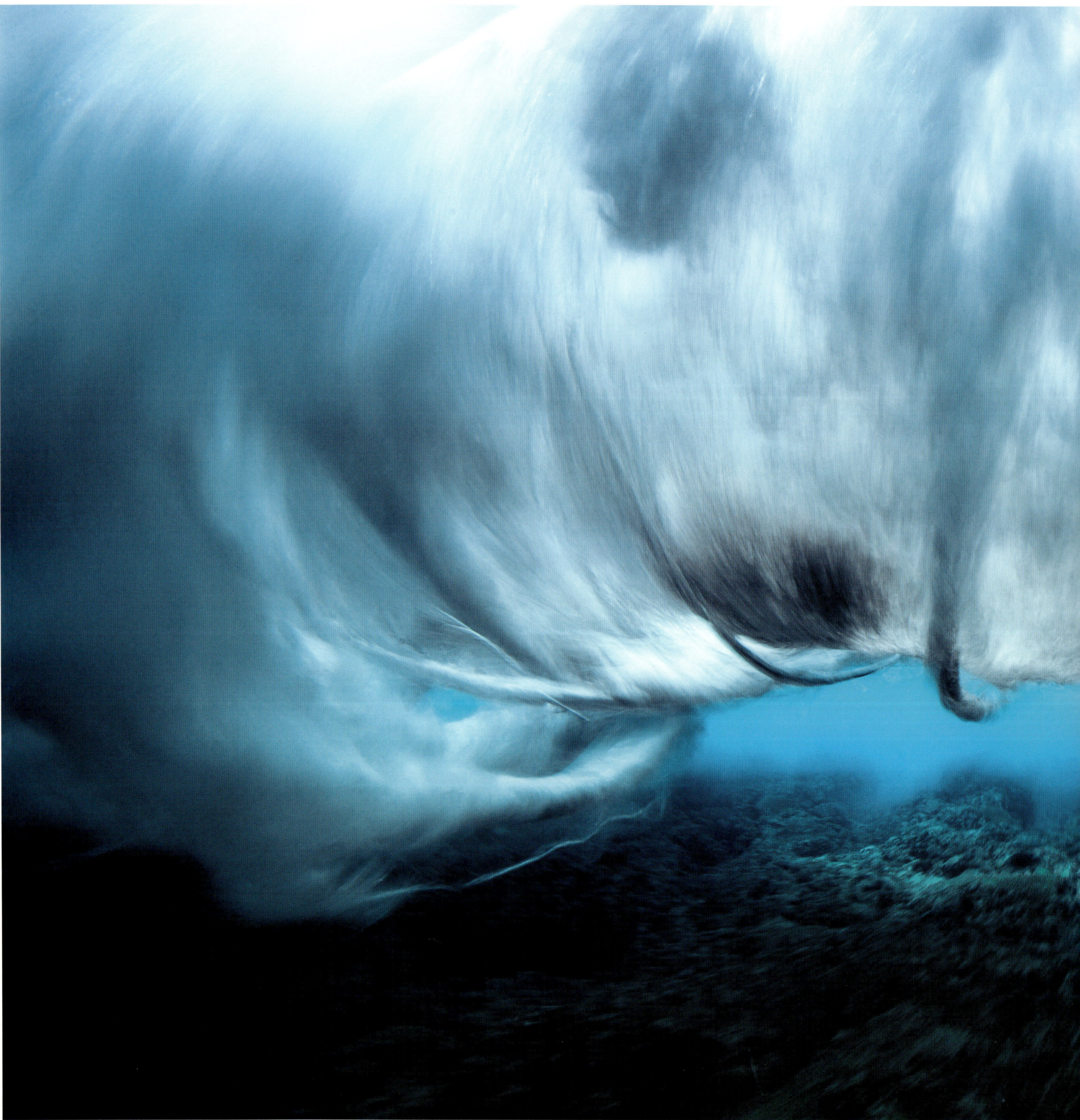

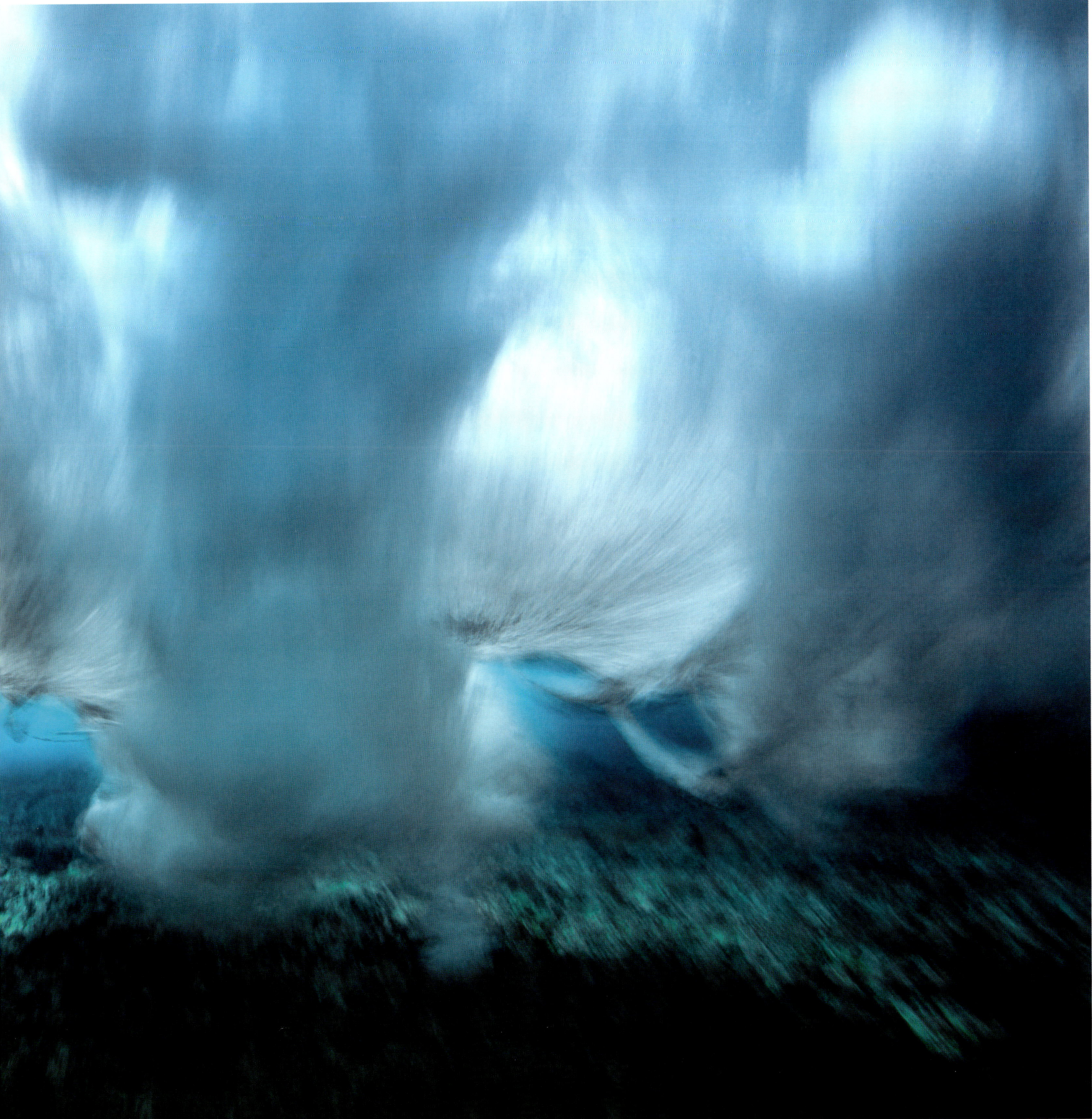

happens on a temperate, rocky coastline is different to a mangrove swamp, a coral reef, or an icy beach inside the Arctic Circle. But the very different inhabitants share in the way they cope with a wide range of opportunities and challenges to survival that contrast with life a few feet either side in the permanent sea or on the dry land.

Wave power is the final, ever-present property of the ocean's waters that we will address in this chapter. While tides are the biggest waves on the planet, we don't tend to think of them as that. And tsunamis, the gigantic, devastating waves that result from seismic activity, are also so irregular in their origin and distinct in their properties that we do not see them as typical waves. We reserve the idea of a wave to the motion given to the surface of the water by the winds and currents. Wind passing over the surface of the sea transfers some of its energy, and this can lead to anything from a ripple to a major swell that might be up to 50 feet (15 meters) in really big seas, with occasional reports of waves more than 100 feet (30 meters). But measuring waves out at sea is, of course, rather difficult and somewhat liable to exaggeration. Some of the largest waves are not simply caused by the wind. Rather, currents that are in some way clashing may combine with added storm power to reach freakish proportions. Extraordinary low- or high-pressure variations somewhere else on the sea may actually begin the process of large waves, which are later experienced a long way from the origin of the swell.

Just occasionally, rogue waves can be seen leaping right out of scale from the surrounding waves—sometimes to devastating effect if they hit a ship unexpectedly. These super-waves are also the stuff of surfing legends when they can be spotted and somewhat predicted near land. Wave-chasing extreme surfers have taken to claiming their place in the record books with videos of derring-do, often being towed into the path of the wave and then risking potential death should they wipe out while riding a board under a collapsing wall of water. In the cases of the more predictable massive waves near land, the size of the wave results from a long "fetch" of wind over the ocean, which then combines with a particular seabed formation offshore that forces up the final flourish of giant surf. Hawaii and California are famous for big surf, but recently the small seaside spot of Praia do Norte, Portugal, one of the most westerly points in Europe, emerged as perhaps the gold spot for such waves (at least until an even more pronounced effect is found). In the autumn, strong winds across the Atlantic combine to force water toward shore, and in part through a massive underwater canyon that runs 125 miles (200 kilometers) out to sea from almost more than half a mile, reaching a depth of 16,000 feet (5 kilometers). The onshore current is channeled toward the coast through the canyon until it reaches the final headwall of the canyon only 160 feet (50 meters) or so below the surface, near the cliffs. As a result, this kicks up a massive force in the water that drives up giant waves, with some of around 100 feet (30 meters) high now being successfully surfed.

Five factors are involved in the creation of any wave: wind speed, the "fetch," the width of the wind, the duration of the wind, and the depth of the sea. These are the elements that shape the amount of effective energy being applied and how that force will be channeled into the wave structure. The wave can then be defined in four dimensions: height (from trough to crest), length (from one crest to the next), period (the time it takes for sequential crests to pass a stationary point), and propagation (the direction of travel).

Vast amounts of energy pass through the seas at any time as a result of this transition of wind power. For a small ripple, the forces of surface tension will equalize and settle the water. For most wind, though, chopping up the sea or producing larger swells, it is gravity that acts as the force leveling out the wind action, pulling down the crest, and thereby creating the movement of the wave and ultimately dissipating it in time. There is an illusion of a great forward movement of water, but it is actually moving up and down; the wave of energy is what passes on, not the water. The water largely stays in the same place, except for a bit of up-and-down movement. If it did not, if it actually moved like the illusion of travel that we see, then all the water would pile up on the shore somewhere, along with all the sea life in it. Fortunately this does not happen. Instead, the energy is what travels and it can be an immense and rapidly moving force, often traveling over thousands of miles of open ocean, spreading the impact of a weather system in one part of the world across to a different continent.

The surface movement of the water with a wave belies the fact that the motion extends downward much farther—7 feet (2 meters) of visible wave may lead to significant movement in the water below that extends 20 feet (6 meters) or more. While this could be destructive if waves get too large, in general the motion is beneficial to life, helping organisms clean themselves or feed, while also playing a vital role in oxygenating the water.

The value of waves to marine life is impossible to quantify as we still know so little about so much of the oceans, and about the waves and all the species that are affected by them. But the value of waves for surfers is easier to calculate. As part of various attempts to preserve great surfing beaches and the conditions that lead to the waves, surfers and their advisors have demonstrated that their transient wave-chasing communities can be worth many millions to an area and need to be respected and protected. There's even an emerging academic discipline called "surfonomics." One PhD student in 2011 estimated that in the United States alone 3.3 million people contribute in the region of $2 billion annually in following the waves. If the movement of the oceans is worth a lot to us just for entertainment, we may expect the evolutionary significance for marine life to be considerably more vital. Priceless, indeed.

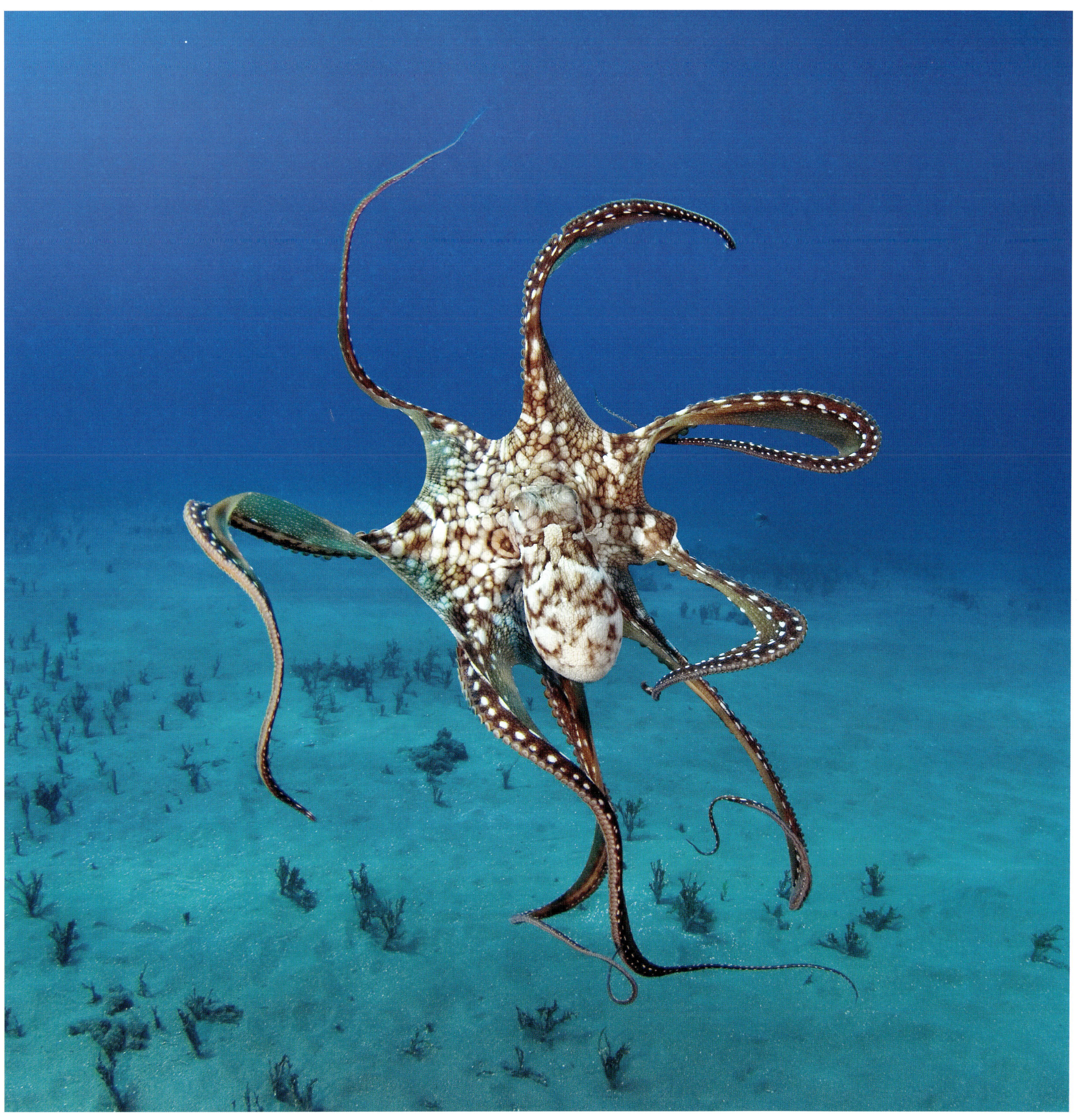

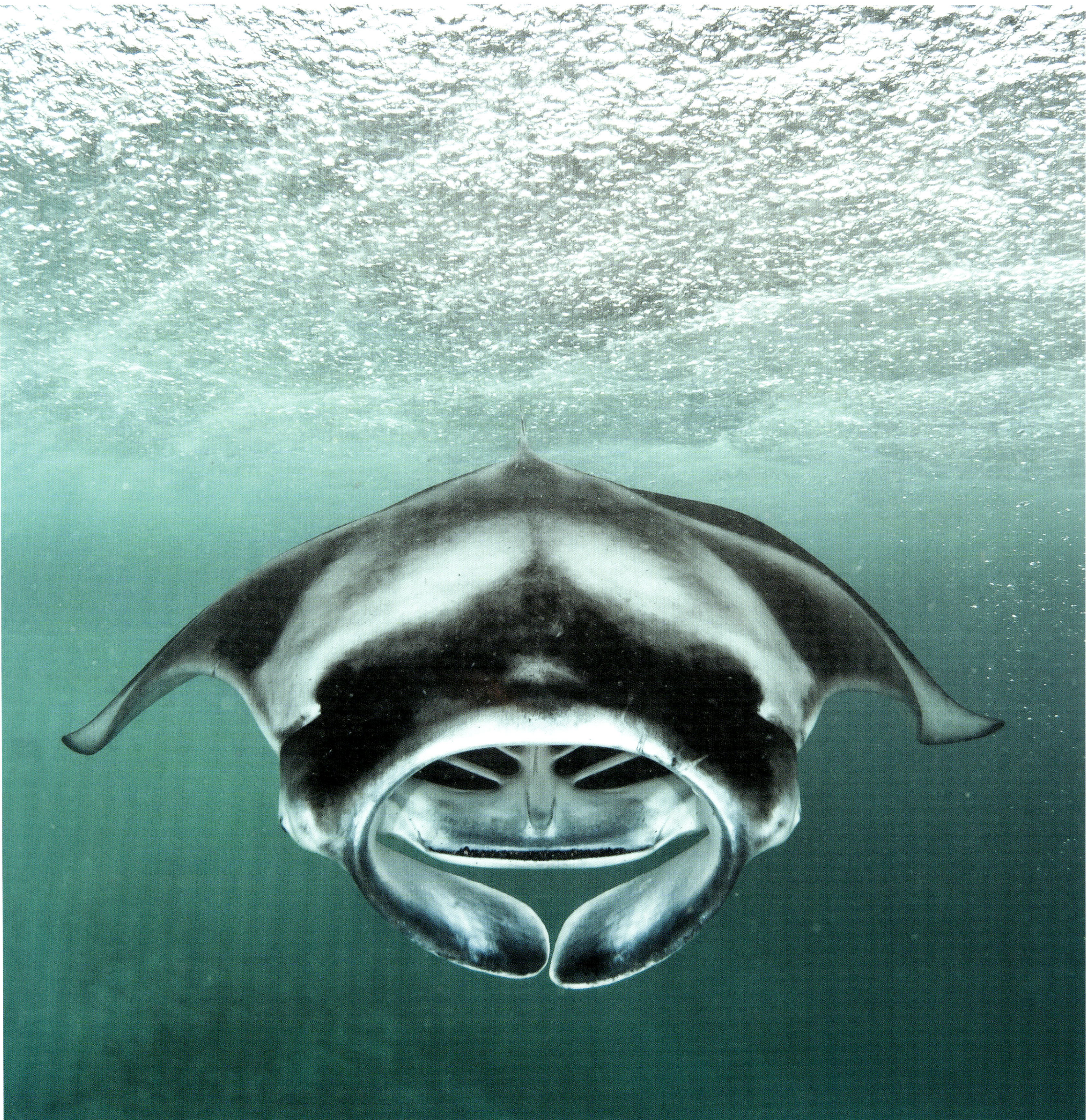

My soul is full of longing for the secret
of the sea, and the heart of the great ocean
sends a thrilling pulse through me.

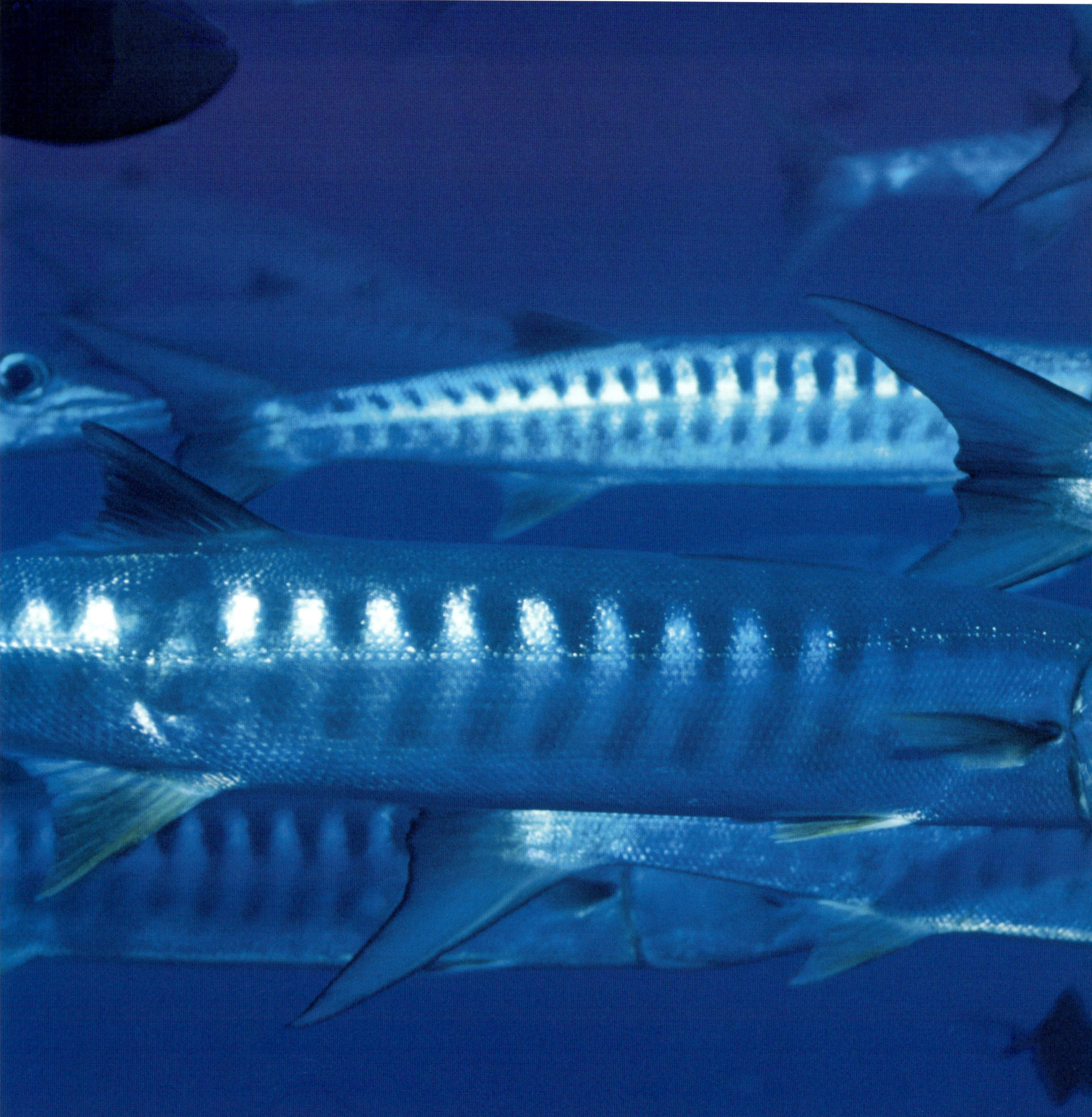

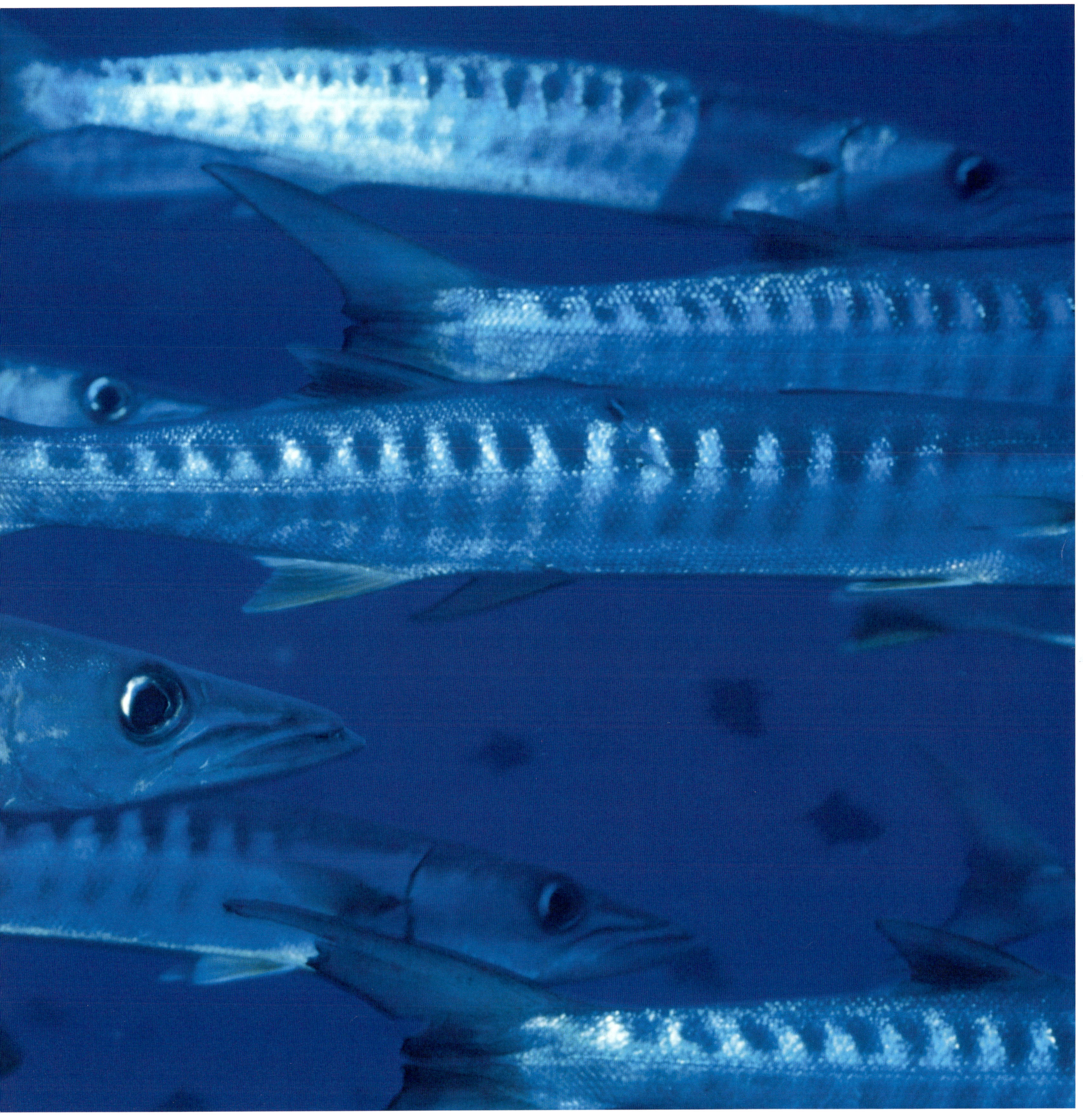

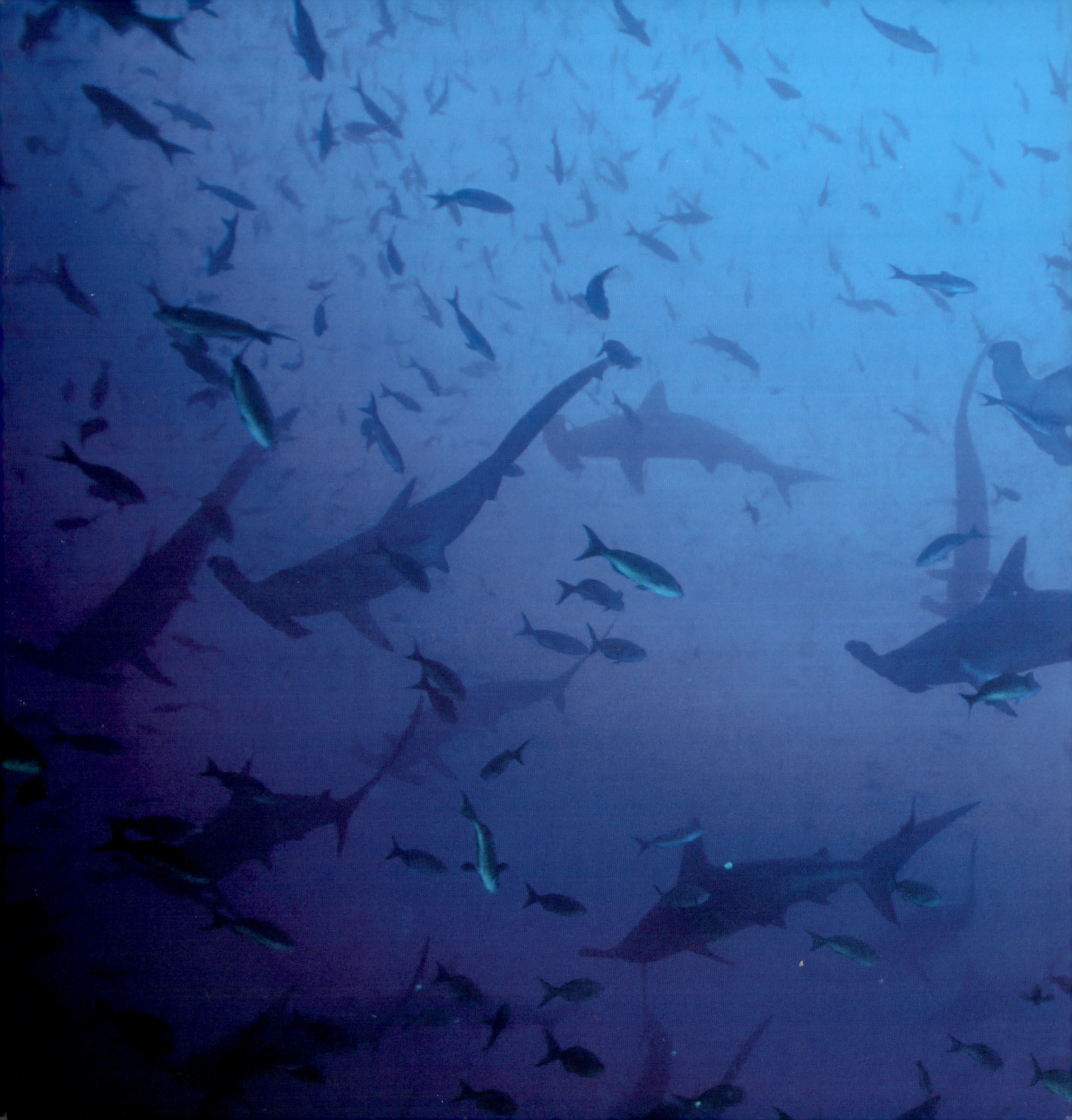

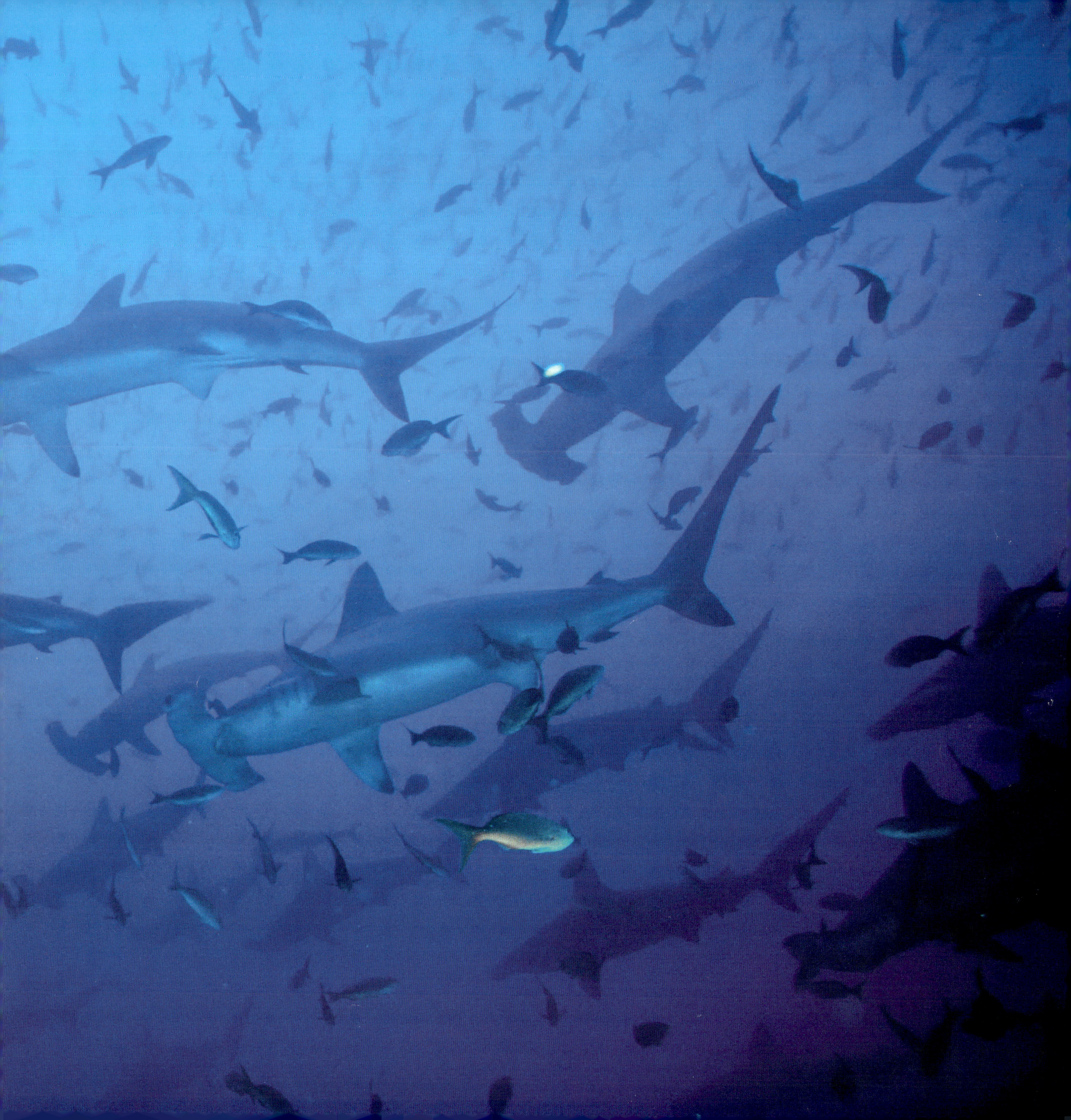

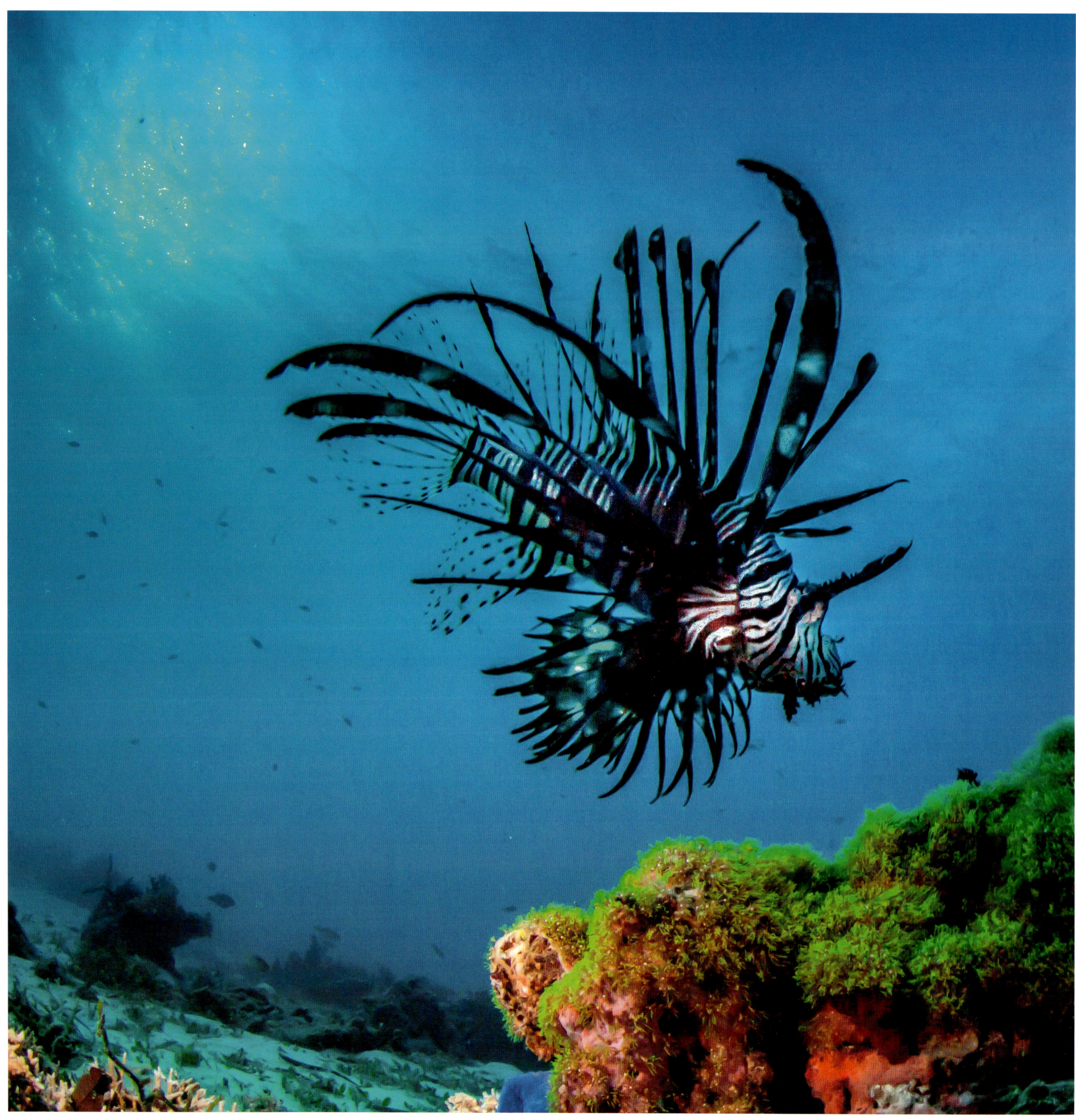

No aquarium, no tank in a marine land,
however spacious it may be, can begin to duplicate
the conditions of the sea.

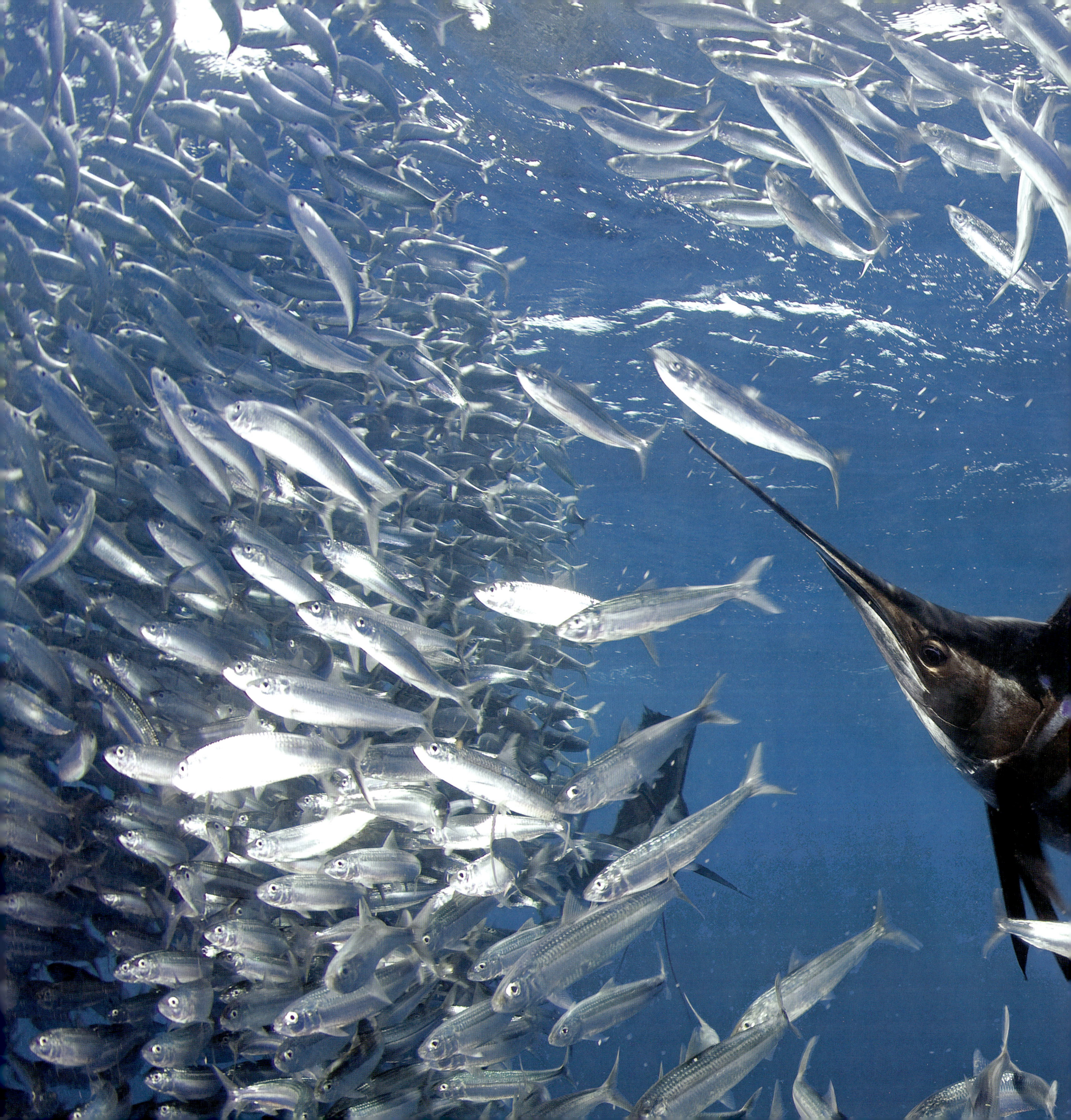

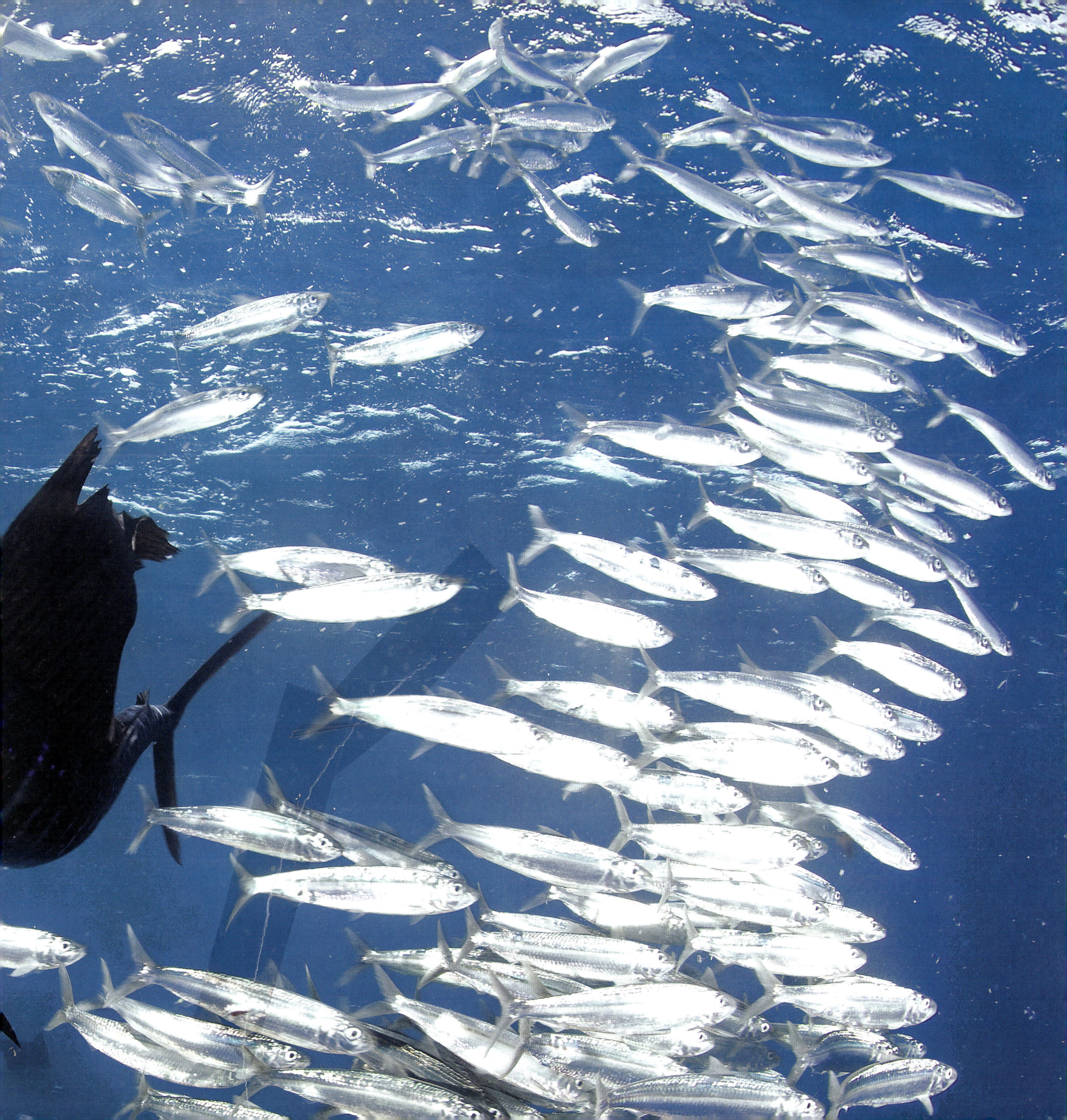

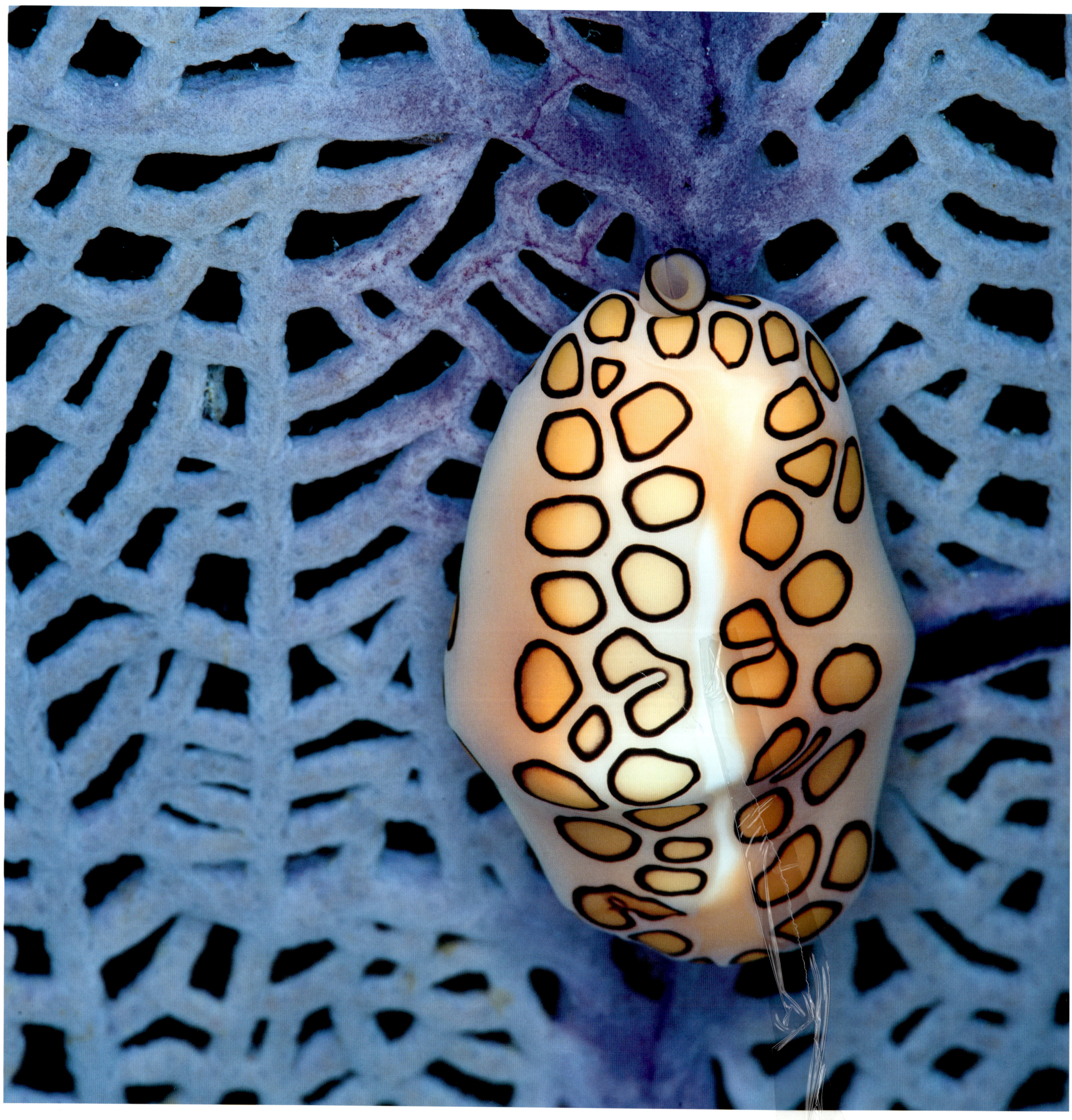

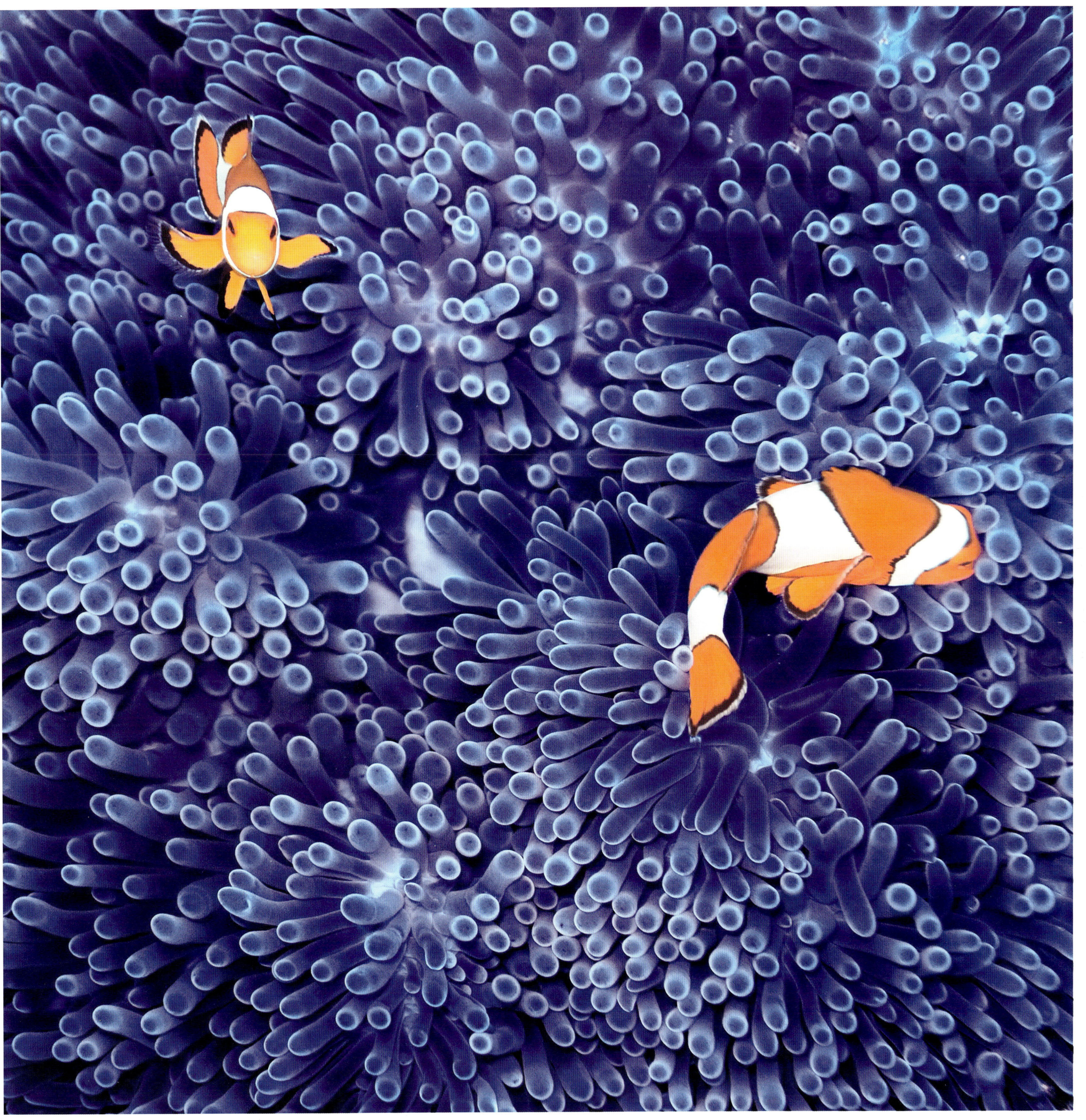

Wild

Our seas are the ultimate wilderness. So wild that they challenge the very notion of "wild"; are the seas untamed, unrestrained, and savage? Yes, they can be all that and yet they are more. We can't conceive of really taming sea creatures or cultivating the waters in the way we do the land and its animals. The oceans operate by their own rules, their own descriptions, that have yet to be parceled up by our language. They are full of the undiscovered in a way that we no longer experience on land. The dark side of the moon is said to be more surveyed than the bottom of the oceans. With that in mind, the oceans seem beyond wild—we need a new word for what is out there, down there. In recent years we have seen an explosion of interest in getting closer to wildlife, even to the point of wanting to "rewild" various countryside. While some parts of the seas need urgent conservation efforts, with pollution an ever-rising concern, most of the earth's surface is a super-wilderness of water life, still impervious, thank goodness, to our intrusion.

This will only add to the allure of the oceans—we are a curious race, our curiosity is crucial to our success as well as a threat to our survival—we will want to know more. People have risked and often lost their lives in attempting to get ever closer to whatever their idea of the true wild on land and at sea might be. For those who want it safer and more comfortable, there are films, television, websites . . . and, of course, books. There is no end of places where we can encounter raw nature recorded and celebrated. However, this coverage is inevitably focused on how one species—our species, *Homo sapiens*—can encounter the others. For this reason, the majority of wilderness that we see is typically land based. If out on the water, the sights are mostly things you can see from a boat or by poking your head below the surface of the water with a snorkel attached.

Whether we are out there encountering it for ourselves, or encountering through media, too often we fail to note that there is a contradiction in the very idea of experiencing the wild. The thing is, to be "wild" means, at its most pure, that the subject is beyond our influence, beyond any adjustment introduced by interacting with humanity. In this the sea may become the true wild, the place that we still don't know, where what we know is that we don't know what is there. The vast quantities of ocean—most of it, in fact—has life moving in it in ways that remain highly mysterious. There are many species still to discover, with most depths never plumbed. There is nowhere wilder. It is untouched by us, and unknown. It is easy to search out images and ideas of this world that present the uncomfortably strange. The "other" is at best unsettling, and often seems dangerous, even terrifying. But seen another way, we are on the outside looking into a near-infinite world of possibilities and wonders.

Most of the creatures and plants that share this planet with us are actually out of sight, for all the apparent congestion in the spots we inhabit. Consider again that figure of 71 percent of the surface of the earth being the seas, and then let us muse on how most of the space in the oceans is both over the horizon and under the surface, going down an *average* depth of about 14,000 feet (4 kilometers). Wow. We quickly get to a sense of, well, trying to comprehend the incomprehensible. These vast, shifting waters are a realm that is alien to our bodies and inexperienced by all but the deep-sea divers among us, and then only in a sampling of random spots. In my outings as a scuba diver, I have never been far from a rock and a sense of the light descending from above. On the occasions when I have looked down into seemingly bottomless depths, there are intimations of vertigo and of something inside my mind warning me to beware. For all the thrill, there is a primal sense that this is not our element. And look at us, why should it be? We have evolved to thrive on land, and only with various artificial devices do we manage to move around for long or far into the seas. We are out of our depth in more ways than one, and somewhere in our subconscious, we always know it.

But this is the nature of wild. Not only is it not for us, but the oceans are also not safe for most of the billions of creatures that occupy them. An emperor penguin lives and feeds off the ice shelf off Antarctica. It has a grand name and flies like a bird underwater, but it makes a meal for a leopard seal. A great white shark would seem the ultimate apex predator, but an orca can disprove that if the shark crosses it. A lobster looks like it has a shell that will protect it from all but humans, and this is almost true. However, it molts the shell to grow. At these points it is vulnerable to strong-jawed fish, such as predators like cod and wolffish, and, for added insecurity, it also has to watch out for other lobsters as they can be cannibalistic. It's a jungle down there. In this, the land and the sea have everything in common, with existence underwritten by the toughest of rules.

To inquire into the workings of the marine wild, to see how this constantly changing environment can thrill, challenge, and question our sense of nature, it is worth looking at key environments up close. Let's begin by considering what we treasure when we seek to enjoy the wild at perhaps the most celebrated of all ocean attractions, the Great Barrier Reef. This is seen as one of the spectacular wonders of the living world, a massive agglomeration of life often compared to the rain forest canopy. It might also be compared to the intense, magnetic growth of life that we foster in our largest cities—where there is a ceaselessly evolving intensity of feeding and productivity. On the reef, marine life is at a density and diversity like nowhere else: Coral reefs are said to be home to at least 25 percent of all marine species and the Great Barrier Reef is the greatest of them all. It has evolved and grown in its current form of living coral for up to 8,000 years, but this sits on a reef structure that is at least 20,000 years old, which itself emerged from numerous earlier phases of coral growth and decline (stimulated by rising and falling sea levels) that go back many millions of years. Coral originated as a life-form at least 350 million years ago. (Of that age is one fossilized reef inland of the Kimberley region, northwestern Australia, which now forms the Napier Range and which was once part of a reef of a scale similar to the Great Barrier Reef.) The name deceives us: The Great Barrier Reef is not one reef but a system of nearly 3,000 separate coral reefs that are closely ranged atop a shallow seabed in which 900 islands are situated along a 1,400-mile (2,300-kilometer) arc off the northeastern Australian coastline. Never mind the works of humanity, this is the single largest structure made by living things, visible from space.

When we look at the life on the reef, we have to start with the coral. Colonies of these small, soft-bodied organisms, called coral polyps, have made, and are making, the very structure of the reef: The hard

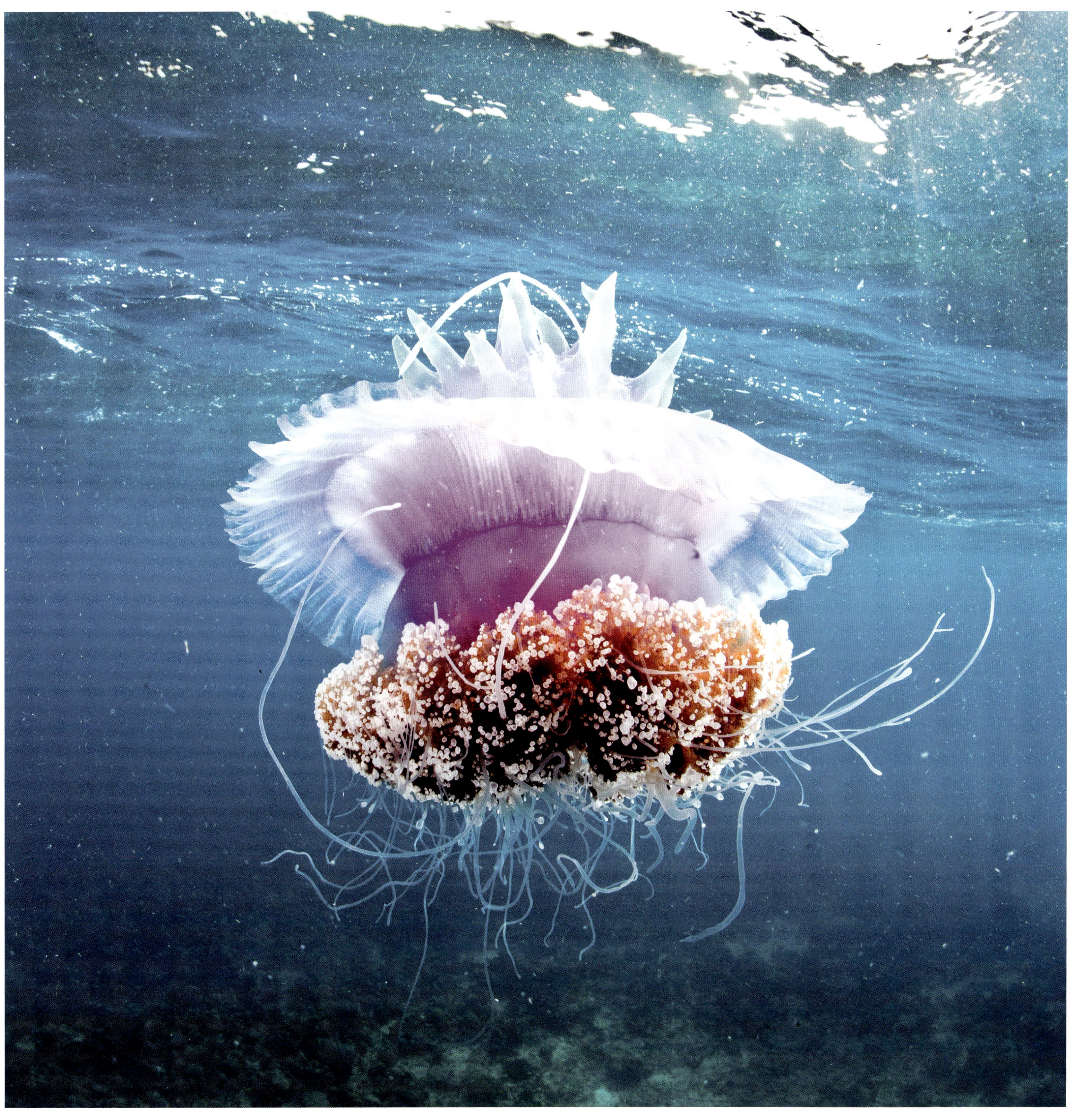

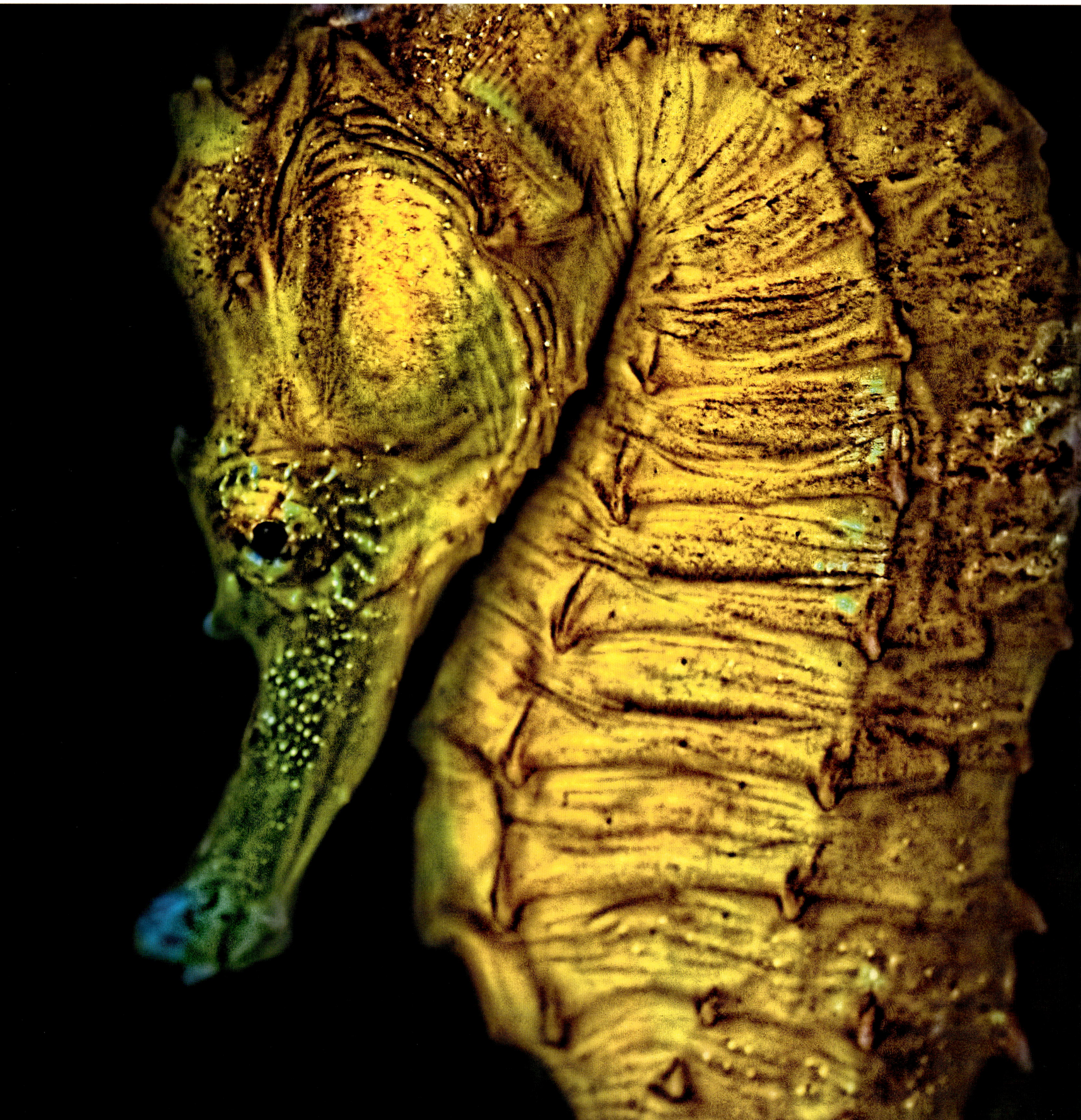

material is a skeletal structure they build as a protection and support. It is a calcium-carbonate deposit made from their processing of the water. Generation after generation of coral have built the reefs. More than 350 kinds of hard coral species are to be found on the Great Barrier Reef (there are also many species of soft coral too, some to be found growing on the hard coral skeletons, but these creatures do not generate a hard deposit). Hard coral needs light and is mainly active near the surface and down to about about 65 feet (20 meters), while being able to live at no more than 500 feet (150 meters) deep. In that top 65 feet (20 meters) of water is an intense mixture of species living on and around the coral, whose dead skeletal deposits provide nooks and crannies that serve as high-rise housing for many creatures. But these are buildings where anybody's home might also be a takeout restaurant for somebody else. Not all the activity is homicidal though: The complex coral structures provide the basis for plant life. Over 500 types of algae are at the core of the plant life of the reef, and perhaps the most fundamental is zooxanthellae, which has a symbiotic relationship with the coral polyps. Zooxanthellae live in the soft tissue of the coral (along with many other sea creatures, such as anemones and sponges) and is why coral needs sunlight. The algae provides food to the coral but is not the only food source. Coral, small and seemingly defenseless as they are, capture small organisms in their stinging tentacles, even very small fish. The tentacles contract to draw this food into the stomach. While zooxanthellae do their work somewhat invisibly, many other algae can be seen covering parts of the reef structure, with red algae being particularly notable. In size, red algae ranges from a film that helps hold the surface of the coral together, preventing it from quickly crumbling apart, to a crusted matting, and then to a bushy seaweed in its most extensive form. Algae take on a range of colors and forms on the reef, with blue and yellow hues and appearances leading to exotic names like "sea lettuce" and "sea grape."

Grasses are another group of plants to be found around the reef, typically in the shallows approaching island shorelines. There are more than fifteen kinds of seagrass on the reef and these are vital in shaping and supporting the environment for other species. Turtles and dugongs—an almost purely herbivorous large mammal—may be found browsing through the grassy beds, feeding directly on the leaves like animals in a field, while the cover provided by these sea meadows nurtures juvenile fish and other small life. The dugong, which is highly threatened around the world, exists on the Great Barrier Reef in relatively high numbers, perhaps ten thousand or so. It is a rather charming creature in a marine world of cut-and-thrust, fish-eat-fish, or even coral-eat-fish. It cruises around at about 6 miles (10 kilometers) an hour and is quite particular about the grass it eats. It doesn't just nibble the leaves but prefers, if possible, to root up the plant and eat it whole, having shaken the sand off. It has been seen to gather up a few choice plants into a pile before tucking in. Of no harm to anybody, grass excepted, it has a life-span that matches ours, if given the chance, and tenderly nurtures its young calves, who bump alongside for company and security (and also because the feeding teats are just behind the front flippers).

One of the most important plants for the reef is, perhaps oddly, not on the reef: It is the mangroves along the Queensland coastline. These shoreline forests with their roots in the sea do a lot to capture sediment and nutrients that would otherwise leach rapidly out to sea and prohibit coral growth. Even though the Great Barrier Reef is 10 miles (16 kilometers) offshore at its nearest point, and is often much farther away from land, it still gains some benefit from the cleaner water (although it also suffers from excessive nutrients and pollutants as a result of farming activities feeding into the river estuaries). The mangrove-reef relationship is symbiotic too, nature's version of communal barter: it turns out the mangroves are able to grow more securely because the distant reef still manages to have a dissipating effect on the wave forces that might reach land.

Beside the algae and grasses, the majority of plant life exists on land, on the nearby islands, where more than two thousand species are to be found, some endemic, with many localized sets of botanical mix, thanks to the considerable distance between the islands from north to south. In the north of the reef system more woody growth has developed, while in the south there tends to be lower-level herbaceous growth only. While these plants are not in the water, they do provide support for species that partly live in, or on, the water, as well as out of the water—for example, amphibians and birds. Birds are also key in having propagated plant life, spreading seeds across the reef islands over time.

The above description of algae and often sparse grasses being the principal plant life of the Great Barrier Reef points to why Charles Darwin described coral reefs as being like oases in deserts. They seem to thrive in areas where the waters are not otherwise supporting rich plant life, due to low levels of nitrogen and phosphorus. These building blocks of life are noticeable by their absence and yet the coral reef seems to prefer this. So how do nutrients find their way up the chain of life if there are few plants feeding herbivores to begin the process? How come the most luxuriantly rich range of life is built on so little? It was only in 2013 that scientists came out with a definitive answer: the silent hero of the piece is the sponge. It turns out that this ancient life-form is still doing fundamental work, recycling vast amounts of organic matter from the water and converting it into a form that can be fed on by sediment-feeding creatures such as snails and crabs. These in turn are fed on, and so the cycle of life begins. Sponges are the super-processors, closing the loop of regeneration on reefs. They do ten times the work of bacteria in recycling organic material, and output as much nutritious material as all the corals and algae combined. The colorful, diverse nature of the reef would simply not be possible if the sponges were not on hand to hoover up the marine fluff and recycle it as a meal for the small guys at the bottom of the food chain. There may be more than two thousand different species of sponge on and around the Great Barrier Reef—new species are still being discovered as the significance of this ancient family keeps rising in the perception of marine scientists. We might reasonably expect to find that they don't all do things the same way and have perhaps remarkable niches of functionality in the life of the reef. While the sponges are vital to the life cycle, they are also in a very careful balance with the coral growth. Sponges and coral are in competition with each other for space to affix themselves. If sponges do too well (which happens if the nutrient levels rise in the sea), then they can start to outperform the coral, leaving it with less space to grow. In such situations a coral reef declines and becomes a more sponge- and seaweed-dominated environment. So, as so often is the case, there is a very

The sea was the cradle of primordial life,
from which the roots of our own existence sprouted.

Billions of years of evolutionary development brought forth
an enchanting variety of forms, colors, lifestyles, and patterns of behavior.

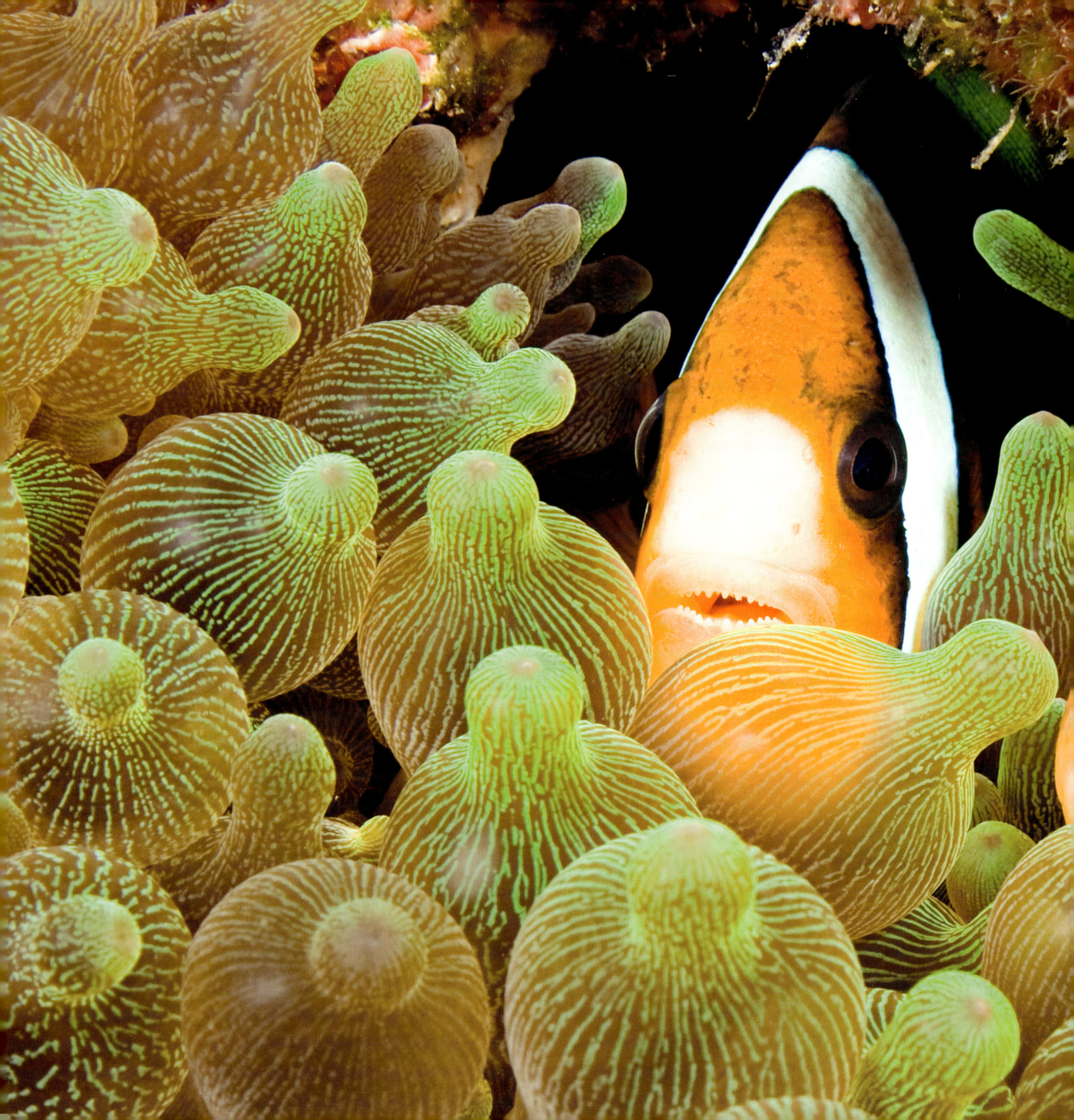

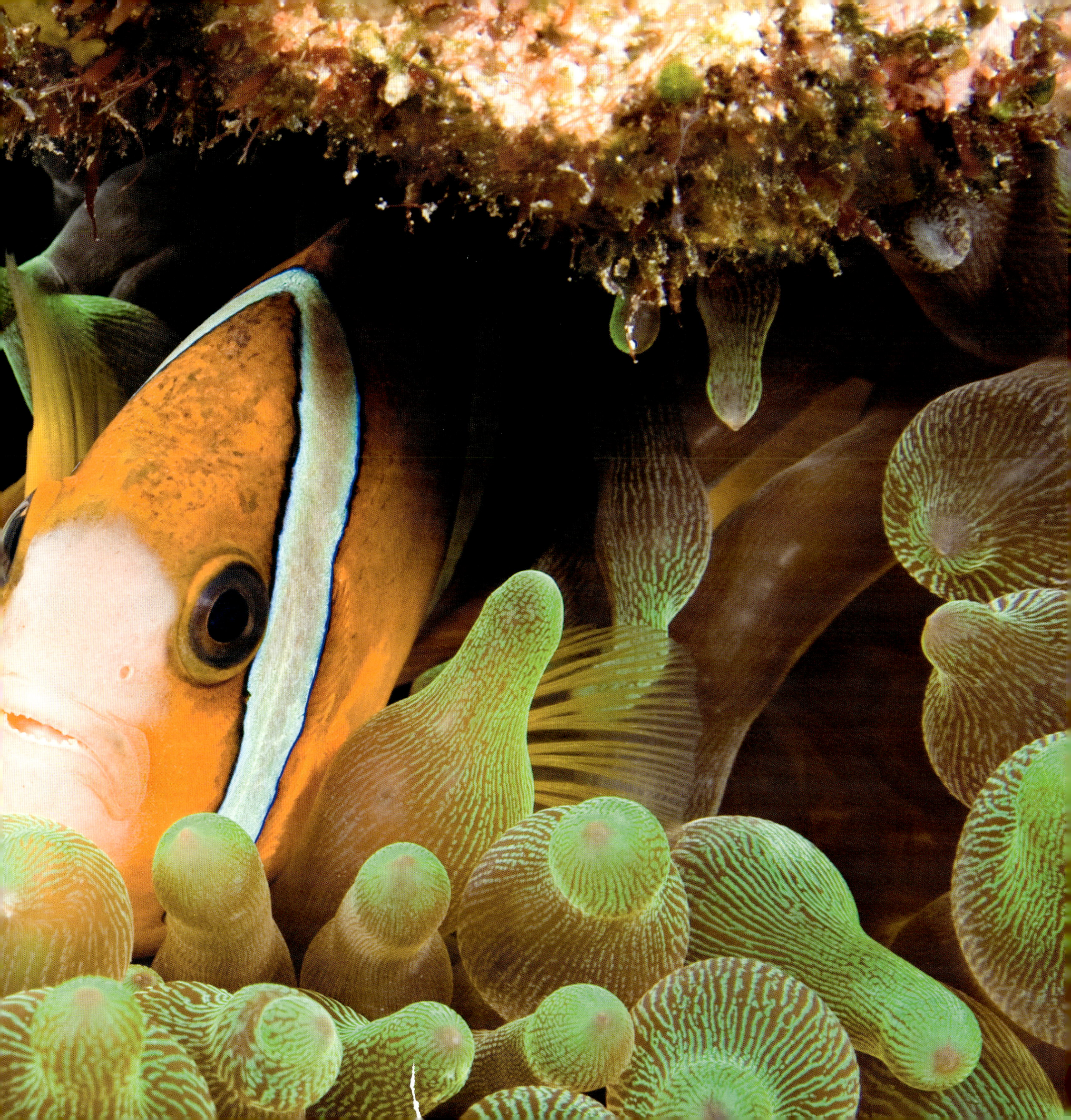

careful balance of relationships between the many life-forms that make up a wonderful biodiverse reef, where if any one key species gets an advantage, it can spell disaster or at least serious alteration to the fate of others.

Now we are at the bottom of the reef, or in its crevices, with the sediment-sifting, low-life of snails and other sifters. The coral and the plant life and the sponges have all done their bit to create the environment that various other forms of faster-moving life will animate. How do we get from these small-fry right up through the pyramid of life to the apex predators of the reef—be they tiger or reef sharks, or sea eagles? This can involve a very convoluted journey or a quick passage. There are more than 1,500 kinds of fish and 3,000 kinds of molluscs and even 130 or so different kinds of shark and ray on the Great Barrier Reef. There are a lot of ways in which, to put it crudely, a small fish can fit inside a medium one and this inside a large one. Not that it happens quite like that. Some very big creatures don't eat big things. For example, humpback whales come to the reef in winter to give birth and to mate. While there they eat very little—they largely live off the body fat they gained from gorging on schools of small fish in the Southern Ocean during the summer, and anyway, they are unable to eat big things—they are baleen whales, filtering the water for large amounts of small fish, crustaceans, and plankton. So at most they may snack a little on the shrimp-like krill or similar while in the area of the reef, albeit in large quantities. This small crustacean, floating around the water on the current, is but a teeny-tiny morsel for a whale, but in bulk, rather like we might greedily scoop a handful of peanuts from a bowl, it can be an appetizer. The krill will have fed on phytoplankton, invisible to our eye but richly plentiful in every drop of reef water. In two steps on the ladder of life we will have gone from the invisible to the very large. More typically, though, many steps are taken in the food chain from bottom to top. But each of those thousands of species mentioned above have their own unique place in the web of life on and around the reef, and they may form a part of many different survival stories—both in what they eat and what eats them. Some have broad omnivorous habits, others are much more niche. Consider the subject of many a colorful image of reef life—the clownfish sheltering in the anemone. This is another example of one of many specific symbiotic relationships, which we have touched on above. It is remarkably mutualistic—each benefits considerably from the other. The first point to note is that most fish would avoid the anemone, or potentially risk becoming its prey, as its tentacles are toxic. This toxin paralyzes some small creatures, enabling the anemone to capture food. However, clownfish (which come in a number of species) either have immunity to the sting or acquire it. It is thought that their skin secretes a mucus that prevents them from being affected by stings, or that the anemone does not sting because the mucus ensures it does not recognize the fish as a target. This allows the two to work together: The movement of the fish aerates the water, bringing in more nutrients, and potentially attracts other small fish into the anemone's tentacles. The clownfish protects itself from predators by hiding in the tentacles while it deters predators and parasites of the anemone. It also nibbles off dead tentacles and feeds off leftovers from the anemone's meals.

Every creature has its place on the reef, in hierarchies of service that may be mutualistic, may be parasitic, may be as predator or prey. When we snorkel around and wonder at the brightly colored fish, we should be also wondering why they take that color or behave in that way. The parrotfish, a medium-sized fish with a distinctive mouth that gives rise to its name, is not just a pretty bluish visitor that comes in variations. It actually serves a vital function in keeping the coral healthy by cleaning the algae off its surface by nibbling it with its teeth. In doing so, it ingests some fragments of coral skeleton, which are then excreted out as sand. In a year, one fish might produce 200 pounds of sand, and so collectively they help build up sandy floors around reefs and may ultimately help develop islands. The fish are also notable in how they adapt their color to suit the environment. And perhaps more remarkable is that almost all species of parrotfish are sequential hermaphrodites, starting out as females and then becoming males.

Even more conspicuous than the parrotfish, and perhaps the most popular subject to spot and photograph, are angelfish. These are seemingly fearless of humans and often demonstrate curiosity around divers. Typically brightly striped, with distinctive, elegant fins, they feed on a variety of sources, from plankton to sponges to small invertebrates. But why they look as they do is a mystery—it doesn't seem to fit as an evolutionary solution at first glance. Indeed, one of the areas where much greater research is still needed is in understanding just why reef life can be so colorful. The fact that we can see the likes of parrotfish, angelfish, and clownfish so easily, and be struck by how brightly colored and patterned they are, may make us wonder how that could possibly work as a way of being discreet around either predators or prey. However, part of the answer may be that fish do not see things as we do, and that these colors can look very different through the alternative workings of a fish eye, or when viewed at low light, or in particular reef locations where these colorful fish might choose to hide when predators are in the area. Sight is not the primary trigger for many fish; instead, sound and smell may be much more significant. It is thought that some fish change how their eyes work when they move from day to night, or from lighter surface areas to deeper areas where colors get filtered out (which is why to our eyes things become very blue quite quickly as we descend). About half of all fish are thought to be able to see ultraviolet light, and quite a few of them may be attuned to read it as a range of colors. For example, damselfish have this ability, and they are consumers of plankton, which reflects a lot of ultraviolet light. Meanwhile, brightly patterned fish, such as parrotfish, are noticeable close up, but the effect created by those patterns can make them disappear quite quickly in flickering light where the color mix starts to work as camouflage. In this respect, they are doing something similar to a herd of zebras on the savanna: they may seem very noticeable in black-and-white stripes, but the effect of this combined with heat haze viewed at a distance creates a pattern that makes it rather difficult for a lion to identify a single creature. And so the brightly colored fish of the reef can suddenly disappear against the splotchy-colored background of various coral, algae, and sponges, their stripes becoming blurred into resembling just another flickering shaft of refracted sunlight filtered through the water. But while the brightly colored fish can disappear into the crowd in that brightly colored environment, their distinctive markings also serve the counterpurpose of identifying them clearly to fellow fish of their own species. No two species have the same coloring and most are quite noticeably different. So the color both merges in, and then helps

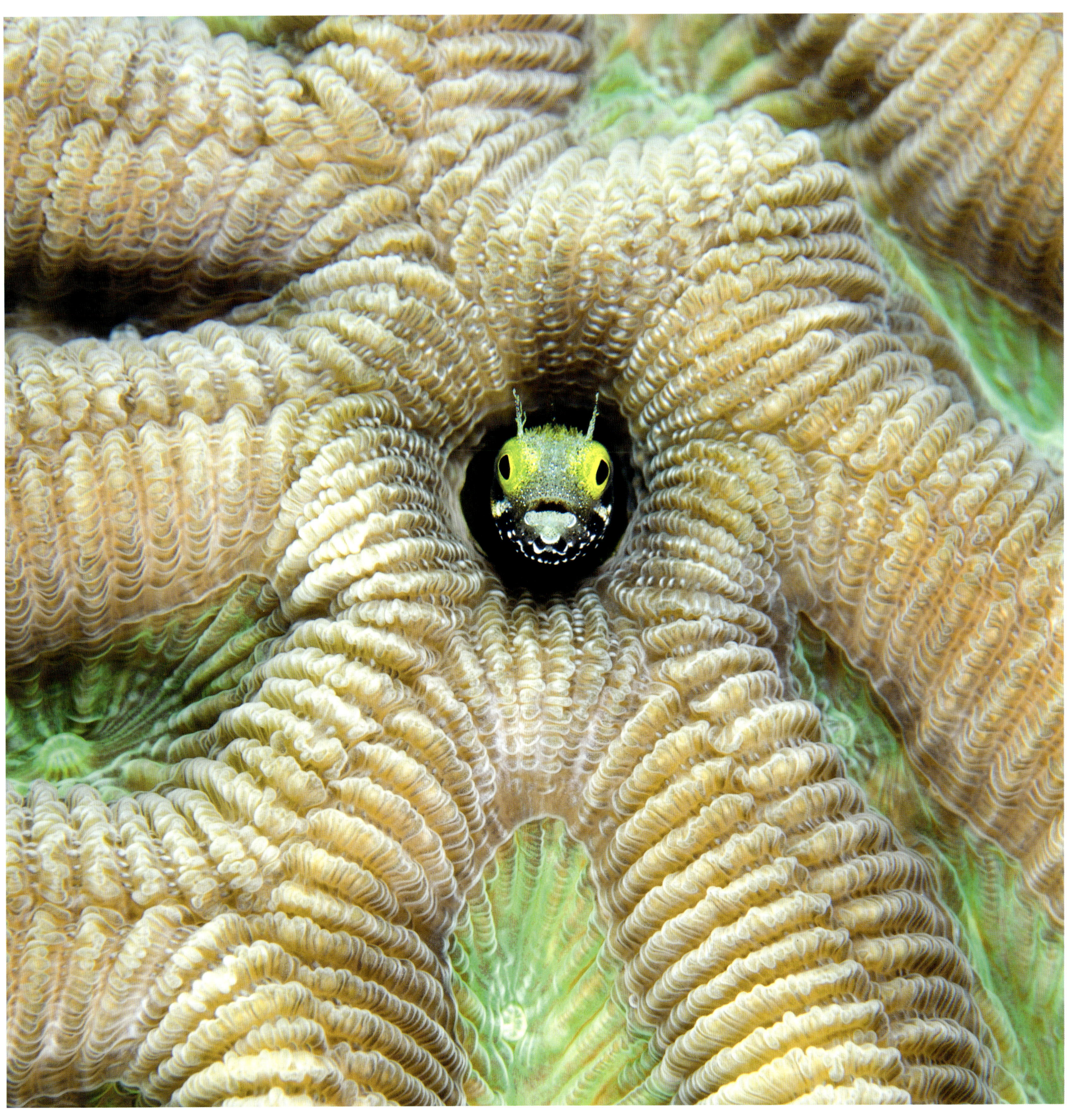

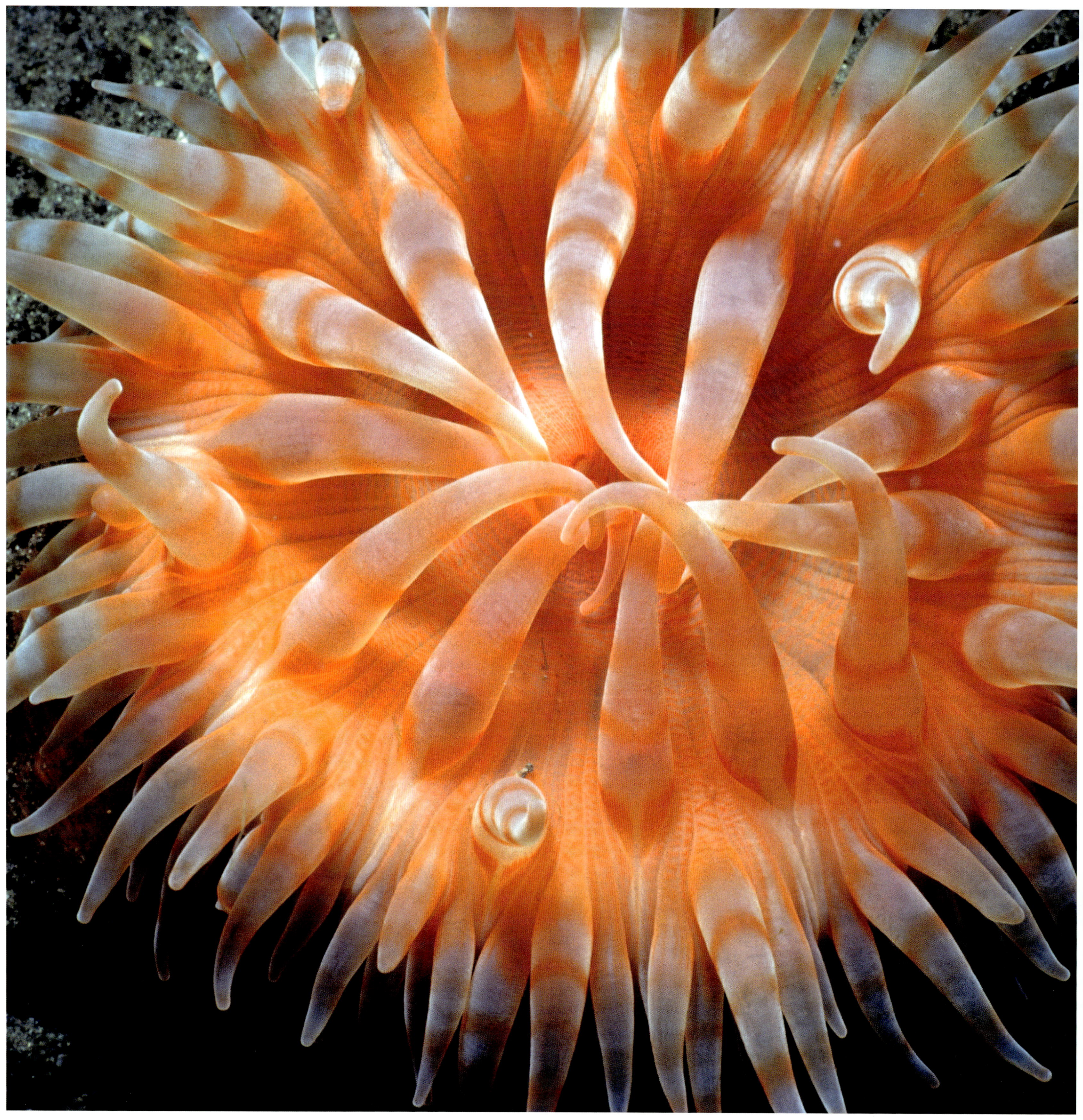

The sea is as near as we come to another world.

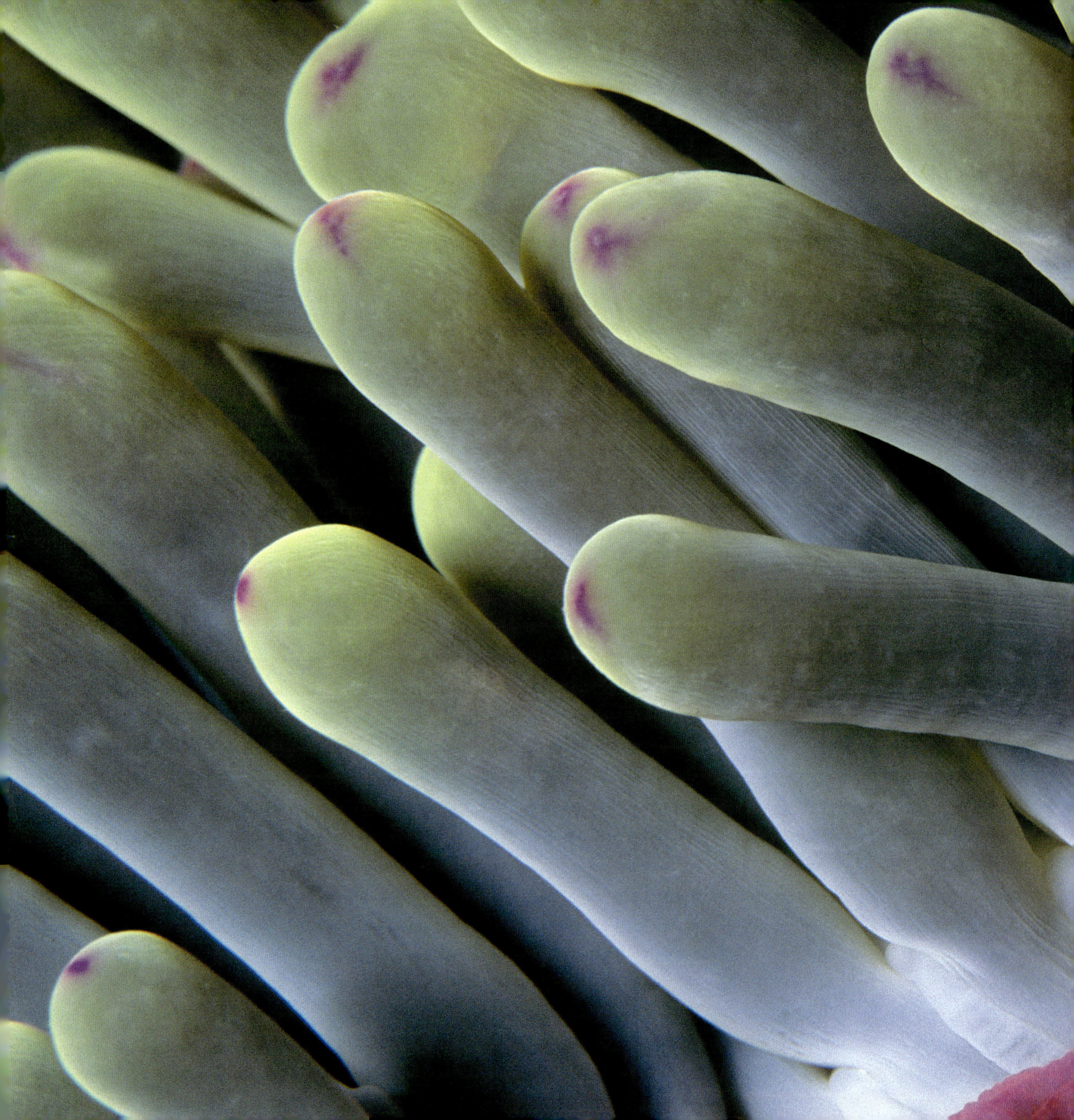

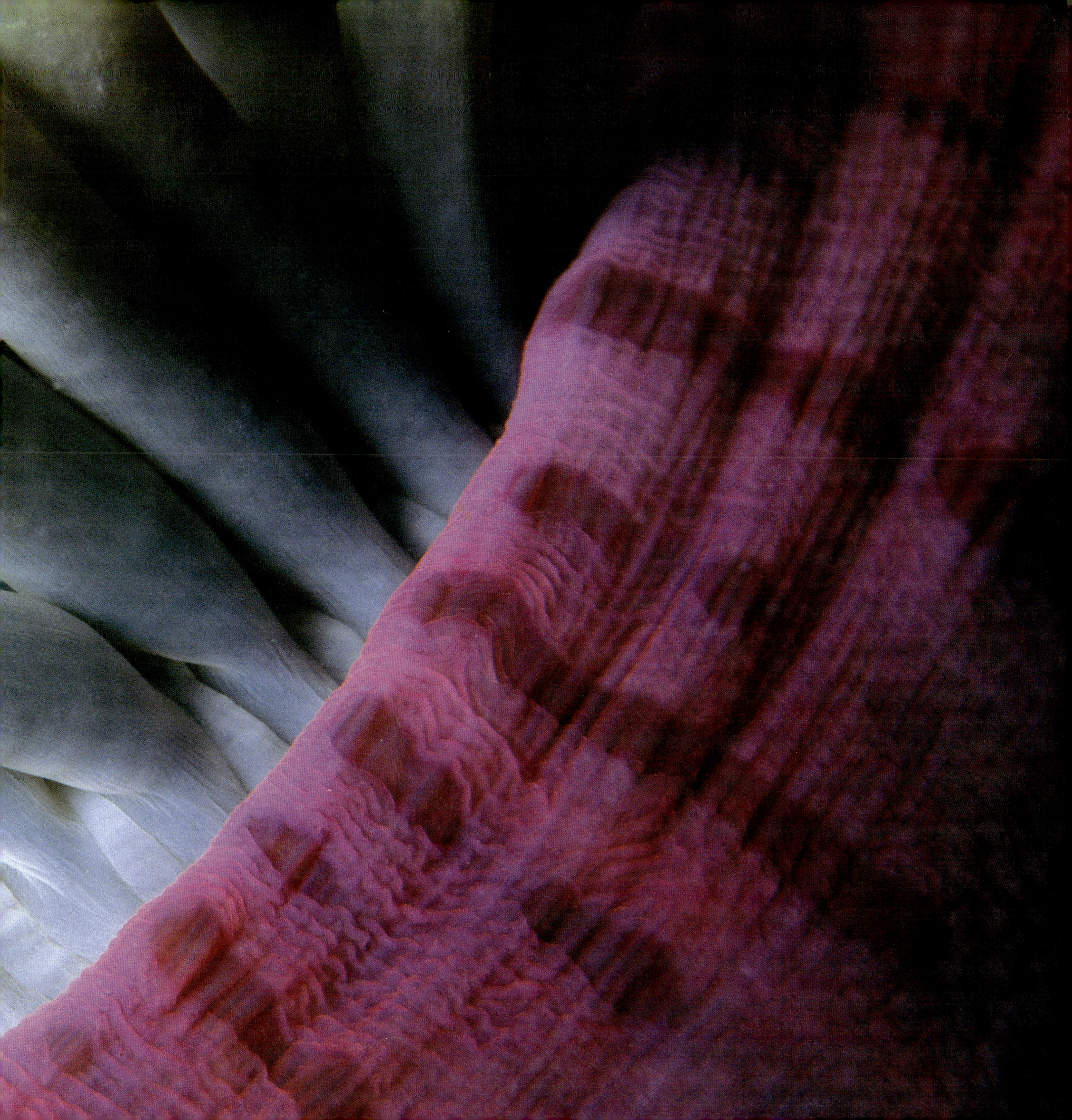

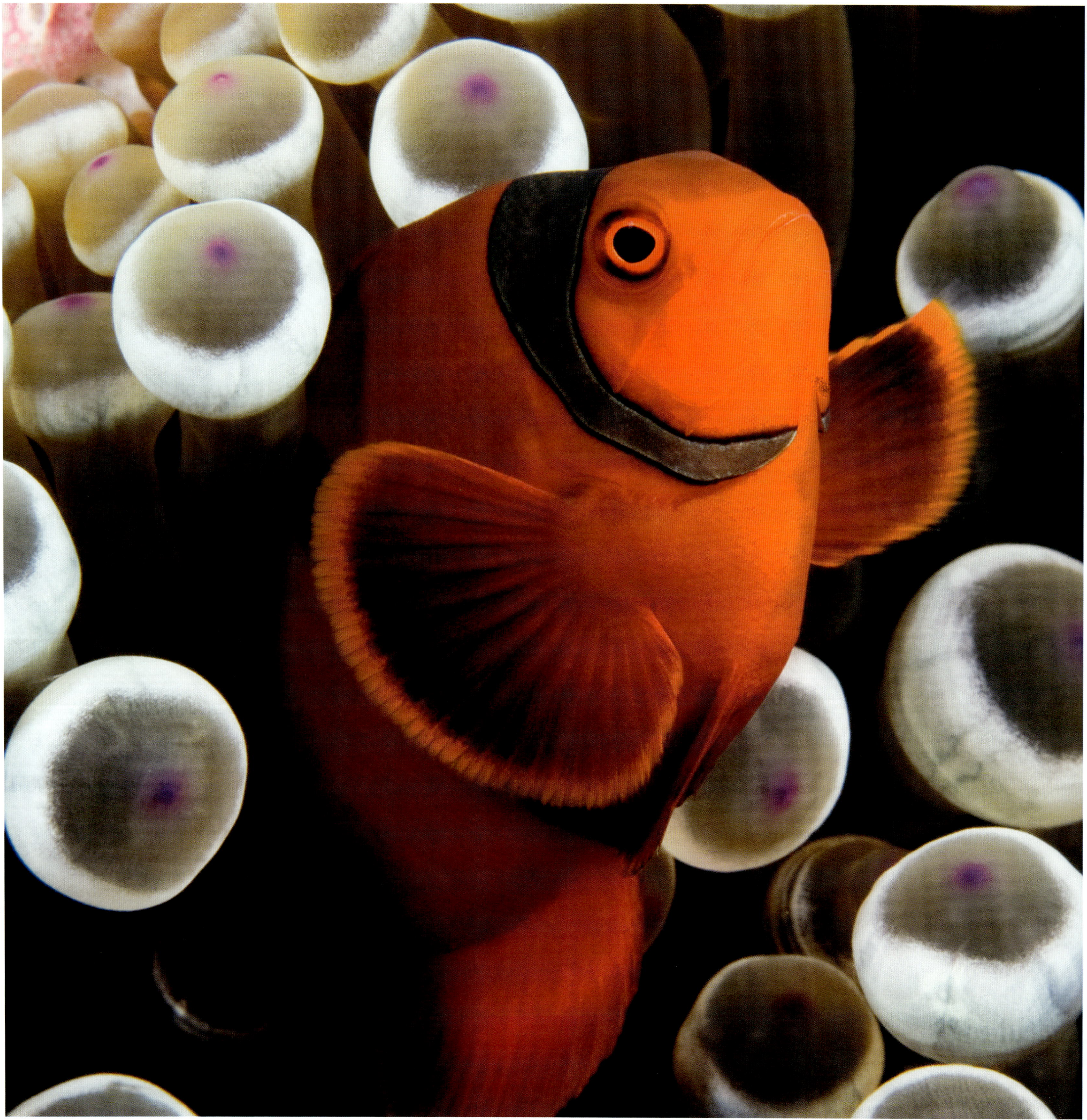

them stand out—it just depends who is looking when and where. In this regard, it may be rather like humans wearing eye-catching clothes to go out for an evening: You may fit in with the general party scene but also stand out for that special other.

One of the attractions of the Great Barrier Reef to humans is that it is a kind of wilderness that is in some ways a Garden of Eden—it can seem that the creatures, for the most part, are non-threatening to us and are not too nasty to each other in front of our snorkel masks. This is an illusion—it's fish eats fish, or fish eats small crustacean and then gets eaten by a big fish, and so on, as indicated above in some detail. But we have good reason for at times finding the reef an almost tame environment, because some creatures display a curiosity and relaxed behavior around us that can almost pass for tame. Among the largest and also very numerous fish on the reef are the groupers and cods. They move slowly, ponderously, and are often social to humans— they will even feed from close to the hand. This might seem strange behavior in a wild creature, a risky strategy . . . but these big fish are confident because they have few threats. That said, they would try to avoid any encounters with the sharks that are typically at the top table of reef life, the apex predator that most would avoid.

Of the many sharks that might be seen near the Great Barrier Reef, it is the tiger shark that stands out as the most common threat to other life (the great white shark is a fearsome predator too, but somewhat less likely to be met, having had its population greatly depleted, as well as being less localized in its behaviors). Aside from its size, teeth, and carnivorous temperament, what makes the tiger shark sit atop the predatory table is that it is up for eating almost anything—from jellyfish and crustaceans through to just about any kind of fish, to mammals such as seals and dolphins, and even birds if they hang around on the surface too long. Sharks are an ancient species and are yet remarkably refined in having effective tools as macropredators: The tiger shark gets its name from its stripes, which help it blend in at a distance in the water, while it has a dark back and a light underside, which also make it hard to see when looking down into darker water or looking up toward the light. Its omnivorous appetites have led to it being called the "garbage shark" on account of what has been found in the stomachs of some that have been caught: all kinds of marine life, including parts of what were dead or dying whales and even other tiger sharks. The indiscriminate feeding means that human pollution can end up in the stomach of the shark, with pieces of metal and plastic being quite commonly found. As this is a creature that is not easily studied or kept in captivity, relatively little is known about the tiger shark. It is known that they live longer than twelve years, but just how long-lived they may be is unknown. It's only in recent years that we have realized just how veteran the great white shark can be. With a life-span that can exceed seventy years and a slow passage to sexual maturity, this is a creature whose population will take a long time to respond if, as has happened, it is on the "at risk" list and needs protection.

It is at the top and the bottom of the pyramid, or perhaps circle, of life on the Great Barrier Reef that we find the most pressing need for greater knowledge on our part, and also greater protection to be afforded to key species. We are just discovering the role of sponges in closing the circle of recycling and fostering life on the reef, and we are still a long way from fully understanding the work of apex predators in controlling population growth and driving diversity. While we are increasingly aware of the vital role of coral in building the reef, and are concerned that threats of warming seas could kill the coral, which would quickly lead to the collapse of the whole ecosystem, we need also to be aware that the whole chain of life can operate to the classic rule: It is only as strong as the weakest link.

It is with this kind of concern in mind that we need to see more broad-ranging approaches to marine preservation and full protection of the environment. The Great Barrier Reef Marine Park Act, enacted by the Australian government in 1975 to protect the reef, was one of the earliest attempts at this, but conservation authorities are struggling to maintain a pristine environment in the face of both tourism and inevitable pollution from nearby land. Worldwide, more than 30 percent of coral reefs are severely damaged, and it is estimated that this will double by 2030 unless we take urgent and extensive actions. We can have some cautious hope: Many governments are working to achieve international commitments to marine protection, and many are increasingly moving to designate large areas as marine reserves or parks. However, many are already well behind their targeted dates. Let's finish this exploration and celebration of the wild by referring to a most recent commitment, yet to fail. In March 2015 a major statement was made in marine protection terms, with the designation of what will be the largest protected area of water in the world—a vast area off the Pitcairn Islands. This group of four islands is one of the most remote in the Pacific Ocean and is the last remaining protectorate of the United Kingdom in the Pacific. The Pitcairns have a certain notoriety, dating right back to the population's origins, as it was where some of the participants in the notorious mutiny on the *Bounty* ended up. This eighteenth-century tale of derring-do was made into a famous film featuring Marlon Brando. It recounts how in 1789 mutineers on the Royal Navy HMS *Bounty* cast adrift in a small launch their unpopular captain, William Bligh, and those seaman who remained loyal to him. They then hid out on the Pitcairn Islands and Tahiti. Perhaps unexpectedly, Bligh survived, reported the desertion, and the Royal Navy ship *Pandora* came out to capture the rebels. They did so on Tahiti, but not on the Pitcairns, and *Pandora* went on to be shipwrecked with loss of life. The Pitcairn rebels survived and the population of the island nation today are the descendants of the mutineers and Tahitians. They continue to get headlines for the wrong reasons, but it is not the people that we are concerned with here, but the fact that there is very little human intervention in this area of spectacularly remote ocean. It is an area of pristine waters and rare seabird life, where humanity seems to have had no significant effect. In an effort to keep it that way, in particular to protect it from factory fishing descending from afar, the British government is assigning marine-reserve status to a contiguous area of 322,138 square miles (834,334 square kilometers). It will patrol the space by satellite: Project Eyes on the Seas will be watching the waters twenty-four hours a day from a research station in Harwell, Oxfordshire, on the other side of the world, but with the British Navy and others at their disposal. Perhaps the age of empire has benefits still to bring us.

Coral reefs represent some of the world's
most spectacular beauty spots,

but they are also the foundation of marine life: without them,
many of the sea's most exquisite species will not survive.

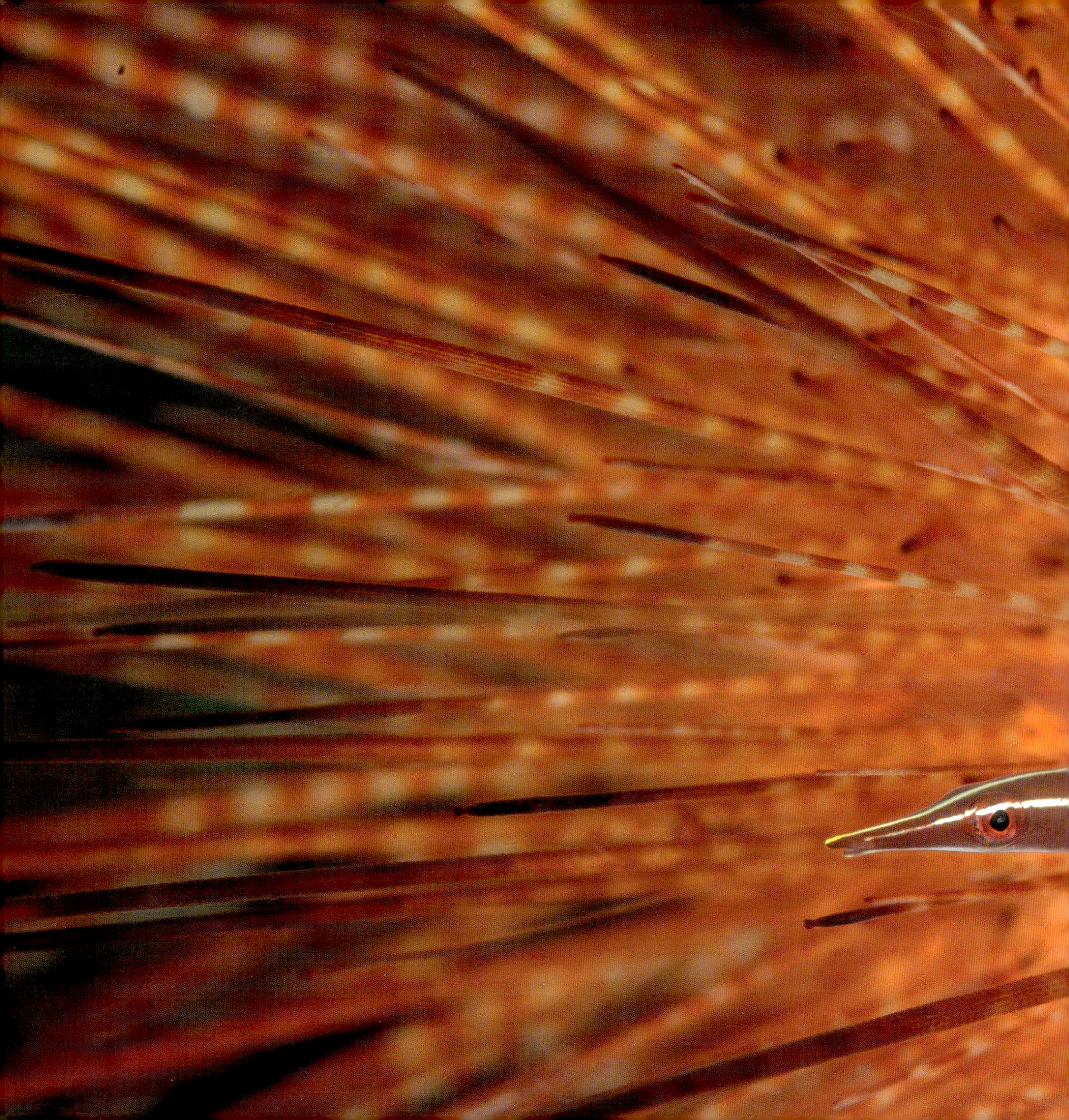

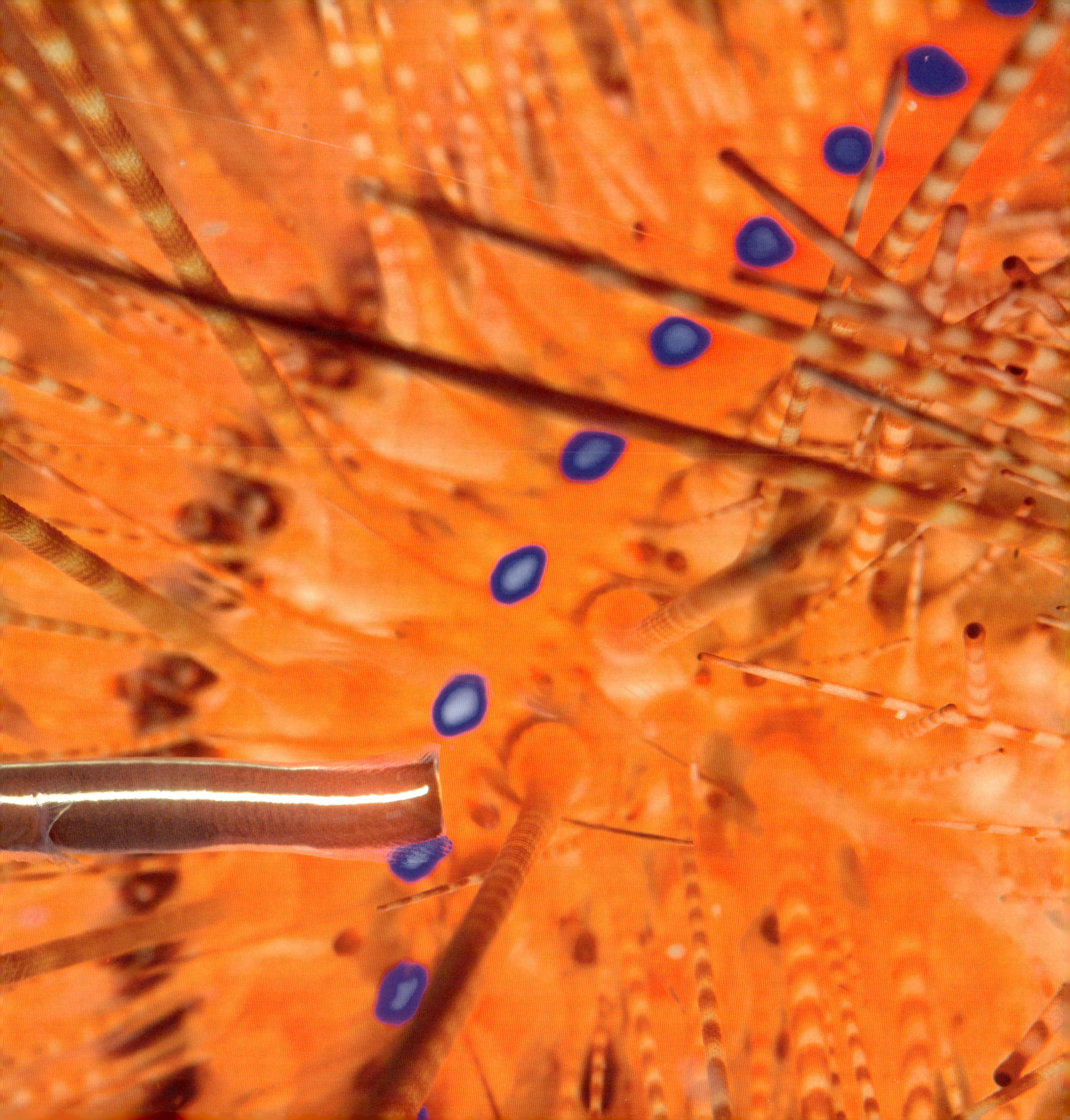

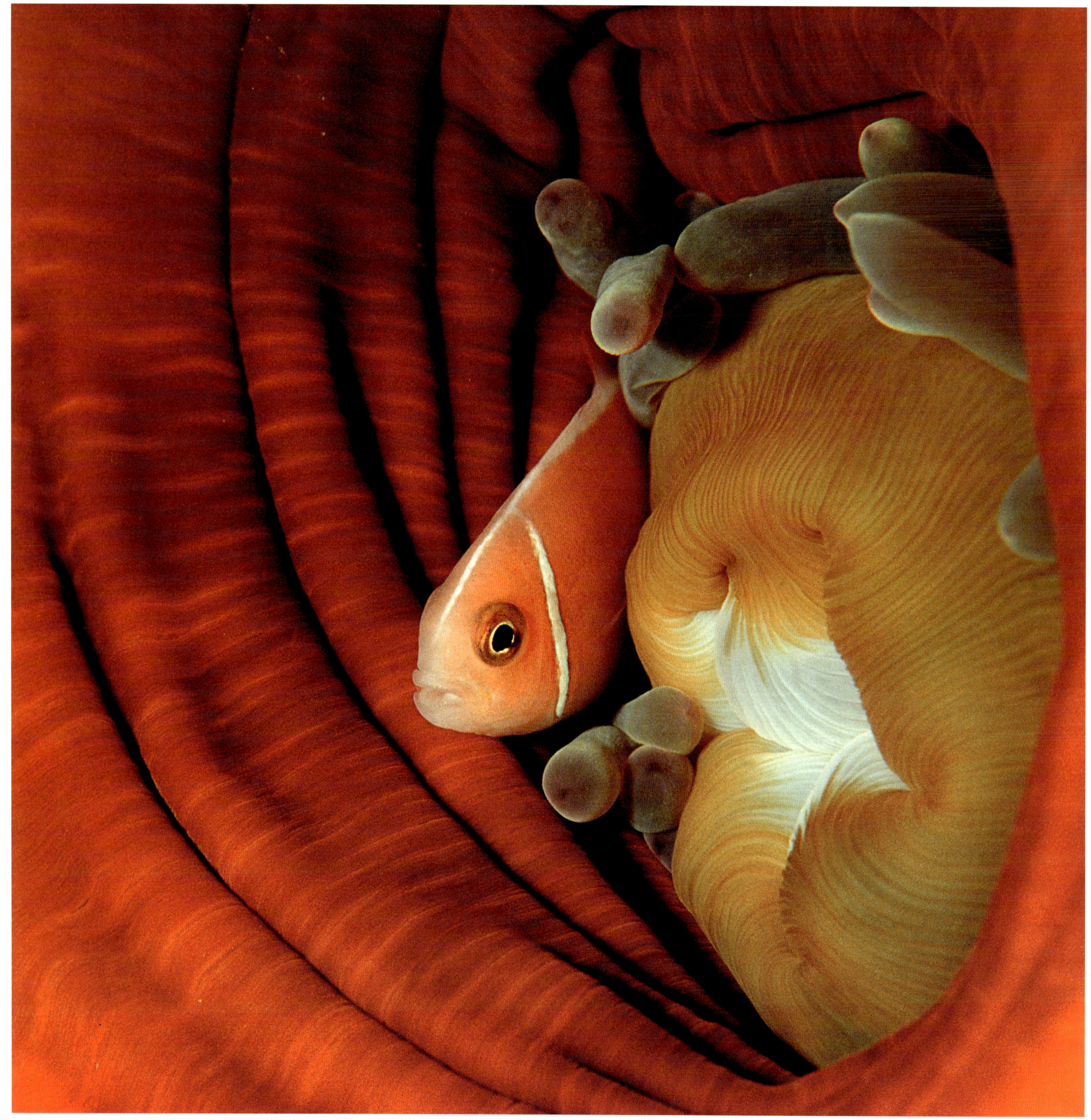

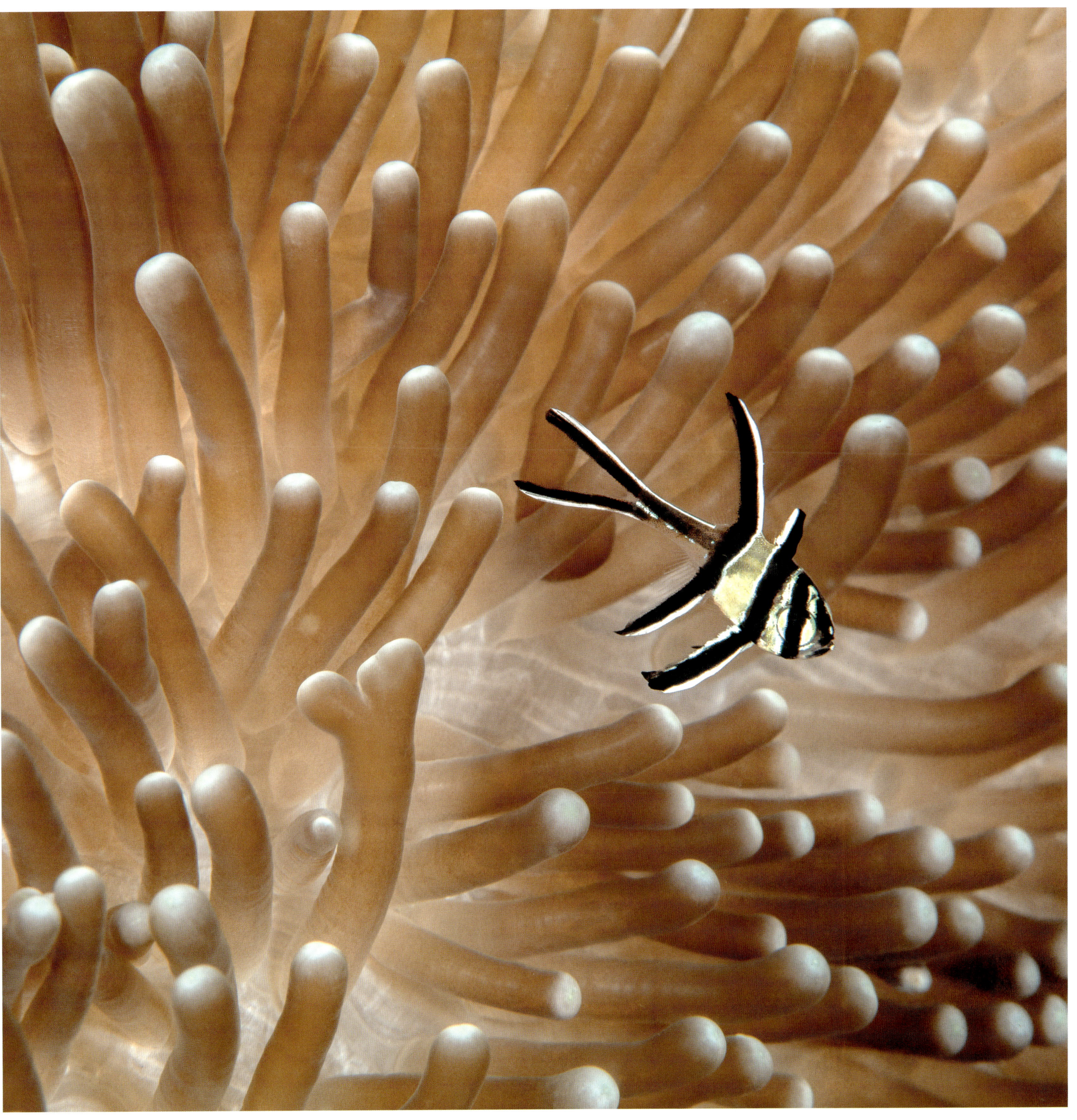

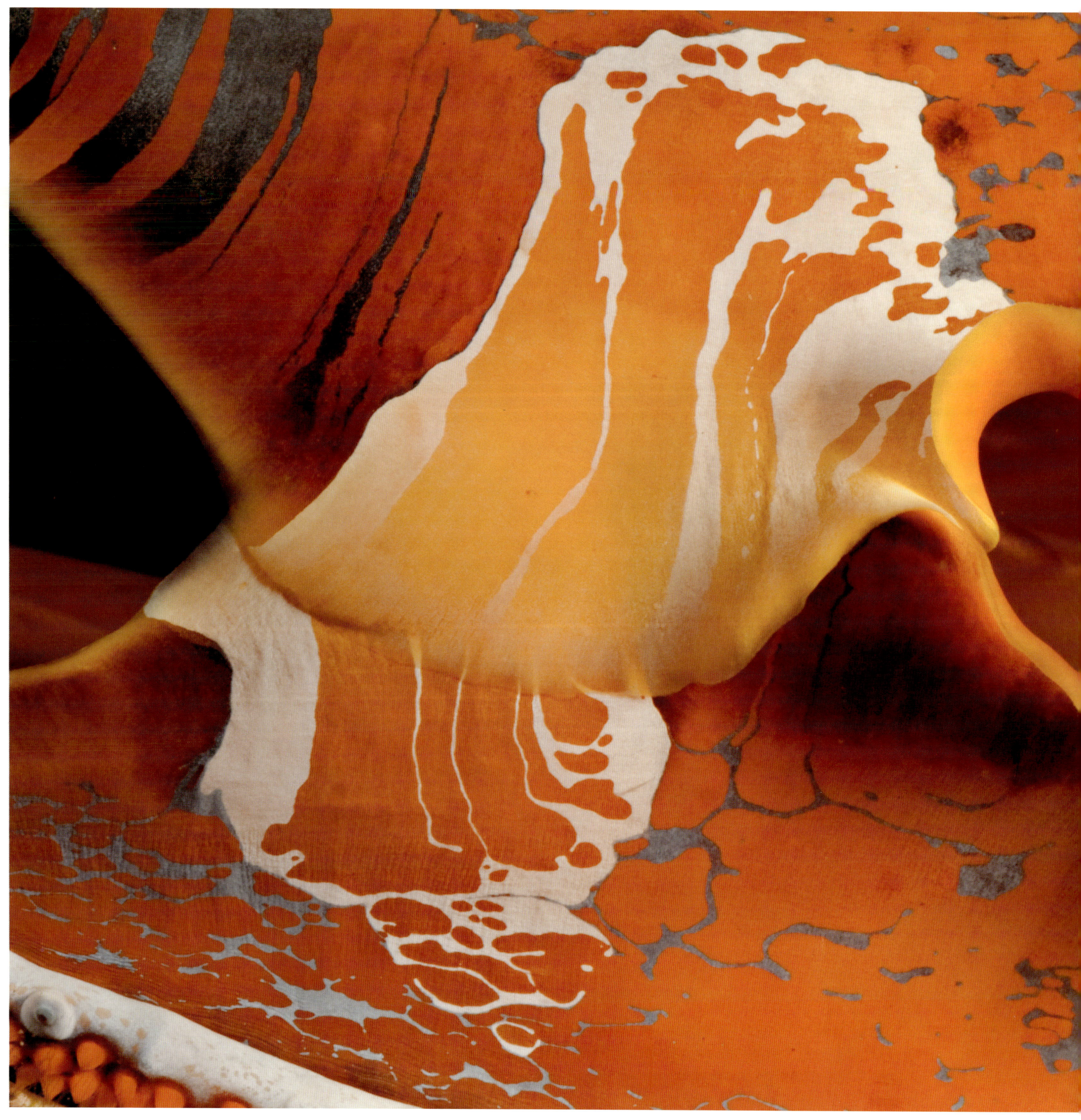

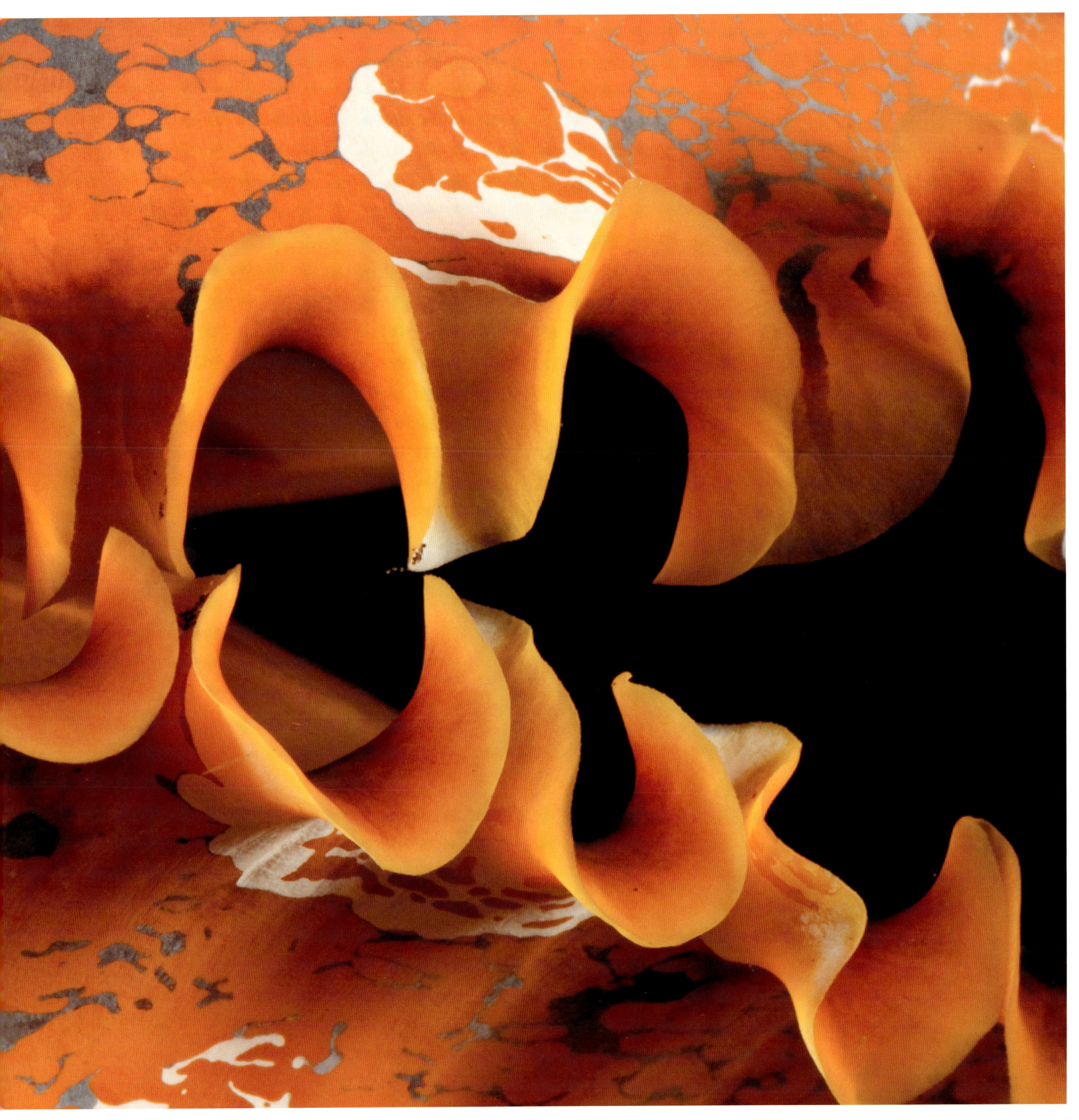

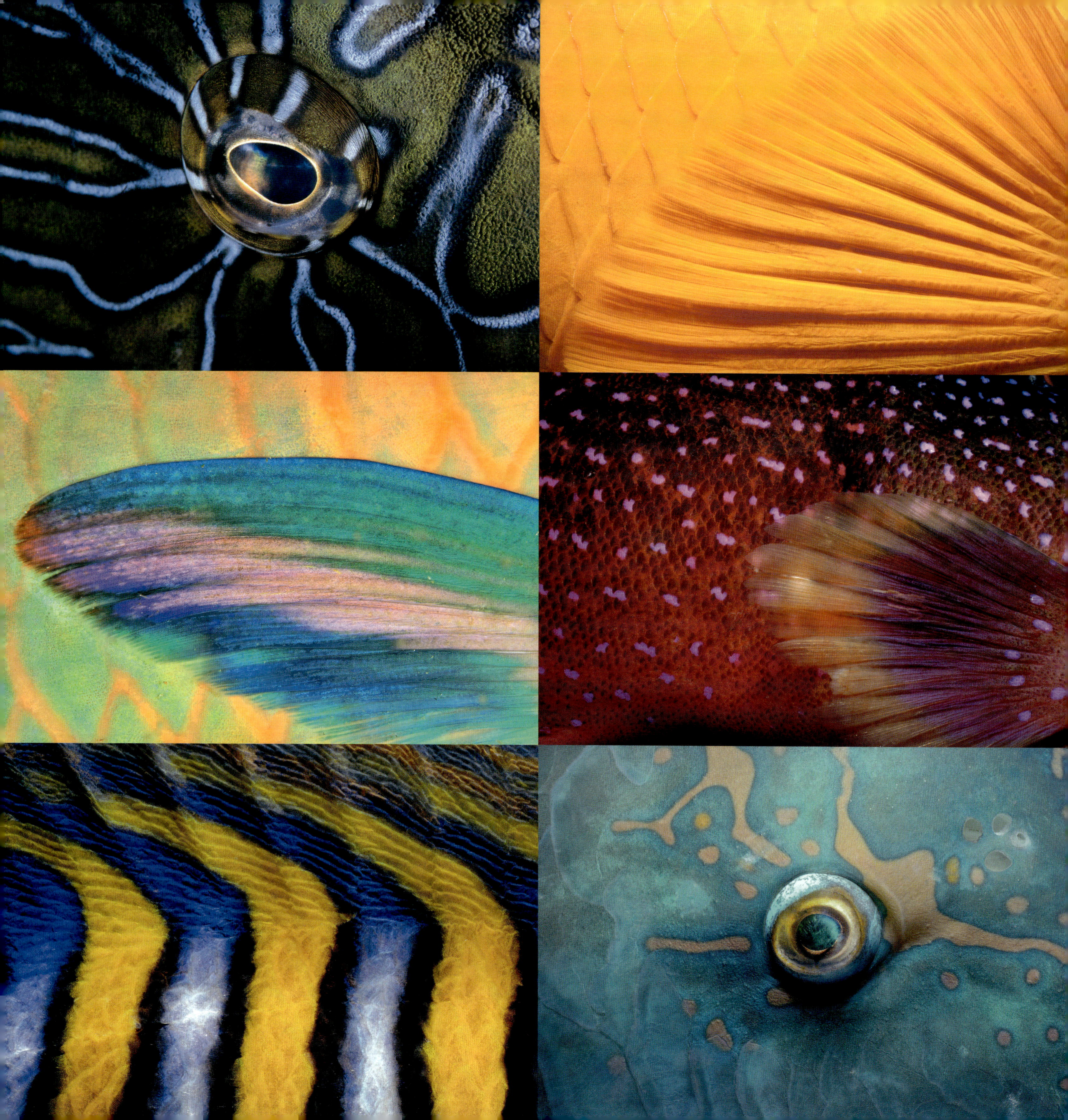

The sea has this contradictory quality,
that the more you see of it, the more it overwhelms
the eye and disappears in its own brightness.

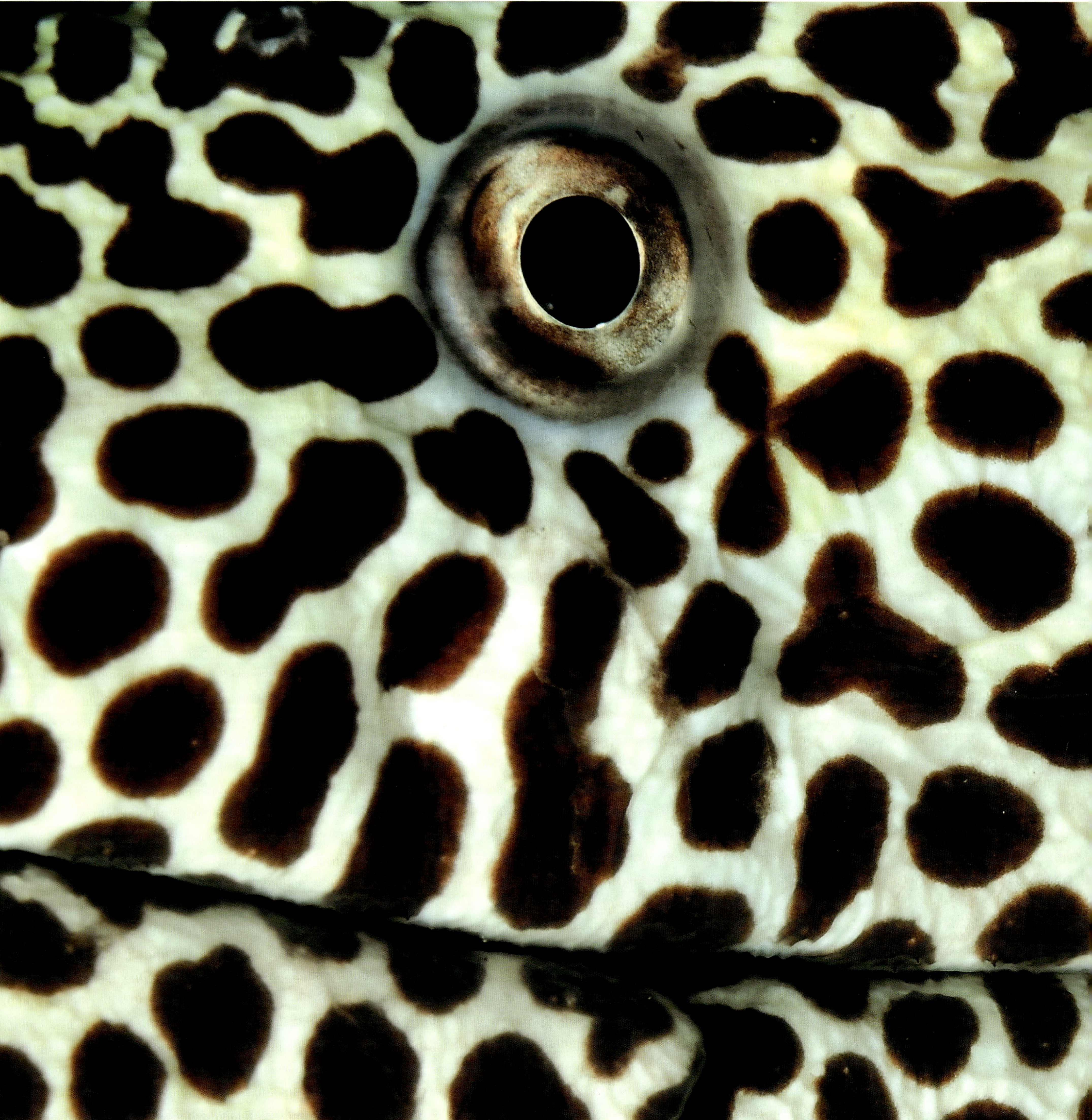

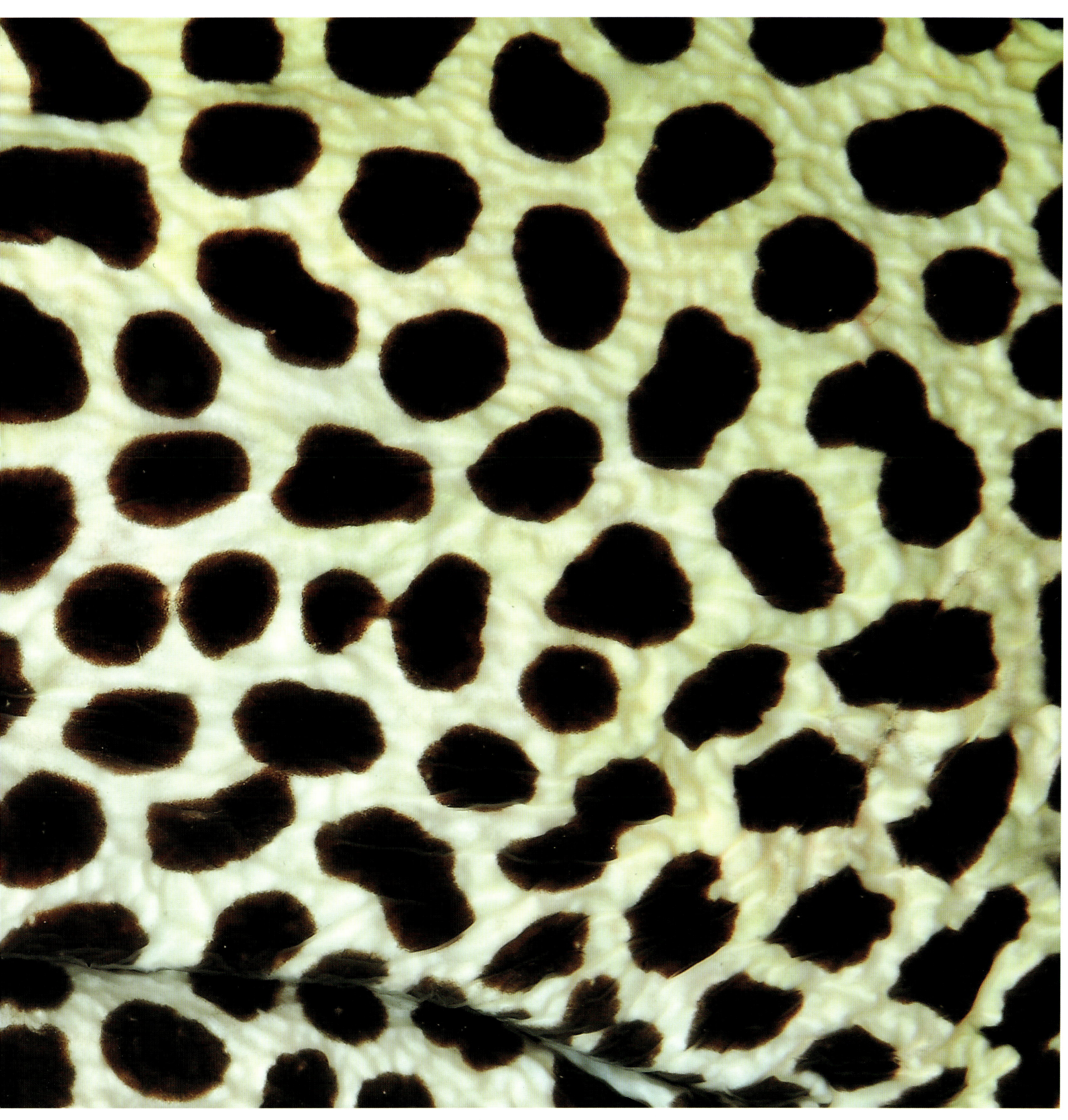

Health to the ocean

means health for us.

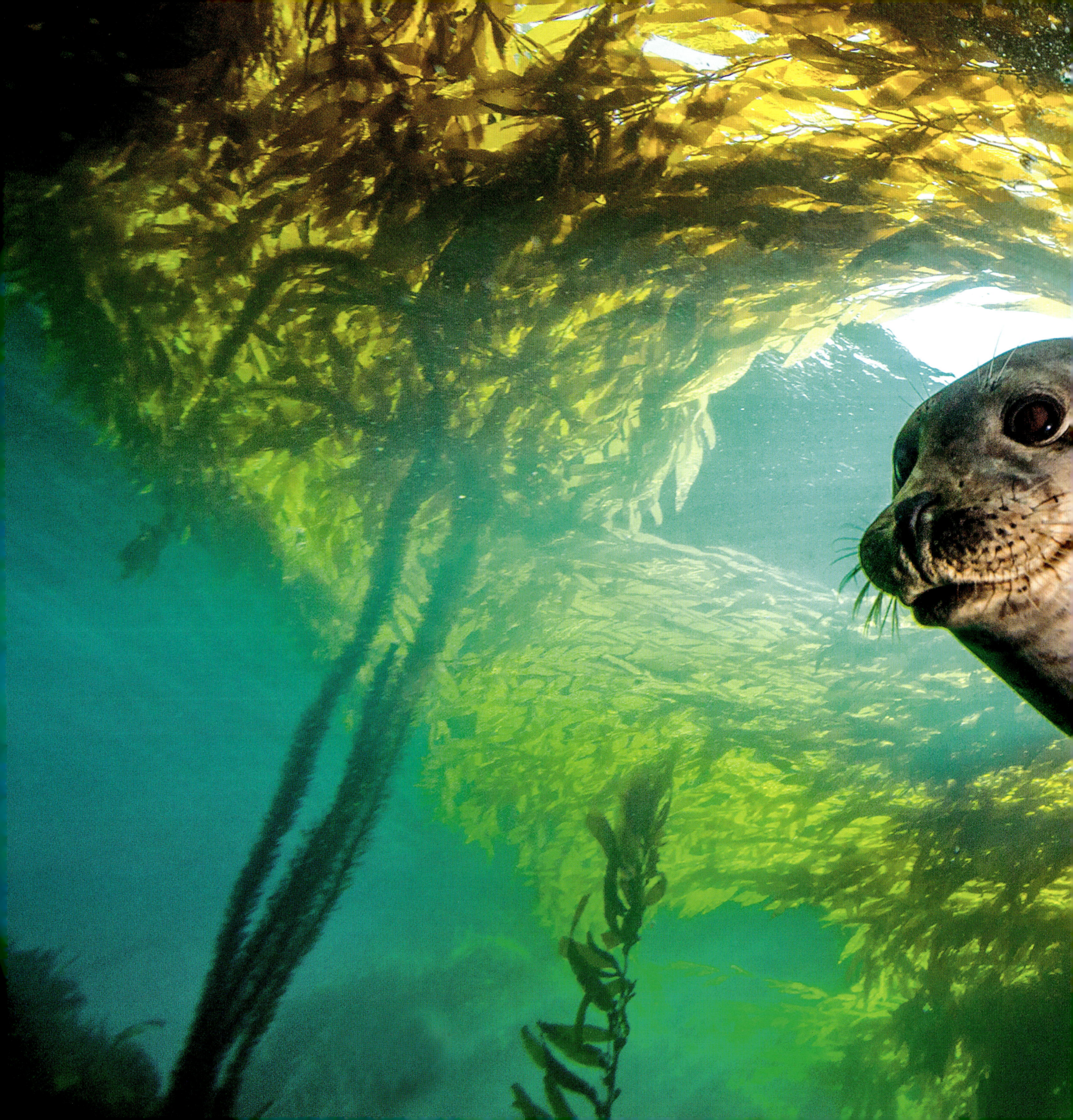

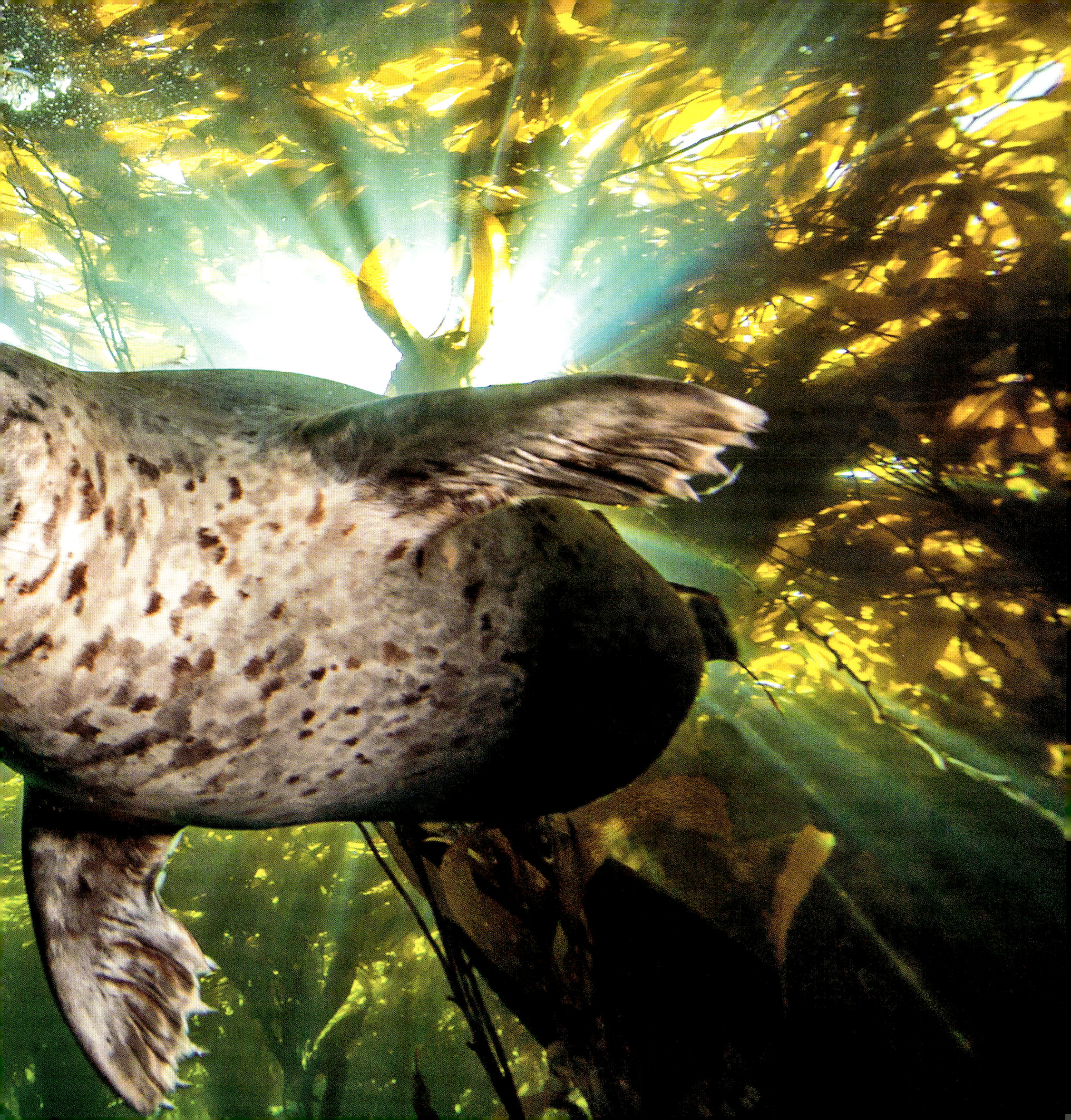

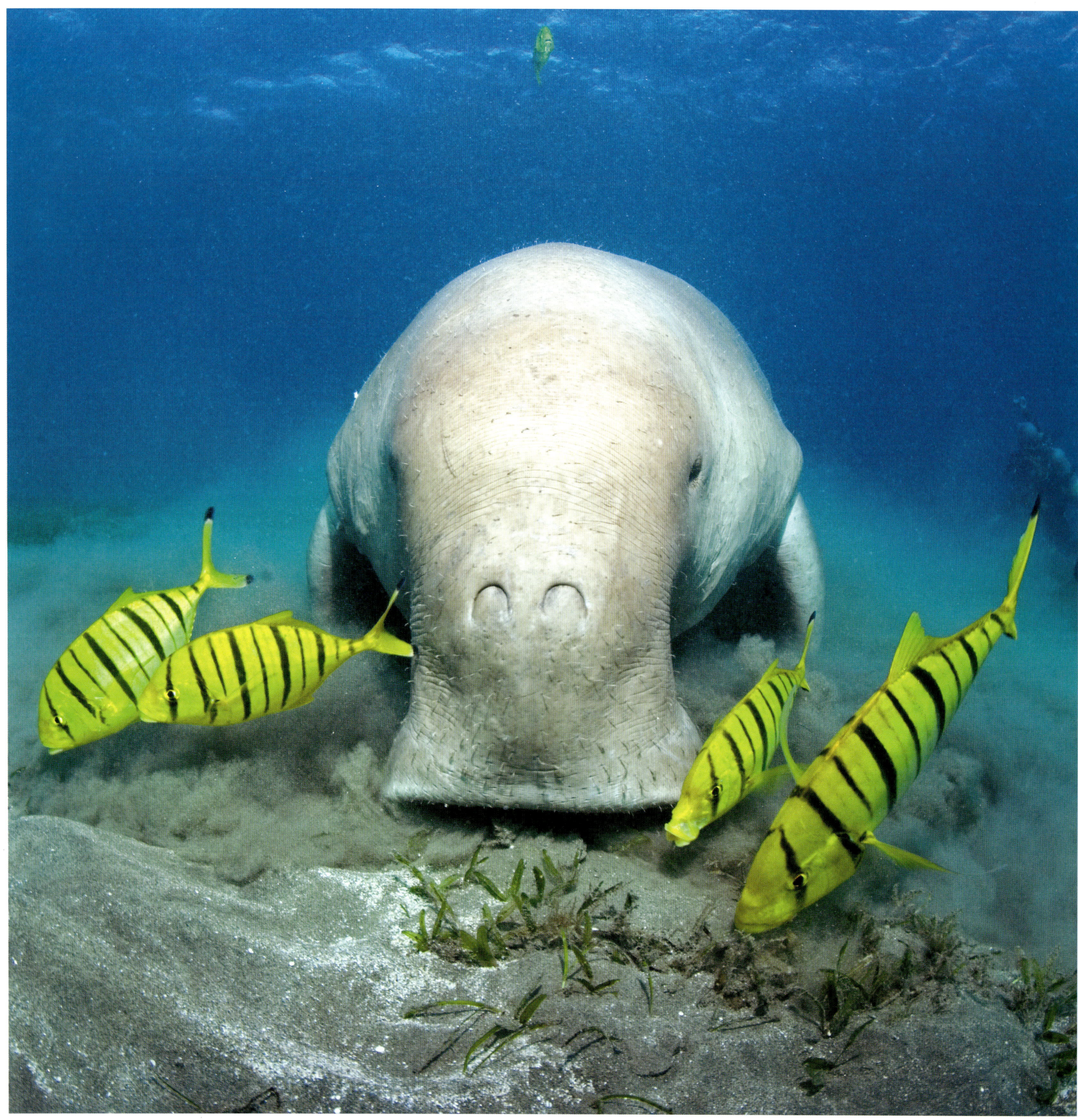

O, wonder!
How many goodly creatures are there here!

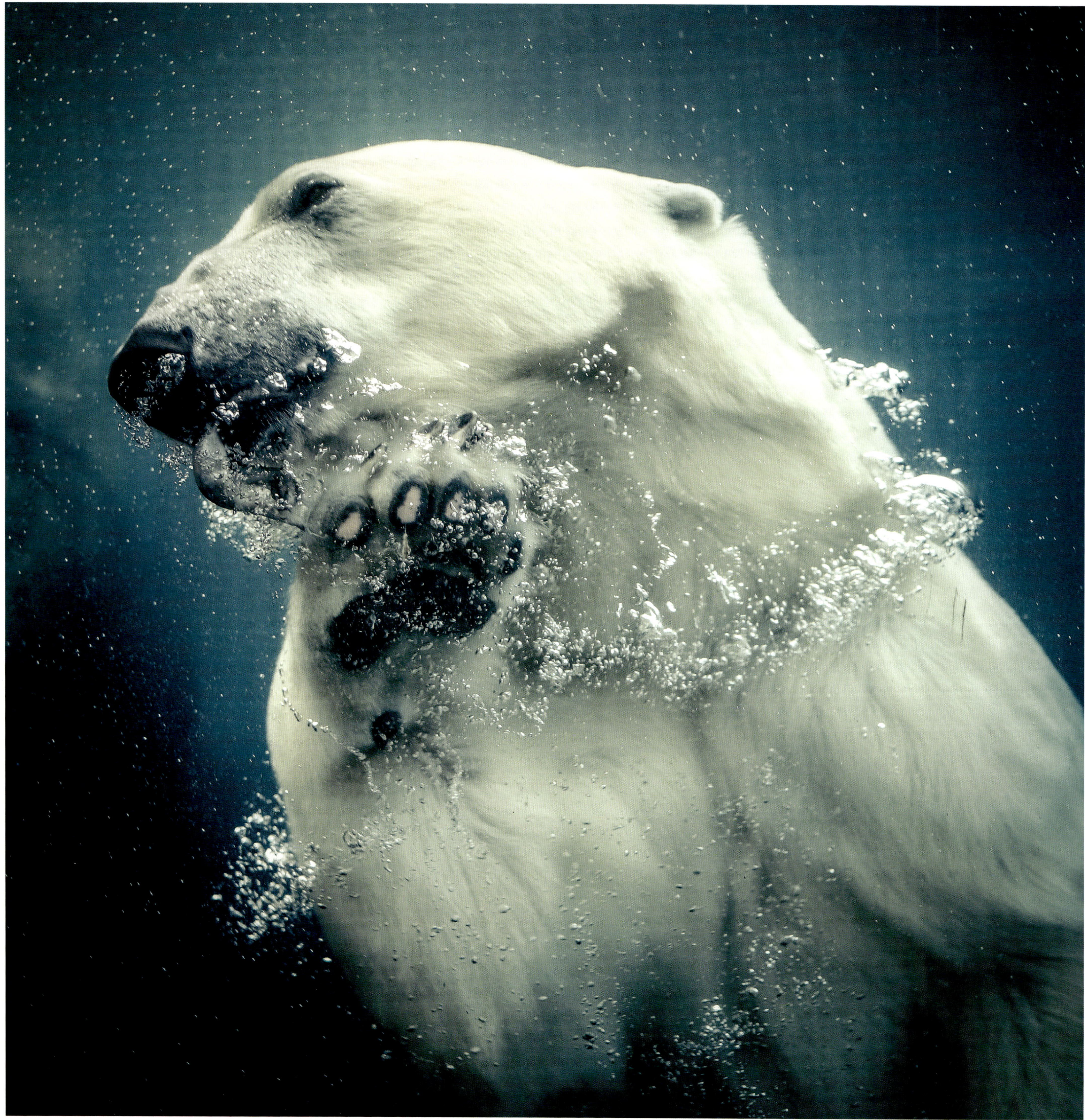

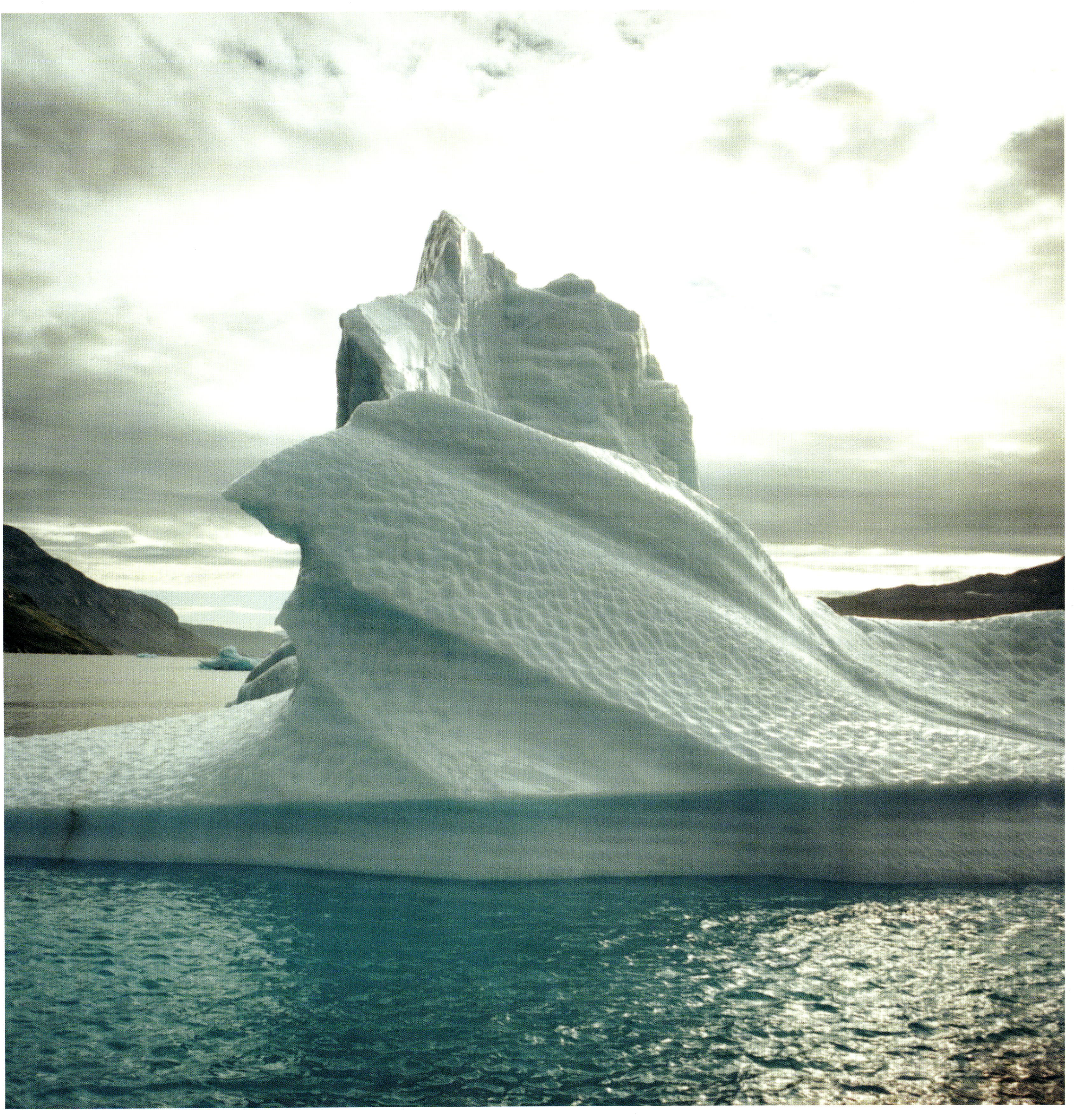

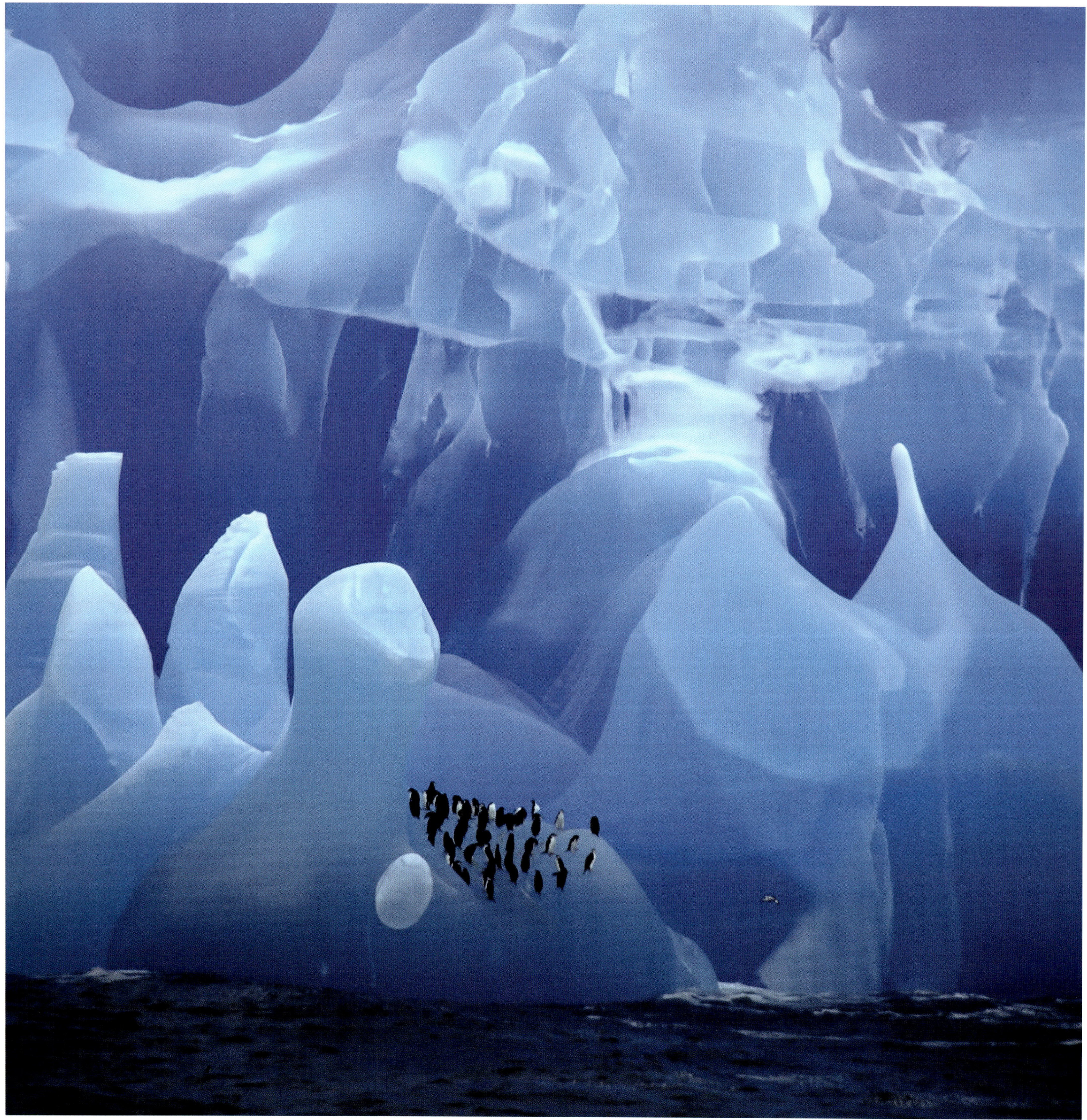

Vast movements of ocean and air currents bring
dramatic change throughout the year. And in a few
special places, these seasonal changes create some
of the greatest wildlife spectacles on Earth.

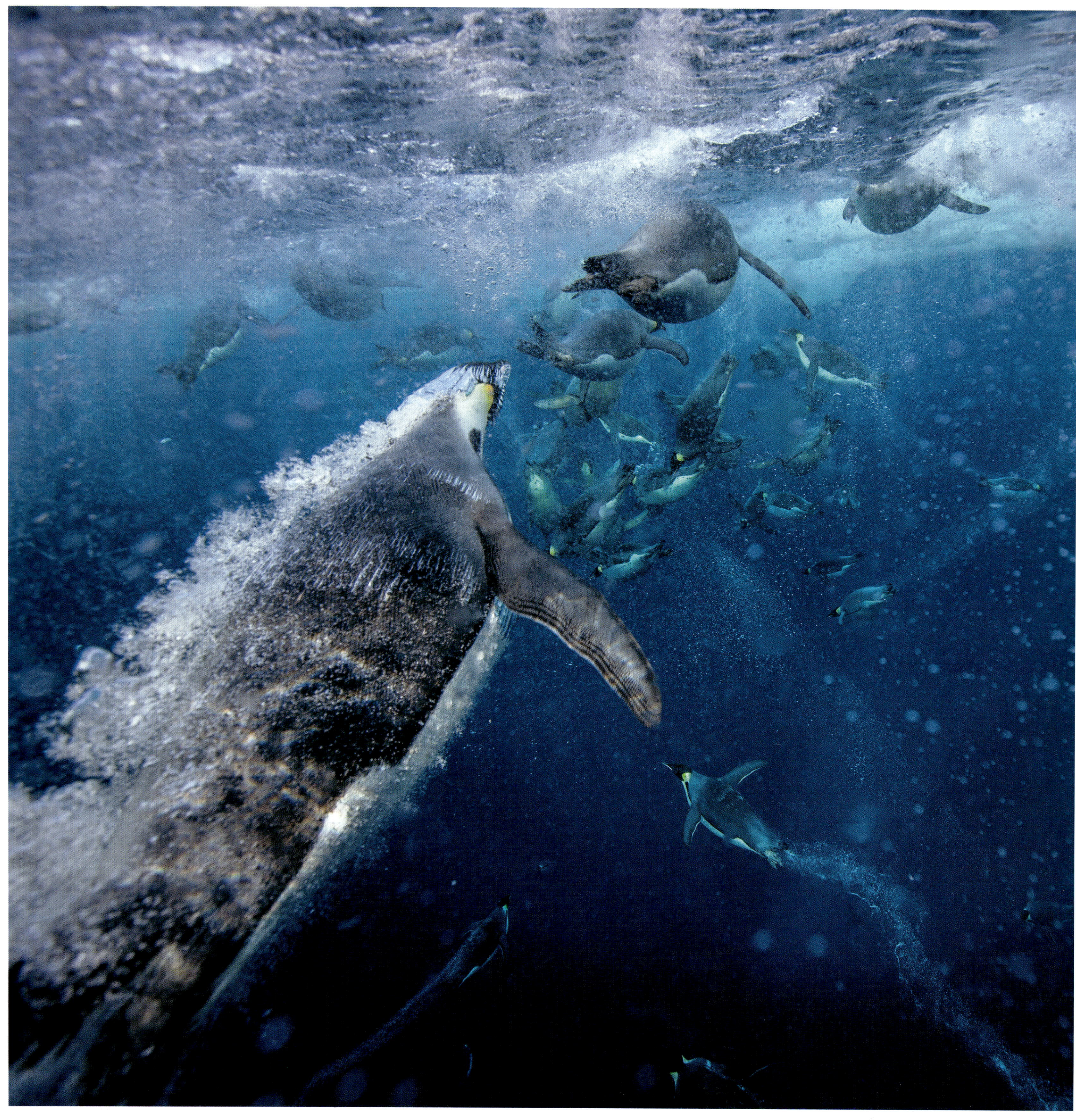

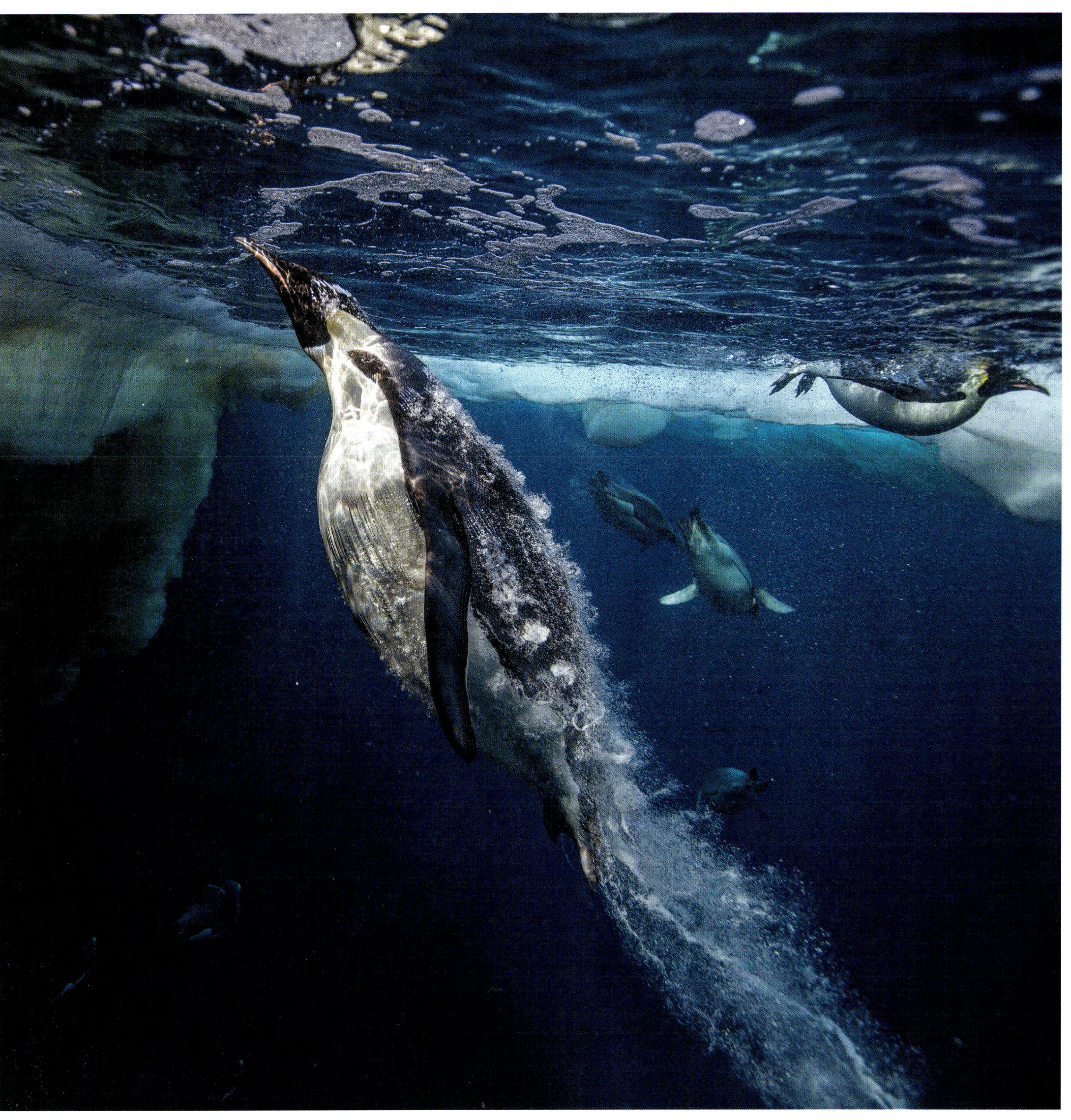

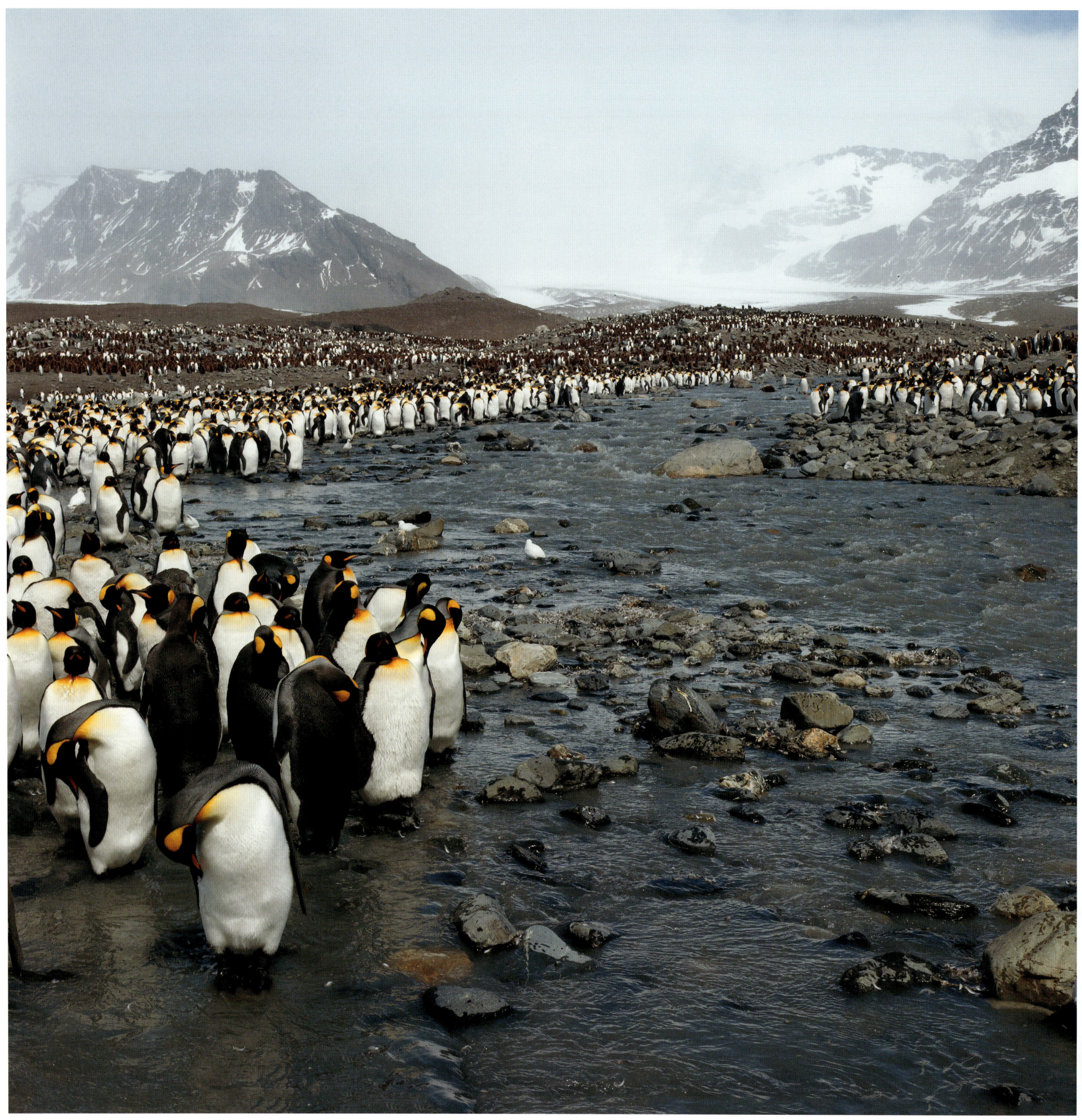

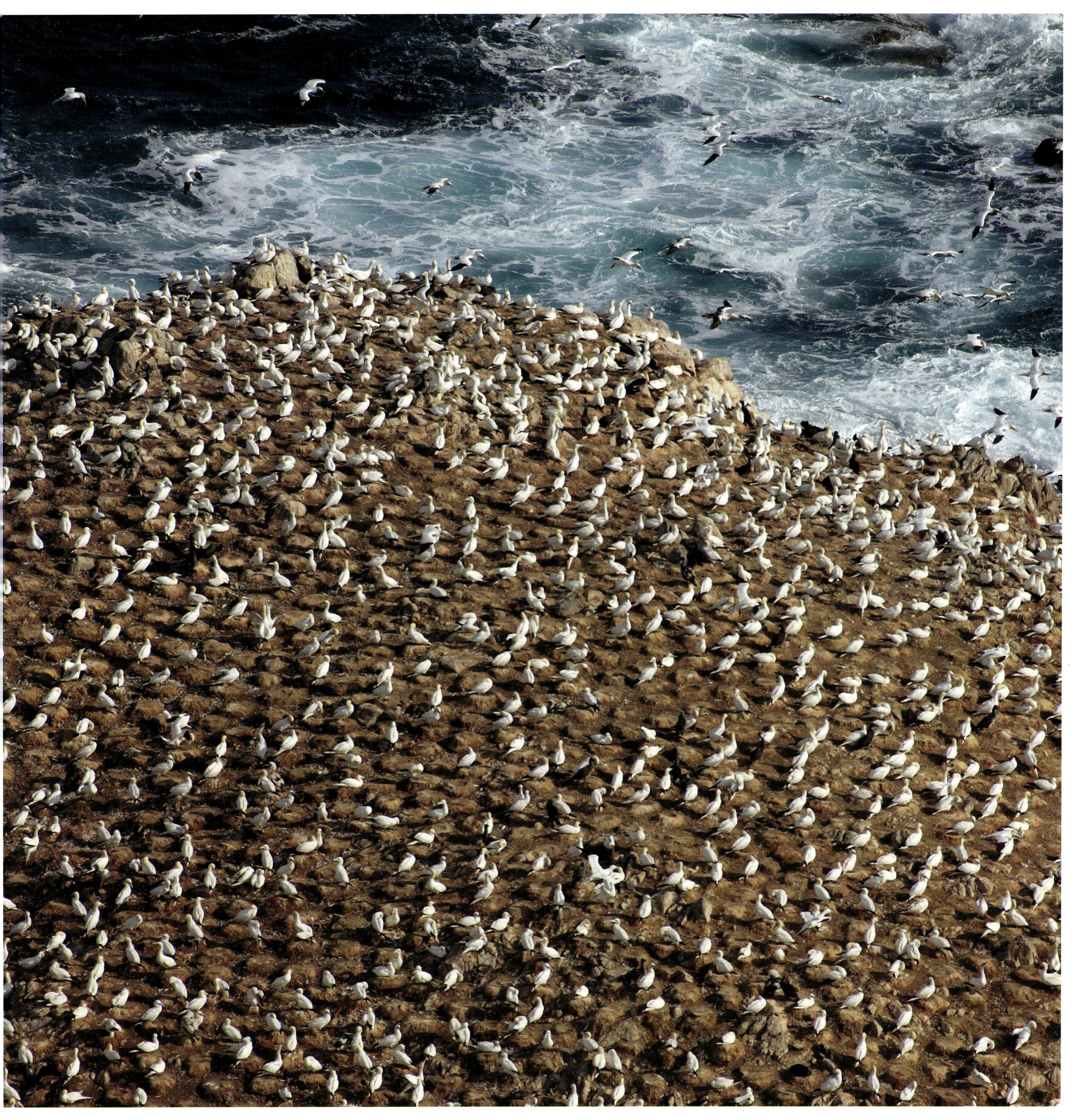

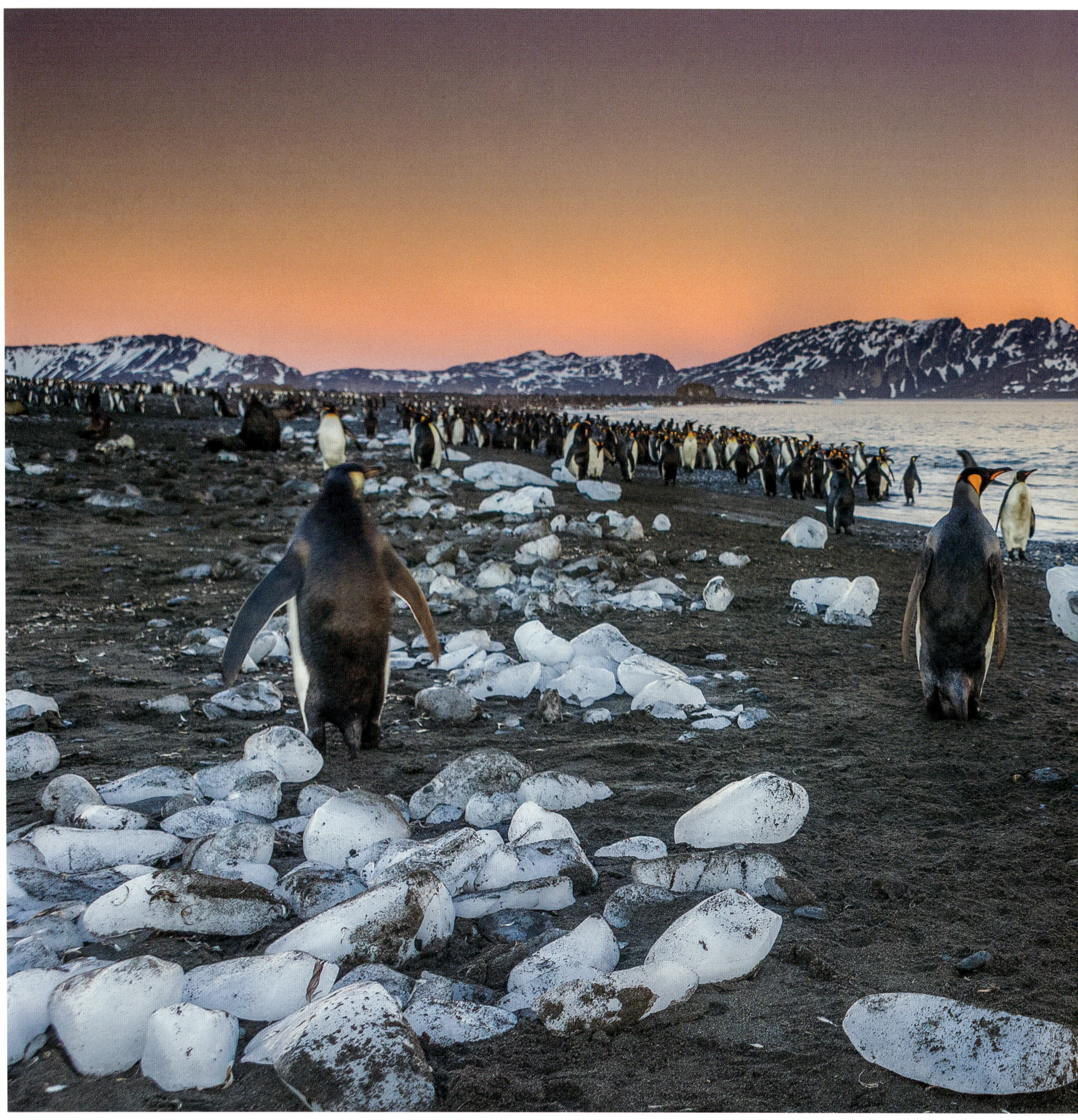

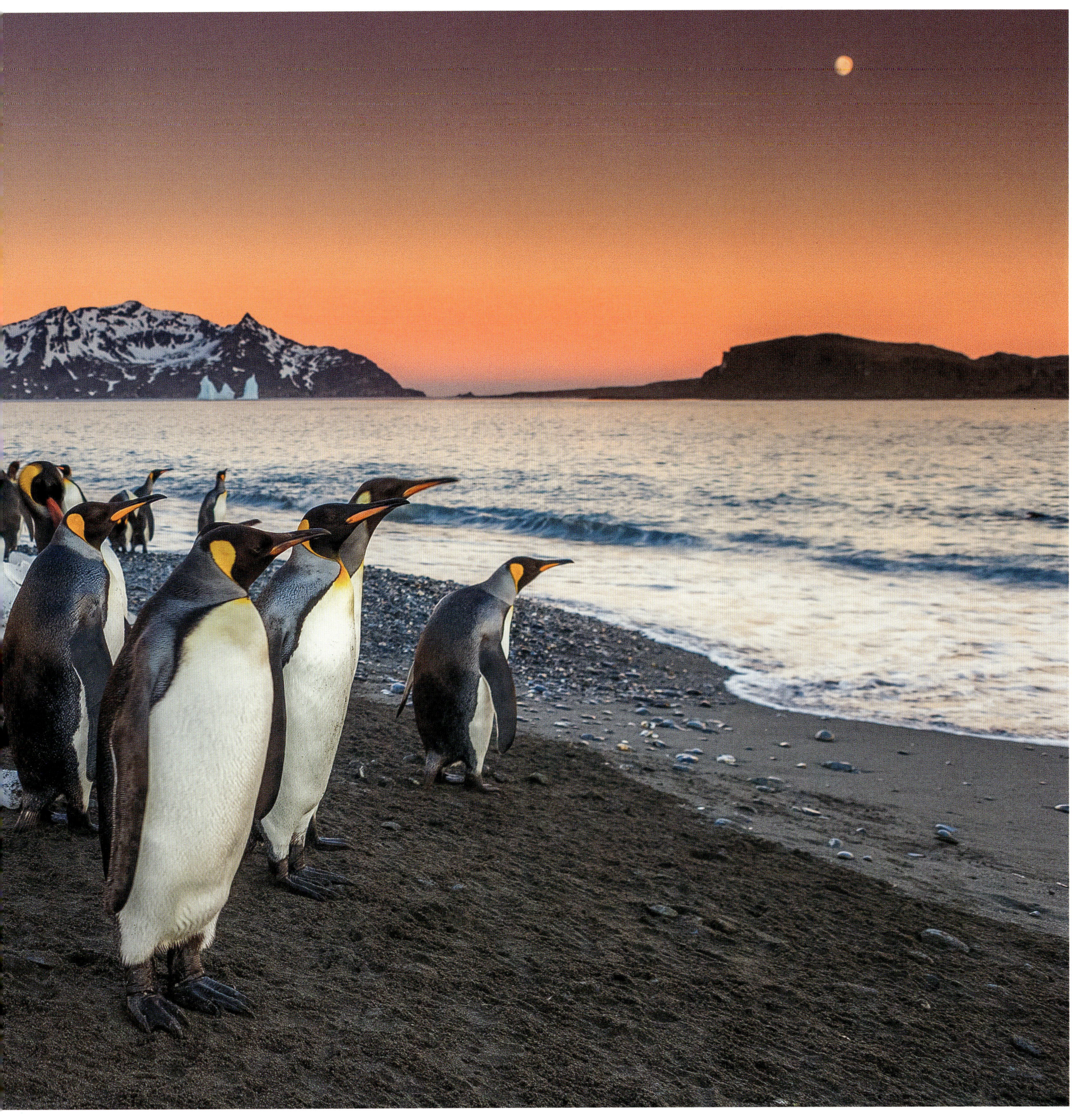

Deep

On January 23, 1960, Jacques Piccard and fellow deep-sea explorer Don Walsh made history by reaching what is thought to be the deepest point of the world's oceans, the Challenger Deep in the Mariana Trench. This 7-mile-long (11-kilometer-long) depression is estimated to be between 35,700 and 35,800 feet (10 and 11 kilometers) deep and sits at a point on the earth's surface that is a long way from anywhere you are likely to have heard of—200 miles (320 kilometers) southwest of the island of Guam, in the western Pacific Ocean. Here is Piccard recalling his first impression:

As we were settling this final fathom, I saw a wonderful thing. Lying on the bottom just beneath us was some type of flatfish, resembling a sole, about one foot long and six inches across. Even as I saw him, his two round eyes on top of his head spied us—a monster of steel—invading his silent realm. Eyes? Why should he have eyes? Merely to see phosphorescence? The floodlight that bathed him was the first real light ever to enter this hadal realm. Here, in an instant, was the answer that biologists had asked for the decades. Could life exist in the greatest depths of the ocean? It could! And not only that, here apparently, was a true, bony teleost fish, not a primitive ray or elasmobranch. Yes, a highly evolved vertebrate, in time's arrow very close to man himself.

It was more than fifty years before anybody returned to those depths. That's far enough to drop Mount Everest in and still have room for Mount Washington on top, and more. Or pick your own combination. Anyway, the next person down was the film director and deep-sea enthusiast James Cameron. First known for the *Terminator* films, then for the Oscar-laden *Titanic* (for which he went down to the bottom of the Atlantic and filmed the wreck of the original ship), Cameron is no dilettante when it comes to ocean exploration. He has used considerable amounts of his film profits to help pioneer new technology for deep-sea diving and related science. And so it was that on March 26, 2012, he descended into Challenger Deep with the hope of new experiences that might extend the story. However, he recorded a significantly different impression than Piccard and Walsh's. During the three hours that his craft explored the bottom of the world—far exceeding the brisk twenty minutes of his predecessors' visit—he reported seeing no fish and nothing alive larger than "an inch." The only things he saw swimming were tiny shrimp-like creatures, scavenging amphipods.

Piccard's tale is now being questioned, as his fish sighting doesn't fit with what has been seen at such depths. However, what Piccard couldn't know, but was perhaps right in spirit about, is that down in the depths there is a lot of life—so we can be grateful that he encouraged our curiosity, even if anything as fishlike as he identified is still to be witnessed by others at the very bottom. In contrast, Cameron had reason to feel disappointed: His own solo journey is so far perhaps the most remarkable aspect of his trip, rather than any amazing revelations. He came back with comparatively thin material compared to his usual gripping narratives, or indeed compared to the growing stories of discovery from deep-ocean environments. We are just starting to see into an almost entirely unexplored realm, one in which danger is ever present, but so, too, is the great chance of finding new species.

Respect is due to anybody prepared to sit all alone in a small, death-defying capsule for many hours, descending and ascending mostly in total darkness, except for the lonely light of the capsule itself, to visit an environment where the water pressure is more than a thousand times what it is at the surface. Malfunctions on Piccard and Walsh's vessel (a cracked outer window prompted a desire to return to the surface pretty quickly) and Cameron's (a broken robot arm obscured the view, which also shortened the trip) indicate the dangerous conditions. Not long after Cameron's trip, the unmanned underwater vehicle *Nereus*, which had successfully descended the Mariana Trench in 2009, was lost—suspected to have imploded under pressure—in the Kermadec Trench off New Zealand. It is suspected that although "only" 33,000 feet (10 kilometers) down, unusually heavy pressure conditions—above those expected at 36,000 feet (11 kilometers), the craft's designed load point—caused the failure.

These obstacles should be no surprise. Fewer people by far have been to the far ocean depths than have visited the moon; less of the place has been seen or sampled; much less is known for sure. While Challenger Deep is thought to be the anti-Everest, the extreme point of depth, we may yet be surprised by further discovery of other depths that challenge it. Year by year, revelations are brought up from the darkness. We have come to realize that there is not one environment down there, but numerous environments, often small and rapidly changing, with each potentially having conditions that nurture new species; each potentially able to revise and extend our knowledge.

Today we see the oceans as mostly having a similar structure, the result of how shifting tectonic plates have come to form the earth's surface, with the addition of billions of years of sedimentary deposits alongside occasional volcanic activities around the edge of the plates and sometimes elsewhere. At the edge of land there is a tidal area, then potentially a degree of shallow water that gives way to the continental shelf, which then drops away at the continental slope—often a very steep descent—and down to a gentler fall, the continental rise, which runs into the abyssal plains that form the bulk of the vast area at the center of oceans. This is how the ocean meets the continents, but there are also many variants for oceanic islands, which are formed as a result of volcanic activity on the seafloor. Here, the drop away from the shore may be dramatic and involve no shallow water or gentler slope. The distances down, and the water spaces involved, are vast. These are worlds that we cannot inhabit in any natural way. For a scuba diver, anything over 100 feet (30 meters) is seen as a deep dive as it brings increased hazards of nitrogen narcosis and decompression sickness ("the bends"), while the most advanced professional diving kit allows highly trained specialist divers to descend to a depth approaching 2,000 feet (600 meters)—pipelines have been fixed by divers working in special suits at beyond 1,600 feet (500 meters). More people have walked on the moon than have achieved a scuba dive below 1,000 feet (300 meters). This is still close to the surface when we think that the average depth of the ocean is more than 12,000 feet (4 kilometers). Deep ocean is described as anything over 650 feet (200 meters), and that accounts for 66 percent of the earth's surface. Furthermore, 80 percent of water volume is below 3,000 feet (1 kilometer) down.

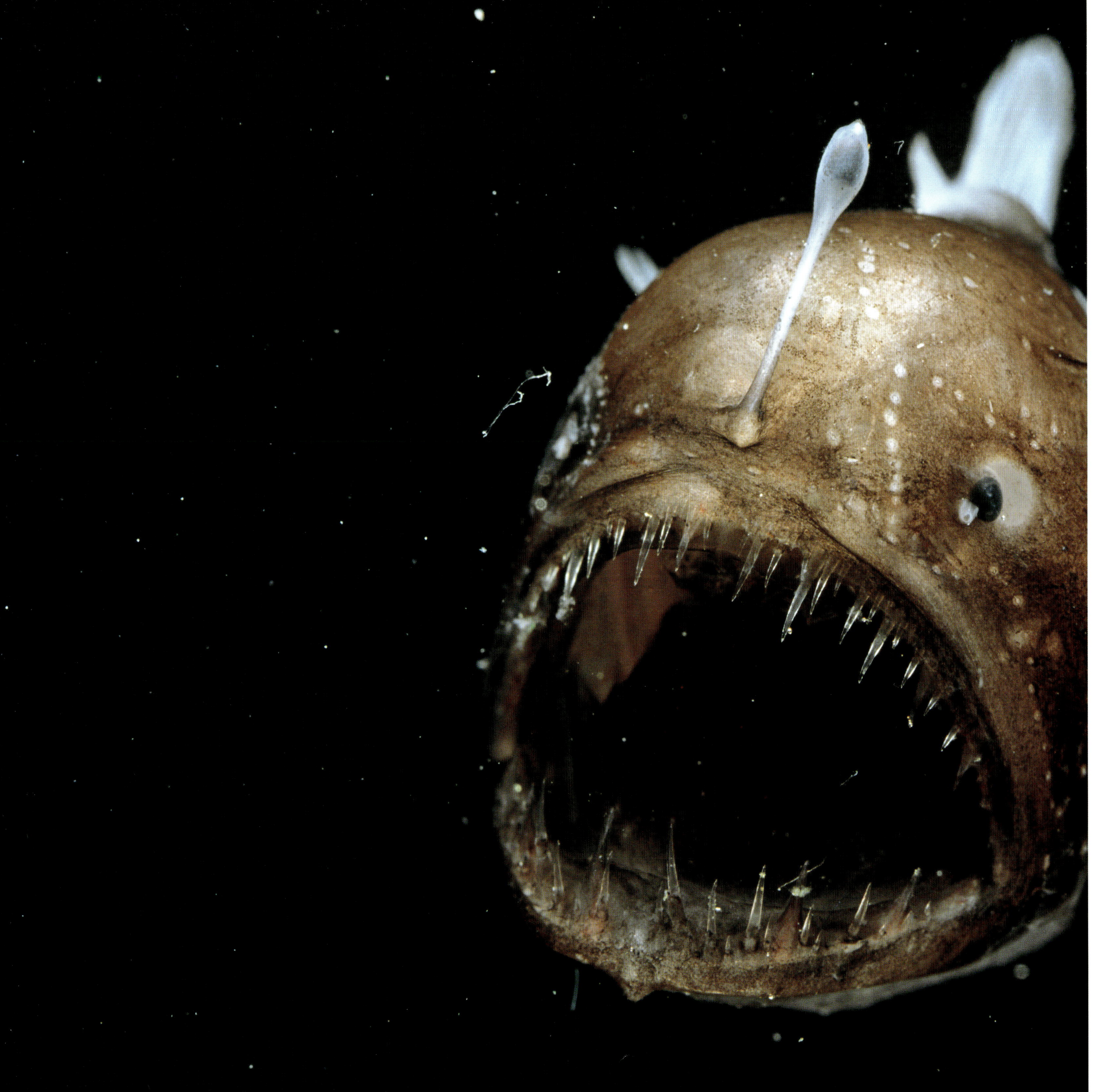

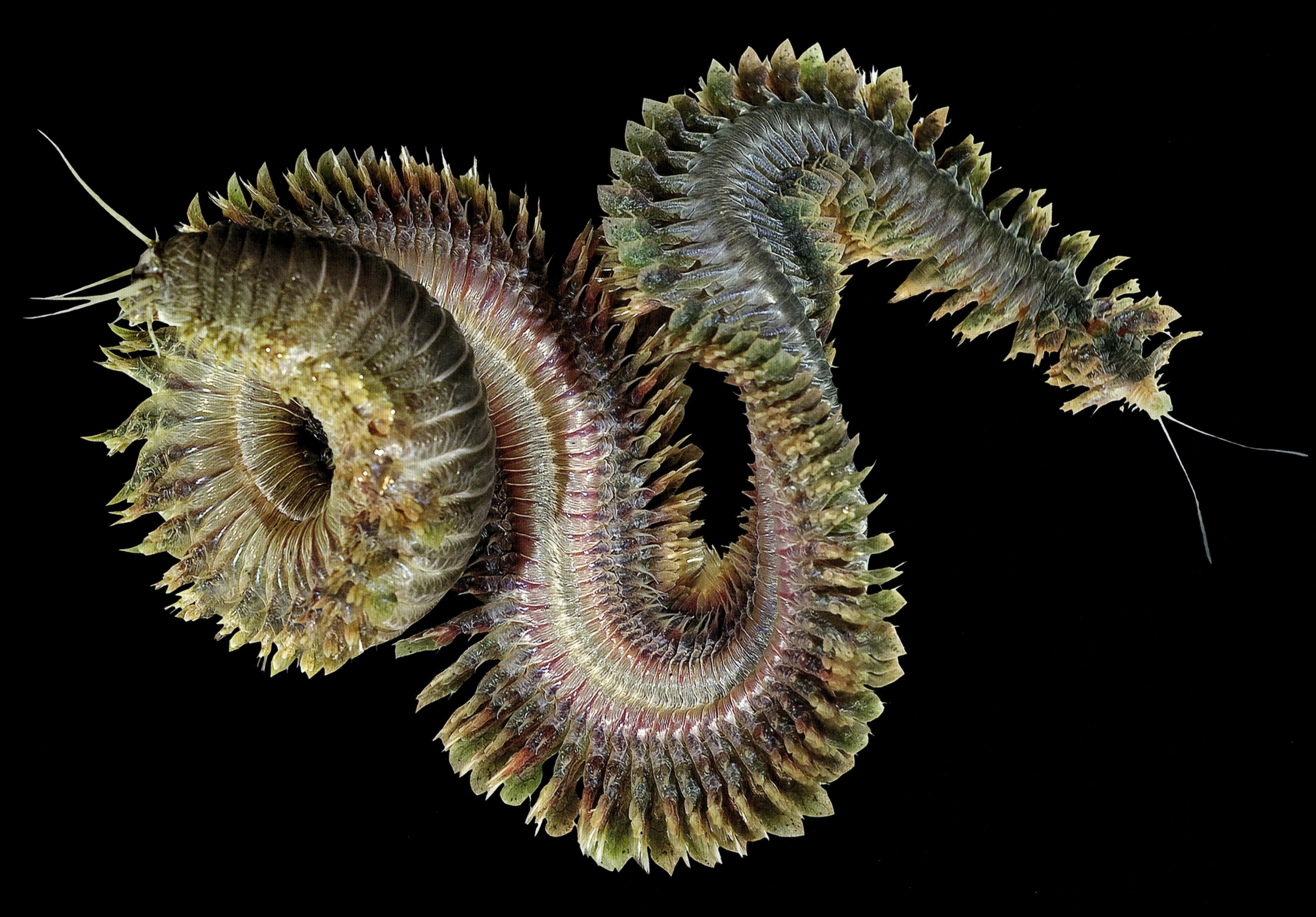

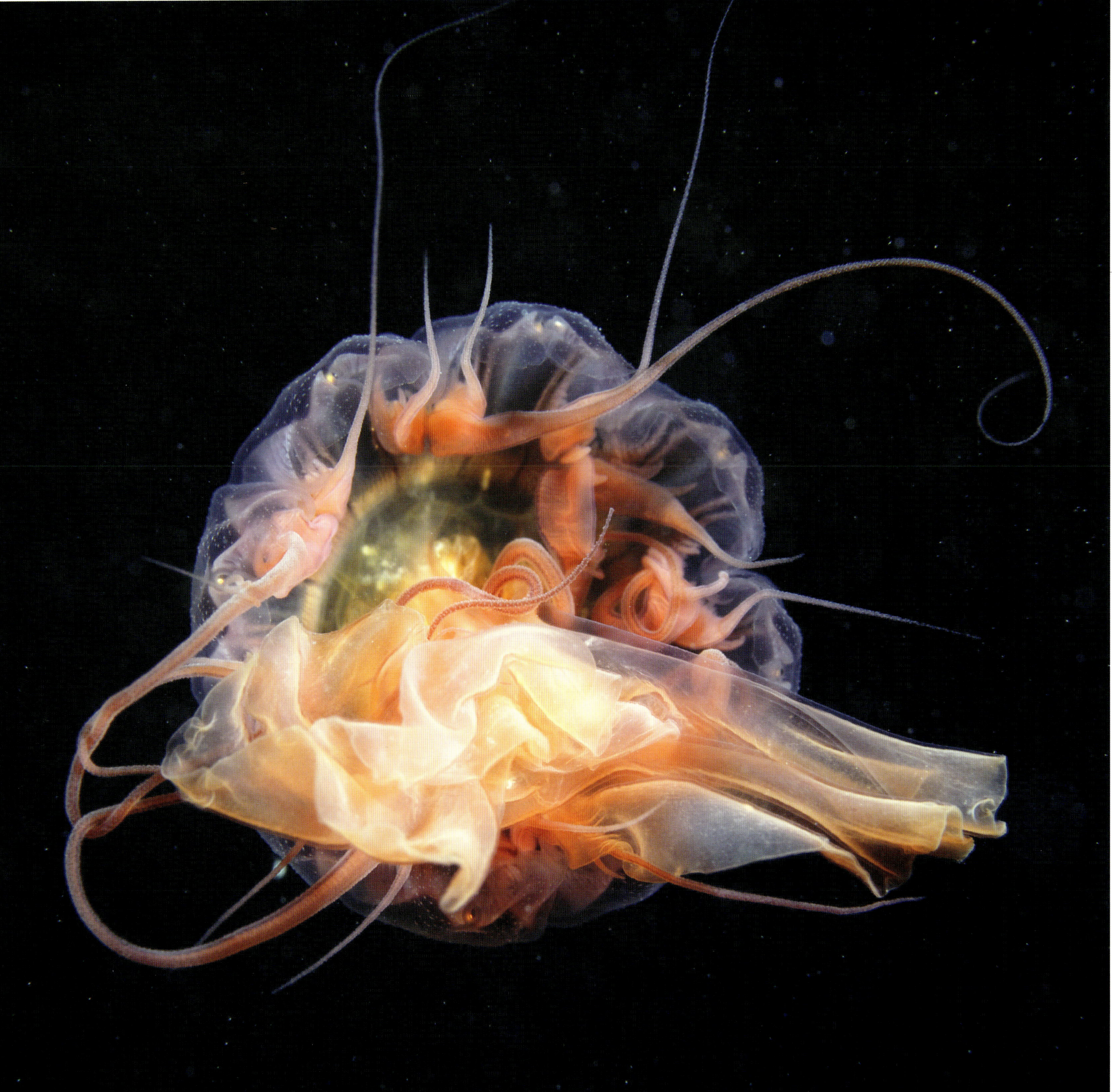

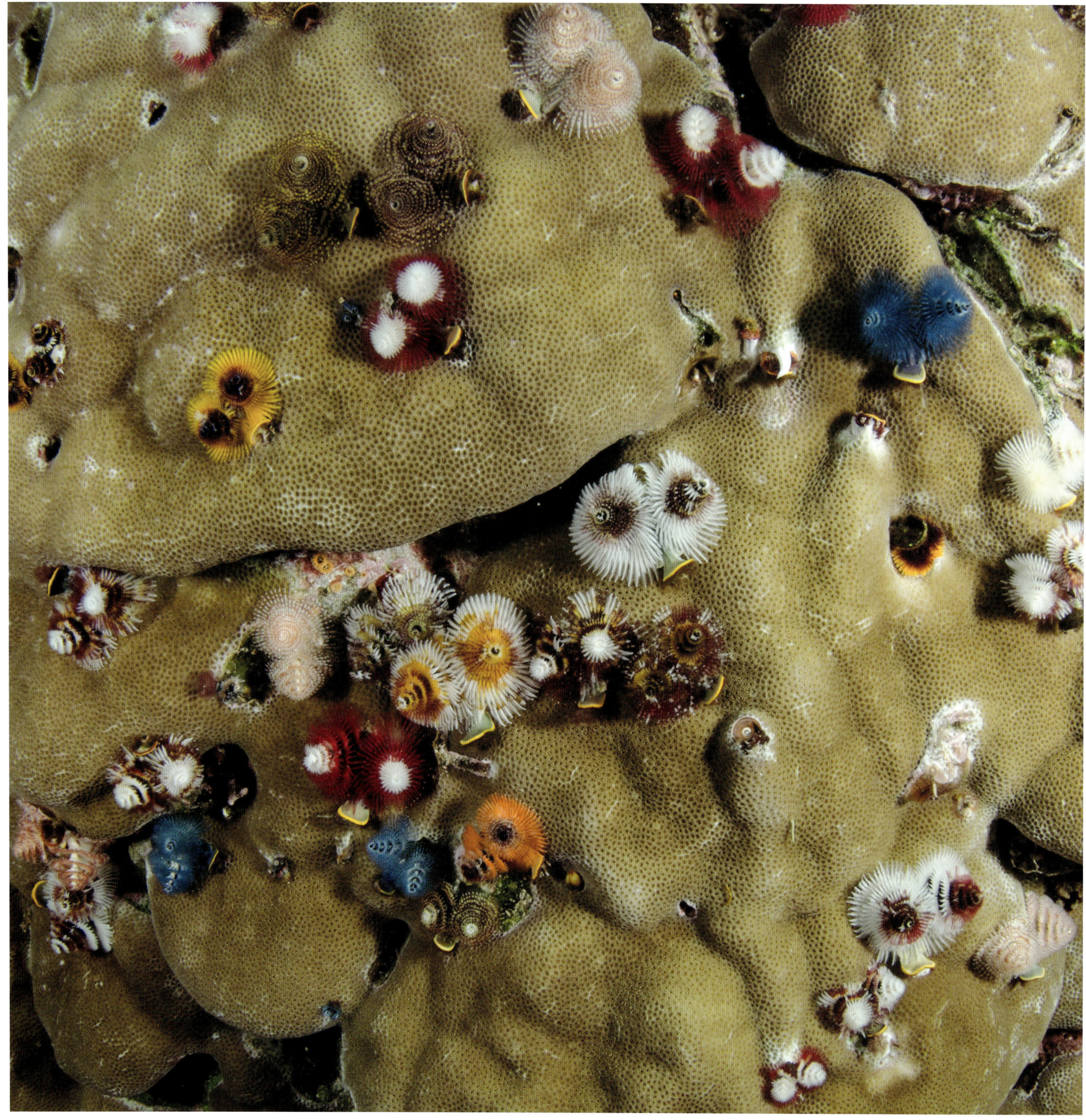

The living world is mostly not as we see it, or can see it. Four-fifths of the water world is dark, cold, and involves crushing conditions more than one hundred times greater than atmospheric pressure.

It is only comparatively recently that we've had some broad sense of the scale of the deep sea and its variety. The name of Cameron's craft, *Deepsea Challenger*, pays homage to HMS *Challenger*, which carried out the first global oceanographic survey in the 1870s, not so far back. This former British warship, converted to carry naturalists and other scientists, circumnavigated the globe while taking samples of water and marine life. During a three-year journey of more than 68,000 nautical miles (125,000 kilometers), the ship's team discovered around 4,700 new species. In showing just how much life might be down in the oceans, the *Challenger* expedition was revolutionary. It disproved the prevailing theory of the time, that the seas were largely dead zones below the point that any light penetrated. This notion was developed by the eminent naturalist Edward Forbes after expeditions in the Mediterranean Sea during the 1840s. He coined the term "azoic zone" (meaning lifeless zone) for water deeper than 300 fathoms (1,800 feet or 540 meters). But the *Challenger* expedition finally refuted that, and the azoic theory is now seen as a classic of misguided thinking, based on inadequate sampling. On *Challenger*'s journey, crisscrossing the Atlantic, Indian, Antarctic, and Pacific Oceans, it made 492 laborious soundings (depth readings) at 360 locations. This involved running out a weighted line with a dredging tool at the end, which was prone to various mechanical problems and accuracy issues and looks archaic compared to our side-scan sonar technology of today. Despite the primitive equipment, *Challenger* discovered what we know now as Challenger Deep. It also discovered the Mid-Atlantic Ridge, the vast sub-ocean mountain range that is like a seam down the side of the world, huge peaks along which runs a crack marking the pulling apart of the plates.

The geological and biological evidence brought back by *Challenger* encouraged notable Victorian scientists to speculate around the case for the lost land of Atlantis, playing on the myth of a great civilization that had disappeared beneath the waves. The evidence for that was not really there, other than that this large under-ocean ridge suggested some different cause and effect for the ocean's evolution than had previously been thought. The scientists on the ship were more inclined to be excited by the range of new species discovered and the similarities and differences of the samples. Their insights into the underwater landscape supported a sense of the oceans as being very old and of a greater size than ever before understood.

Within a few years the new insights into the undersea mountains and abysses, and the connections between species around the world, were interpreted in a radical new theory by a German meteorologist and explorer, Alfred Wegener. He proposed the idea that continents drifted around, as if floating over the surface of the globe, and that they'd been doing this for millions of years, hence the fossil and geological evidence showing that different continents were once connected. This idea was widely criticized and still lacked much support when Wegener died on an exploration in the snowy wastes of Greenland in 1930. Now it is taken as central to our understanding of how life is connected, as well as influencing further interpretation of plate tectonics and their effect on the surface of the earth. With just a few clues from the *Challenger* expedition and from other research under the oceans, he came up with the basis for much of what we believe happened over the course of Earth's long history. We now reject Wegener's theory of floating continents because we know that continents are driven apart and together as fixtures on the plates of the earth's crust, which are either moving apart (as in the Mid-Atlantic Ridge) or under and over each other (like the Mariana Trench). But, in general, Wegener made a huge breakthrough in seeing the earth as one giant, round Rubik's Cube of sorts, with parts moving into place and thereby affecting other parts.

Future oceanic surveying was affected by some of the major stories of the beginning of the twentieth century. The sinking in 1912 of the ocean liner *Titanic* after it hit an iceberg, and then, in World War I, of the HMS *Lusitania* by a U-boat, pointed out the need to better understand and look into the mysteries of the ocean. The pioneering of echolocation as a way of spotting underwater objects was in part driven by the concerns raised from those disasters. A breakthrough point was a week in June 1922 when a US Navy physicist, Harvey C. Hayes, took around nine hundred soundings on an Atlantic crossing using his new Hayes Sonic Depth Finder, massively increasing the information available on the shape of the ocean floor.

Between 1925 and 1927 the German ship *Meteor* built on that work, with an extensive survey across the Atlantic, its chief purpose being to see if ocean water could provide gold in an economical manner, and in a quantity, to deal with Germany's crippling war-debt repayments. The *Meteor*'s crew succeeded in advancing our understanding of ocean science while coming up short on the main task. Gold may be a trace material in seawater, but it was in no way accessible enough or available in sufficient enough quantities to pay off the debt that would later be seen as a fundamental root cause of World War II. What the expedition did confirm, using an echo sounder, was the path of the Mid-Atlantic Ridge.

It was only in the 1950s that studies led by American marine researchers Marie Tharp and Bruce Heezen—who used data from sonar readings to publish the first physiographic map of the North Atlantic in 1957, and later published the first map of the entire ocean floor in 1977—gave a firm interpretation of the profile of the Mid-Atlantic Ridge as confirmation of Wegener's theory, with the added appreciation that the crack in the ridge (which Tharp and Heezen incidentally discovered) was a volcanic fissure, evidence of plates moving apart, and that this was the driving force for continental drift. This spreading—at approximately 1 inch (2.5 centimeters) per year per year for the Mid-Atlantic Ridge—is also seen in the East African Rift, a series of rift valleys that is splitting the African plate in two. Eventually the rift will become so deep that it will be flooded by the Gulf of Aden, turning the Horn of Africa into a continental island.

Since then, more sophisticated sonar devices have given us ways of producing detailed mapping of the depth of the oceans and the shape of the seabed. And yet we still don't really know what is down there without having a robotic device or actual person down in the depths looking. We can see things 238,000 miles (384,000 kilometers) away on the moon, but not a few miles down in our oceans.

It's worth reminding ourselves of what happens when we just look at the oceans: We look, but we can't see far, despite water usually being transparent. We can easily look through a small amount of water inside a glass, even with some distortions, but in the sea, water exists with key variations that affect our sight. First, the surface reflects light so that our vision through it is compromised or may be obscured to a point at which we see nothing beneath the waves. This is because the more the surface ripples or swells, the more light is reflected, in effect producing something akin to a mirror effect back to us. If we duck underneath wearing a snorkeling mask we will see much more clearly, but in deep water our vision becomes highly restricted by new factors that come into play, causing everything to peter out into an impenetrable blue. In every drop we have the filtering effect of chlorophyll and organic matter (all those millions of creatures), which cumulatively give color to the water, but this is then affected by how light waves are filtered, so that ultimately only the blue part of the spectrum is left to penetrate the furthest and color our overall perception. It's an illusion—if we drew a glass of seawater out we'd find that the water is of similar clarity at all points—but this light-filtering distortion is a powerful one that has shaped our view of the sea from earliest times. We talk about the "ocean blue" or sometimes describe it as green (an effect of more turbulent water filtering out the light spectrum differently). But imagine if this light filtration was as it is with the air: If we can see up to the moon and the stars, or see satellites drifting across the night sky, could we not see down to the ocean depths? No, not given all the other life that is in the sea, every drop full of living matter, much more than in the air. Even without the light filtering, looking through seawater would always make for a thickly foggy view.

So we keep grasping for clues. And while we know we still know so little, our vision of the ocean is very different from what the men aboard HMS *Challenger* discovered. But they laid the groundwork. Their samples of water from around the world showed just how much life was still to be discovered. The scientific value of the specimens brought back is evidenced by the fifty volumes of reports produced over nineteen years after the *Challenger* docked. Even today, some of the specimens, held at the Natural History Museum in London, are used in research, as many of the *Challenger*'s finds established "type species" that are used to classify new discoveries.

Besides plumbing for depth, the *Challenger*'s scientists pioneered ideas about measuring temperature and currents, as well as trying to sample water and species from different levels in the sea. Most of the processes were not pioneering, which was good, as it was better that the focus was on the finds rather than the methods. As dredge after dredge brought up new life, it must have been an eye-opening, career-shaping moment for the naturalists on board. For the ordinary crew, the memoirs of one seaman, Joseph Matkin, indicate that initial excitement on the wonders of the deep became outweighed by thinking about what was for dinner, or if the ship's rations of Madeira wine might be shared. There was clearly a class division: "The Scientifics," wrote Matkin in a letter home, "keep all their information and specimens to themselves and I dare say you will know as much about the Expedition from the papers, by and bye, as we do who are in the ship." What perhaps remains remarkable is that in the years since, although technology has moved forward dramatically, there is so much still to do. This is partly because researching the undersea world remains a relatively resource-poor area of study (and that dates back to *Challenger*, because despite all the crew's discoveries, there was no inclination to fund a successor mission). Compared to what we spend on sending space probes off to Mars and elsewhere, it seems astounding that we have never invested anything like that amount in looking closely and extensively on a regular basis to find what is down in the sea.

The challenge of knowing the ocean depths is that we are dealing with a four-dimensional world of shape-shifting. Here, everything is changing all the time as water moves and millions of living things move with and through it, both up and down and across in any direction. Currents can quickly change the environment of the bed, scouring the sediments and rearranging whatever is down there. It's even harder to get a fixed idea about the many layers of temperature and salinity and the related life that exist down through the depths. If a probe goes and looks at a place in the water or on the seabed one day, we can be certain that it will be different on another day, or even moments later, when currents and the movements of animals deliver a new experience. A key reason for the movement of creatures is similar to why our species keeps moving around: we go where the money is, and deep-sea life goes where the food is. As the food is typically falling from above it, in often unpredictable ways, the configuration of species below is itself unpredictable. Even at the bottom, sediment is rarely static or the same: It may have different life growing on it as plants, or living in it as creatures, or it may be moved by inconsistent currents. This may influence what food is on offer, and when, and that will influence what life is present. Waves of researchers have refined this model of biodiverse patches since the 1970s. The phrase "rainforests of the deep" has increasingly been used to explain how the unique conditions of often small patches can spawn great biodiversity in the deep sea, with species differentiation and specialization occurring in ways that bear comparison with how the rainforest canopy supports unique ecosystems in just one tree. In the same way that unique beetle species may be found to live in the micro-environment atop a single trunk, so might deep-sea species adapt to survive the immensely specific conditions they find in a deep sea trench, or by a "black smoker" of underwater volcanic venting.

In order to understand how life literally "goes with the flow" underwater, it is time to discuss the weather. Not the weather as we see it, but the kind of deep-water precipitation experienced down through the depths. The first 660 feet (200 meters) of water is what is known as the photic zone, and it is where photosynthesis can occur and thereby support the creation of plant life, which supports the creation of the chain of other life. Below this level, food is what descends from above, or is about hunting down and eating your neighbors. Organic matter is constantly falling down through the water, a varying precipitation of dead or dying creatures and the rejects from messy eating up above. It changes as you go down; each layer is affected by the life and the conditions that apply there

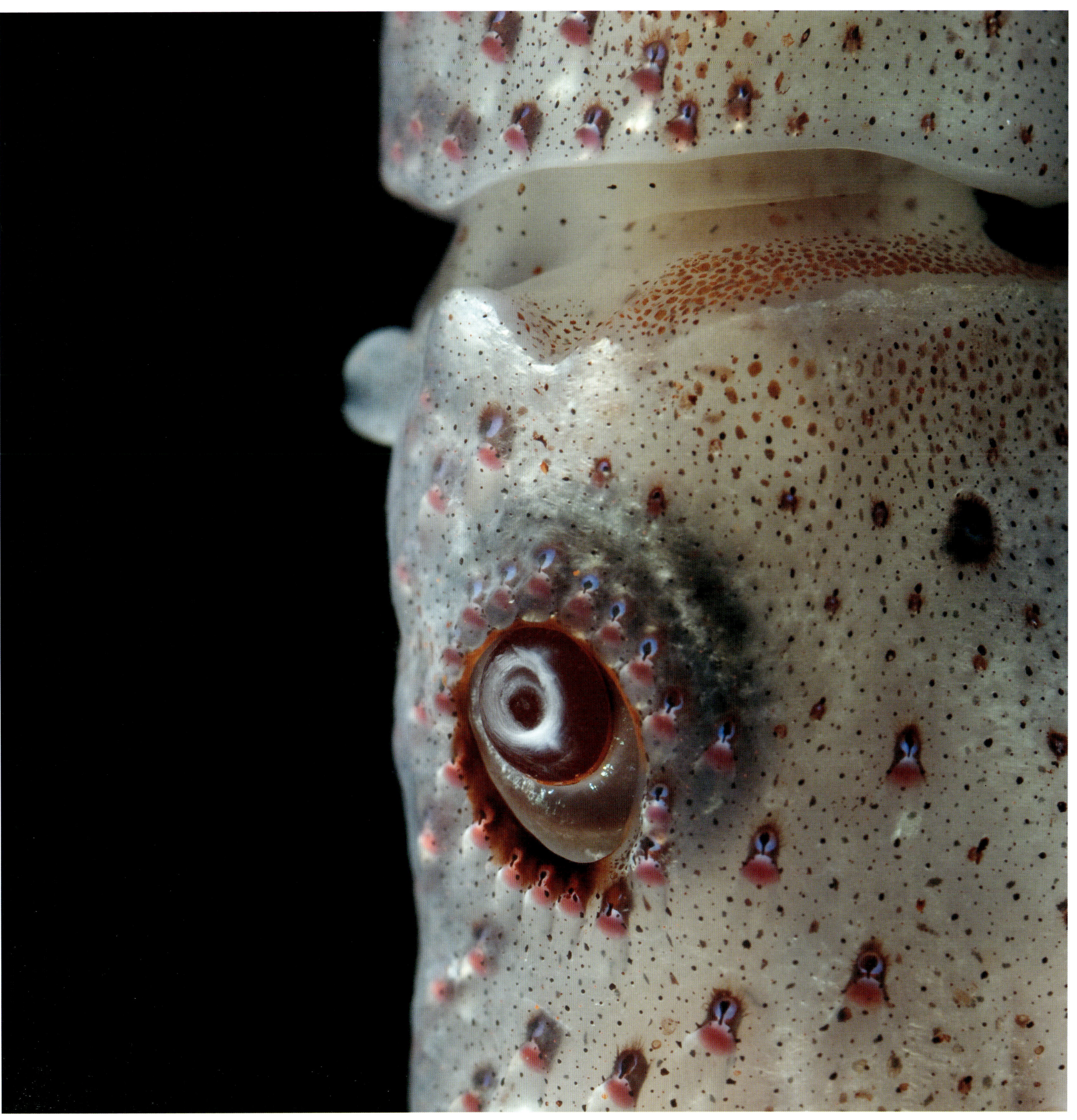

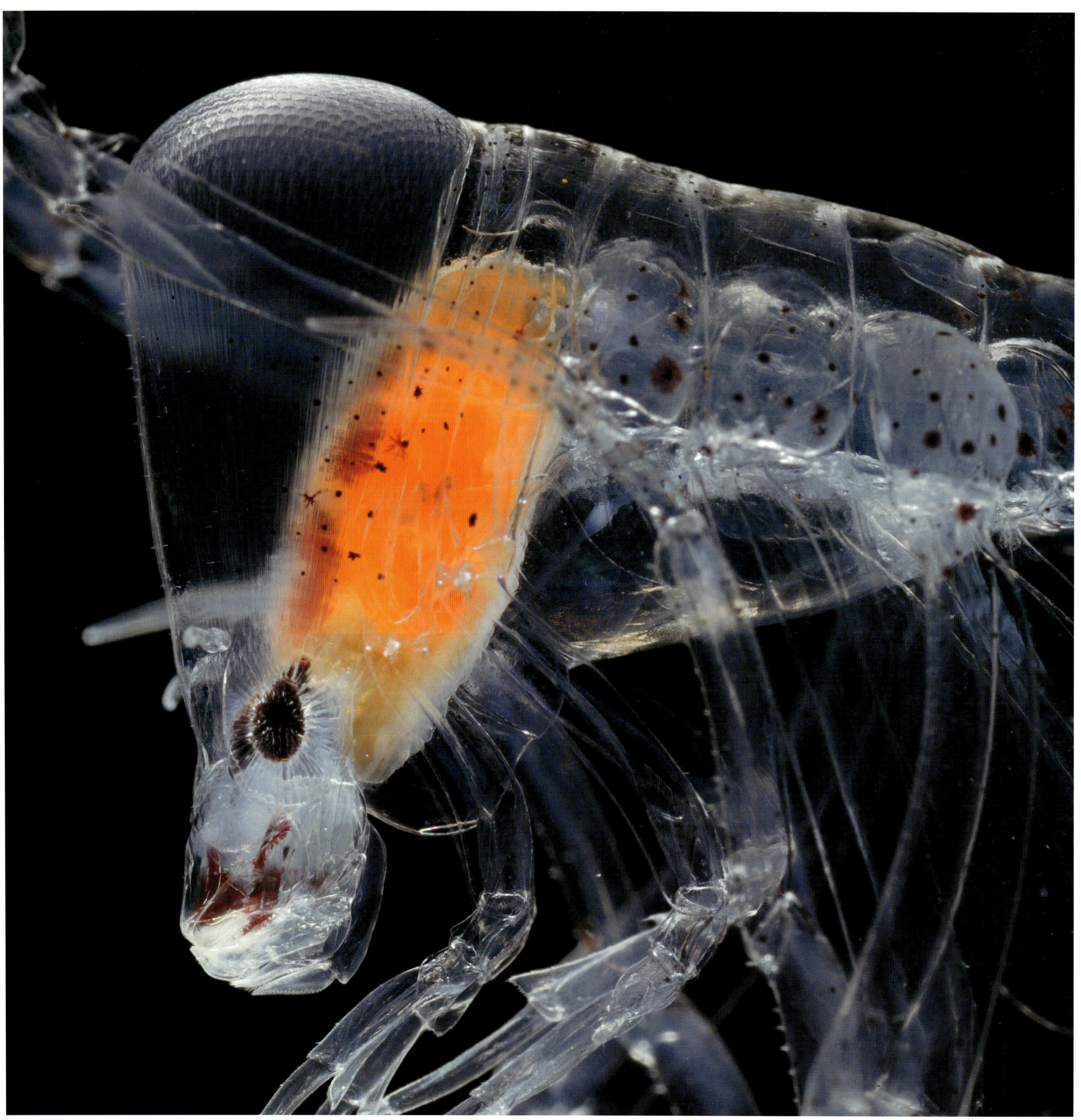

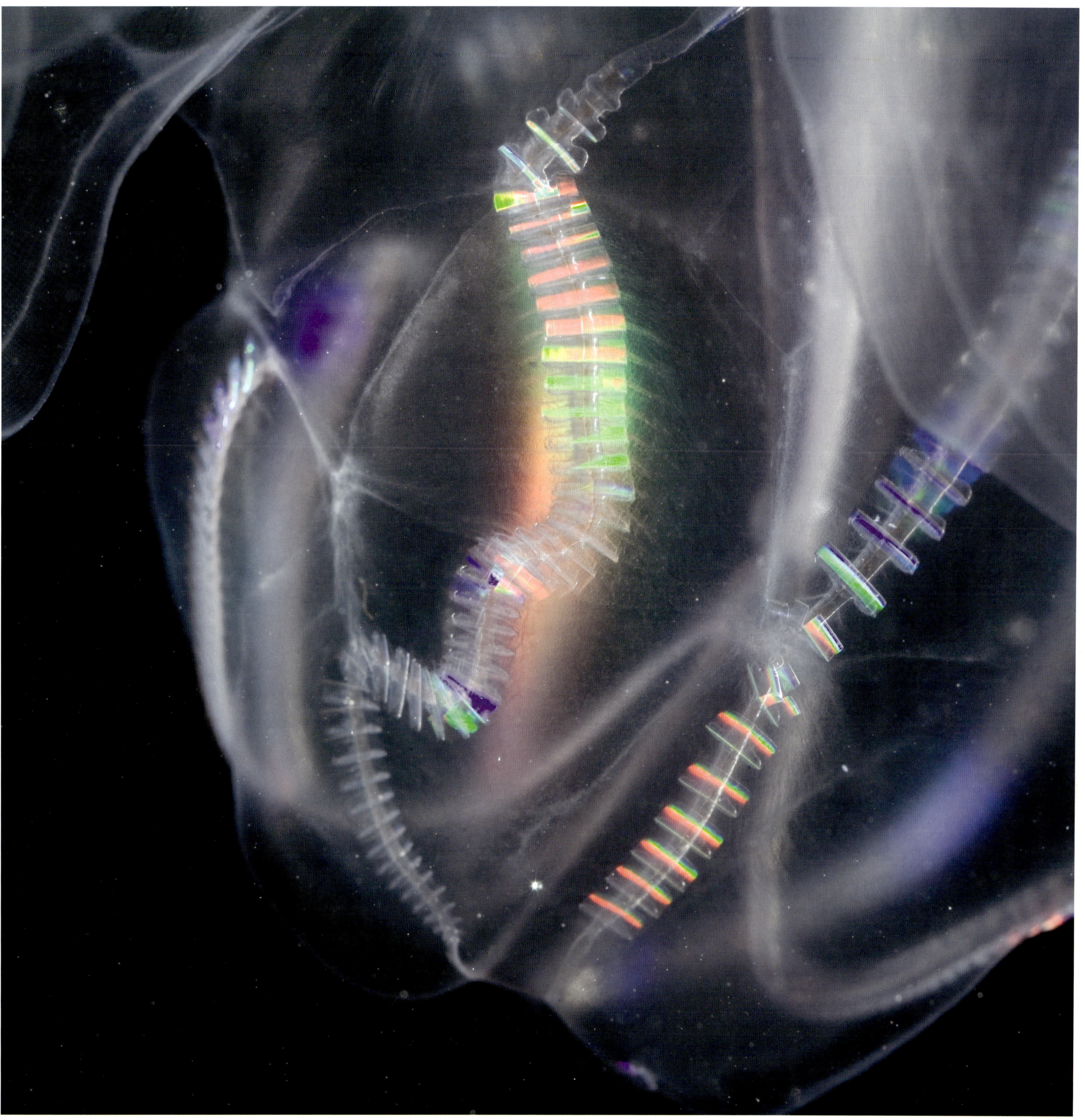

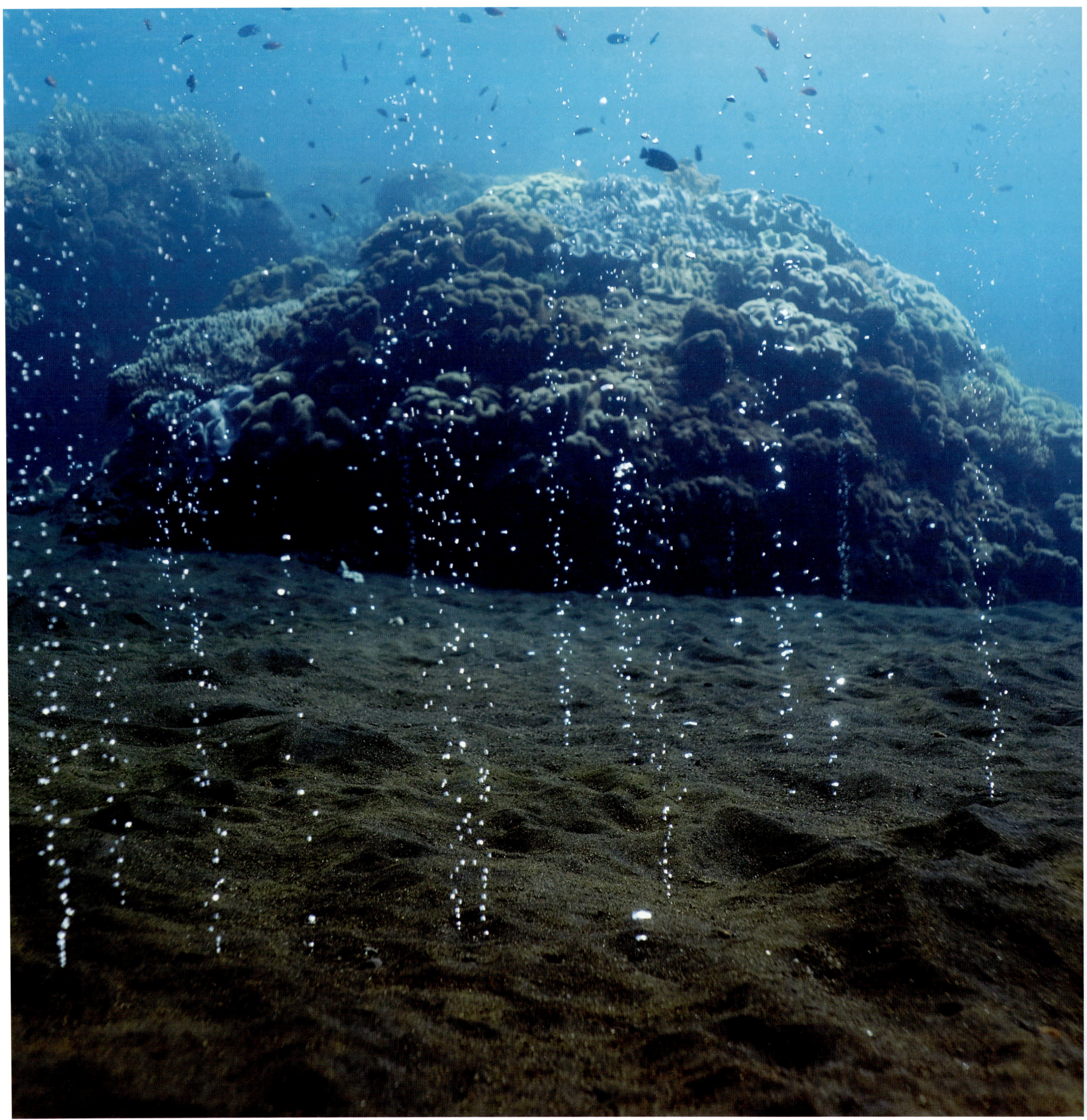

We know more about the surface of the moon
and about Mars than we do about the deep sea floor.

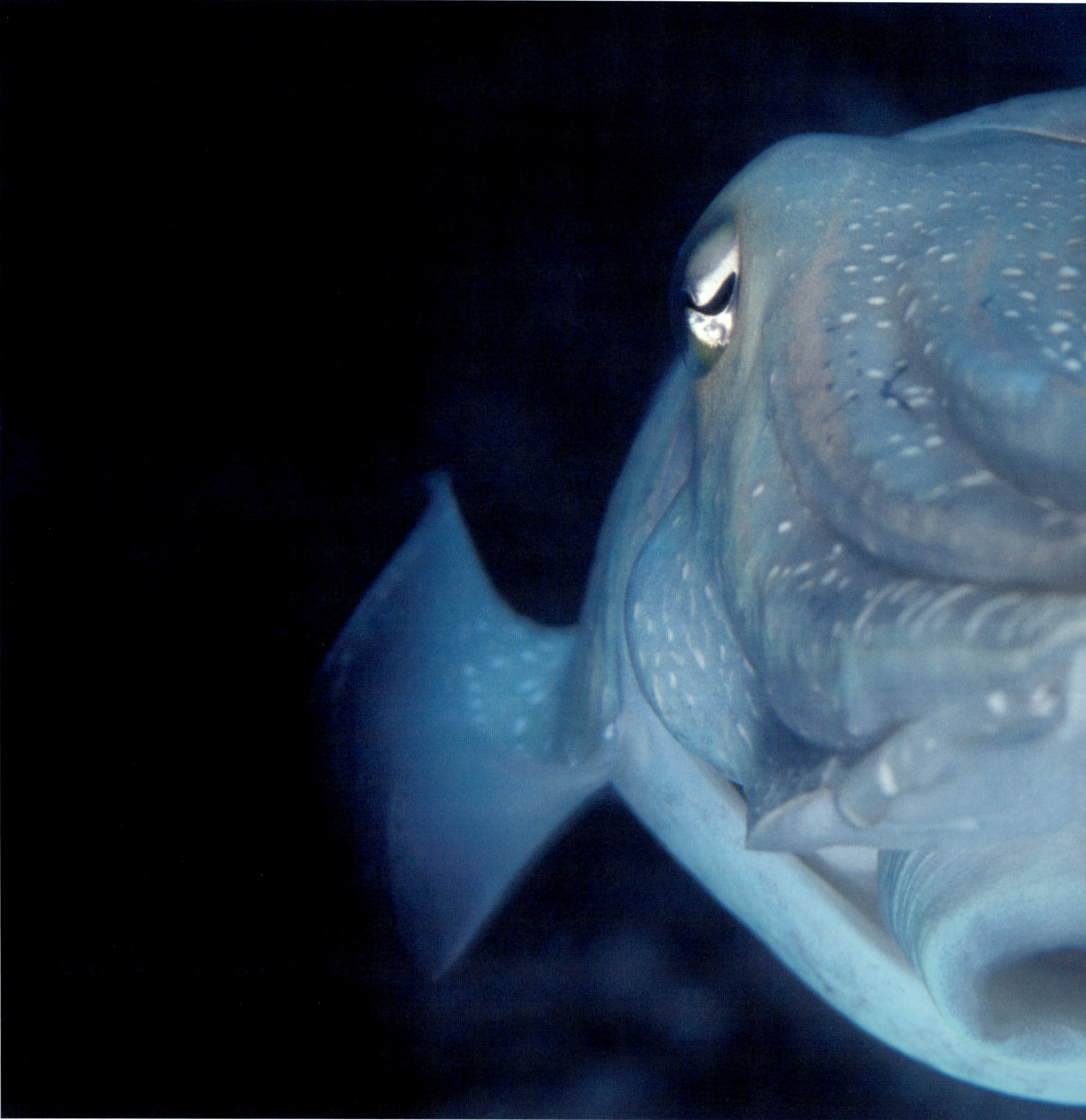

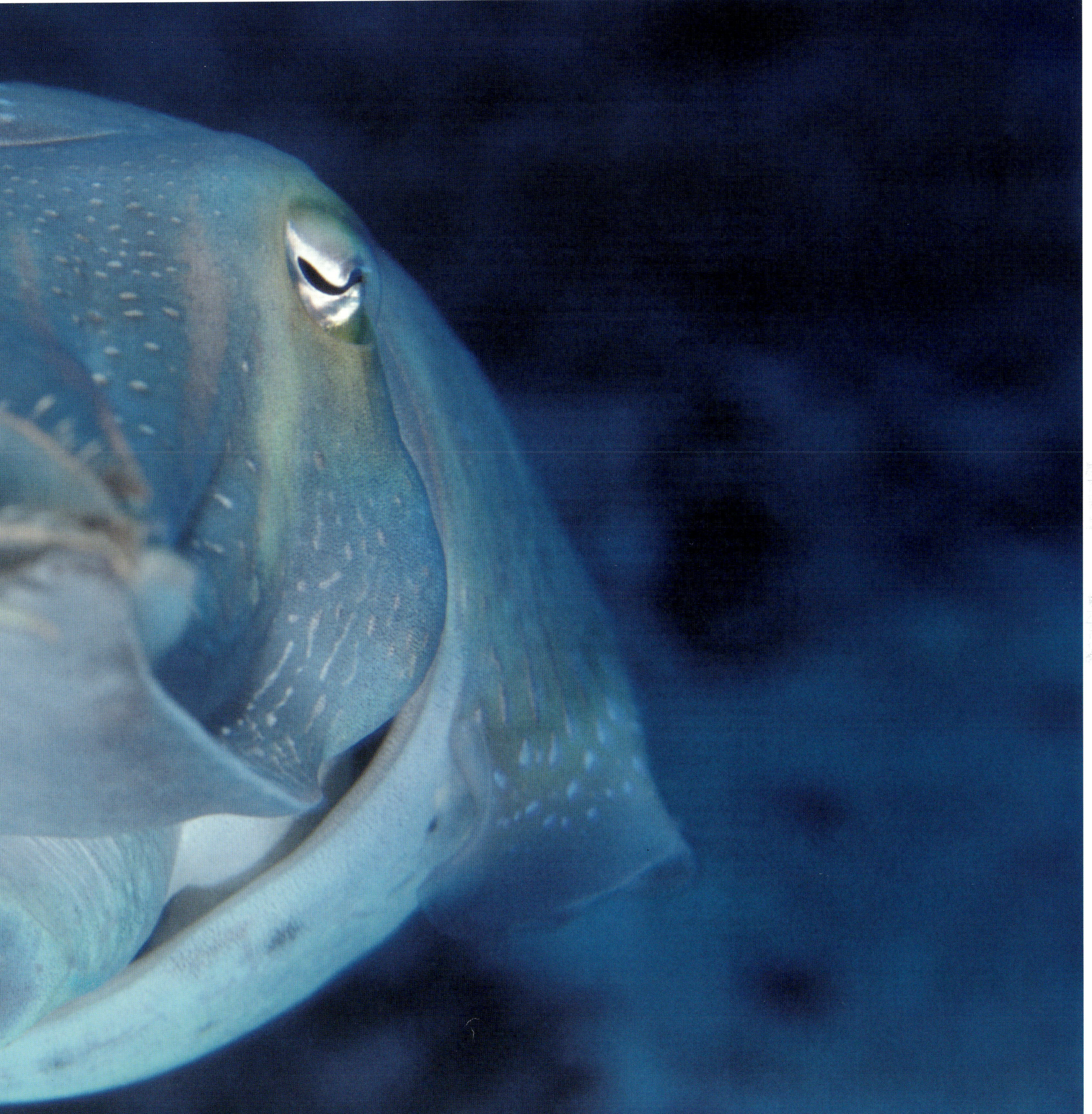

as well as the residue of life above. While "marine snow" is constantly falling, it varies in intensity and content. When there are vast blooms of algae in the spring, shortly thereafter the overproduction that is not consumed descends down through the ocean. And when there are migrating bands of carnivorous creatures, anything from cod to orcas, then that will lead to a trail of offcuts. That influences what's for supper at various levels below, and will also lead to movements of life as it follows the food. On top of this is the influence of major and minor currents, which can be like a conveyor belt for both food and diners. The result is that what you see in one place at one time won't be the same a moment later. All this confirms a notion about life that Heraclitus first defined around 500 BCE, the idea that "you can't step in the same river twice." He also said that "all entities move and nothing remains still." He was talking philosophically about existence in general, but it works as a precise description of all life in the ocean.

These changing conditions in part explain how difficult it is to track deep marine life and describe the ecosystems. They are not fixed, and estimating what is to be found there is an exercise in intelligent guessing. The Census of Marine Life, a $650 million, ten-year project that reported its findings in 2010, came up with the conclusion that there are around 230,000 identified marine species (having cross-checked the numerous multiple names for a single species) and yet noted that this is only 20 percent of the total number of species in the oceans. By a complex reasoning based on the known samples, and on the "known unknowns" (as Donald Rumsfeld once tried to describe a challenging situation), the census concluded that there were more than one million species, at least, yet to be discovered. Others have estimated that there may be as many as 10 million species that are still unknown in the ocean. Most of these will be in the deep sea.

So how do we get to look closer at this world not made for humans? There are some ingenious solutions being invented. One of the most successful series of devices is celebrated in a series of enormously popular YouTube videos that has garnered millions of views. It is based on highlights from footage captured at various depths in the Mariana Trench on a pioneering tool created at the University of Aberdeen. The film includes images of a new type of snailfish living at the greatest depth at which any fish has ever been observed and recorded, 26,700 feet (8 kilometers) deep. That is still well above Piccard's claim of seeing a fish, but does suggest that the possibility is not so improbable. These films were made as part of the Hadal Eco System Studies (HADES) program, a collaboration between nine institutions to investigate the final 16,000 feet (5 kilometers) of the hadal zone, which is the greatest depth beyond 20,000 feet (6 kilometers), areas confined to the trenches that descend below the abyssal floor of the ocean and are largely around the Pacific. The Oceanlab team at the University of Aberdeen built a series of unmanned "autonomous landers"—known as Hadal-Landers—durable and fairly simple contraptions that can withstand the tremendous pressures at depth. Around 9 feet (3 meters) high, with a pyramid shape rising out of a set of splayed, weighted legs, the structure is reminiscent of *Apollo 11*, the first craft to land on the moon. It carries a camera and bait, and is simply dropped

overboard and allowed to sink to the bottom. After filming what happens around the bait (mackerel is a favorite morsel to offer up), the device is retrieved by sending a signal to release ballast, and as it is designed to be buoyant, this permits the craft to rise back to the surface. Dr. Alan Jamieson, a leading scientist behind the design and research, said of the record-breaking snailfish sighting: "This really deep fish did not look like anything we had seen before, nor does it look like anything we know of. It is unbelievably fragile, with large wing-like fins and a head resembling a cartoon dog."

One particular oddity about the weird and wonderful fish and other creatures that gather around the bait is that they live in an area where there is relatively little marine snow falling. At the surface above the Mariana, Kermadec, and Tonga Trenches, where Oceanlab has focused its investigations, there is typically clear blue water. This may be picture-postcard tropical, but it also means that there is very little falling detritus, nor are there the carcasses that the bottom feeders depend upon. The mackerel package is designed to be manna from heaven and has proved highly effective. What is perhaps surprising is that given the general prevailing conditions, snailfish and other creatures of size are actually hanging around on the off chance of such a rare meal. What do they eat when the Hadal-Lander is not delivering? It is thought that their metabolisms are designed to work pretty slowly down in that cold, oxygen-poor water. Time passes slowly at these depths. The creatures there burn calories carefully and live for a long time.

But while these animals seem to obey the general rule that the chain of life is in general hanging off the photosynthetic activity that begins near the surface with plankton plants, and then goes up in size before falling down into the depths, not all deep-water life depends on this. There are increasing signs that other rules can apply. It can be an alternative universe down there, where life is simply not as we know it. This became apparent in the 1970s when a team of researchers working around underwater volcanic activity off the Galapagos archipelago found strangely large patches of life where the tough volcanic rock at depth in oxygen-deprived waters led to an expectation for little or no life. At 8,000 feet (2.5 kilometers) down on the Galapagos Rift, researchers John Corliss and John Edmond, working in the cramped conditions of the relatively primitive submarine *Alvin*, looked out to see "a fabulous scene . . . an oasis. Reefs of mussels and fields of giant clams were bathed in the shimmering water, along with crabs, anemones, and large pink fish." What they later realized they were seeing was life built around not photosynthesis but chemosynthesis. Instead of seeing a chain of life pinned on bacteria that fed on sunlight, they were seeing an entire ecosystem that was dependent on bacteria that fed off hydrogen sulfide, which was in rich supply from volcanic activity. Water running into the volcanic vents around the ridge was broken down to generate hydrogen sulfide. The waters shimmered with the rising heat and gases coming through from the seismic activity and in this, new life prospered. The penny dropped on the nature of this life when the scientists opened up their samples back on the ship and detected the rotten-egg smell of sulfur. Over many millions of years a whole different chain of life has been developing that operates on similar but vitally different rules.

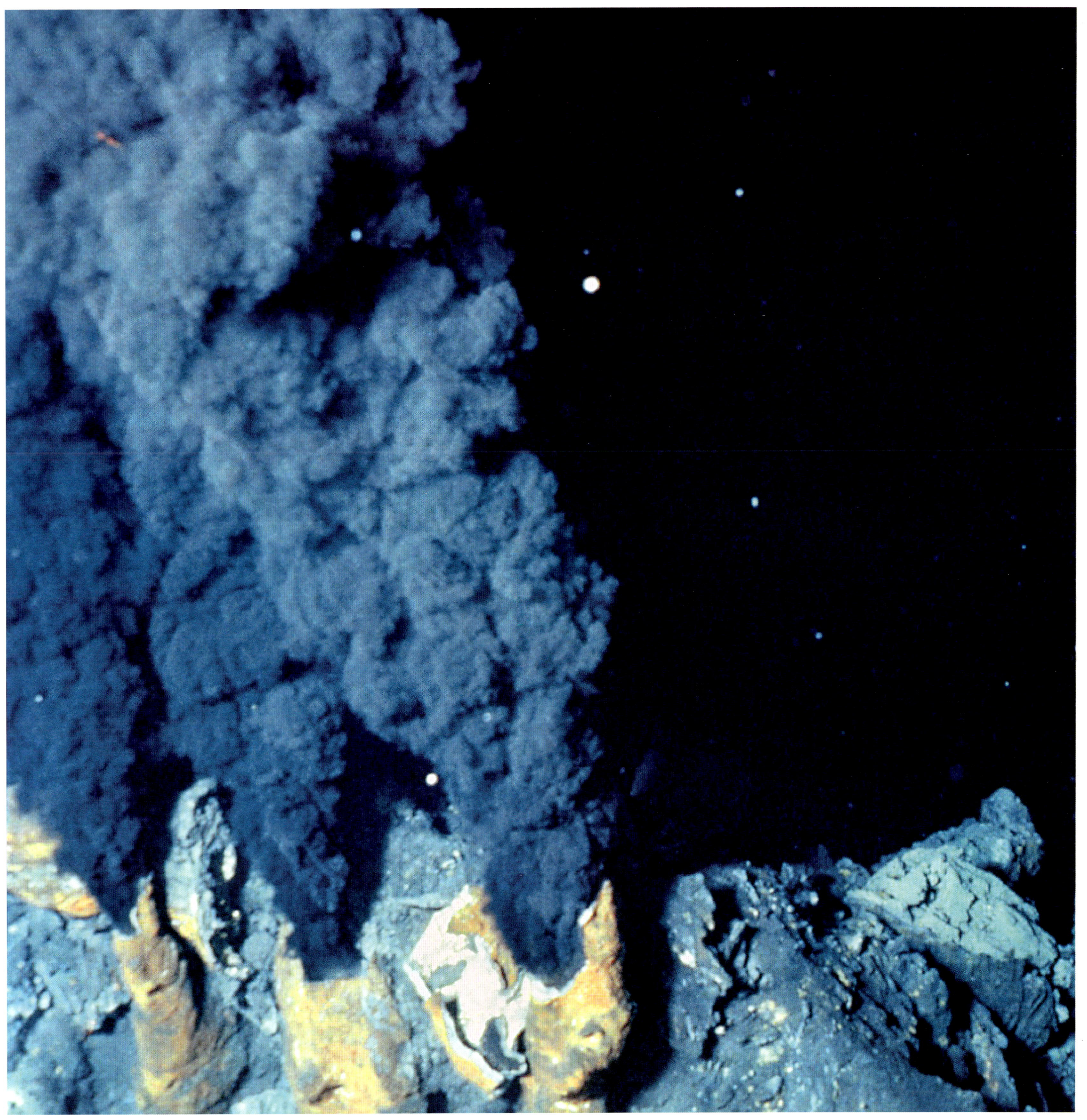

Consider the subtleness of the sea;
how its most dreaded creatures glide under water,

unapparent for the most part, and treacherously
hidden beneath the loveliest tints of azure.

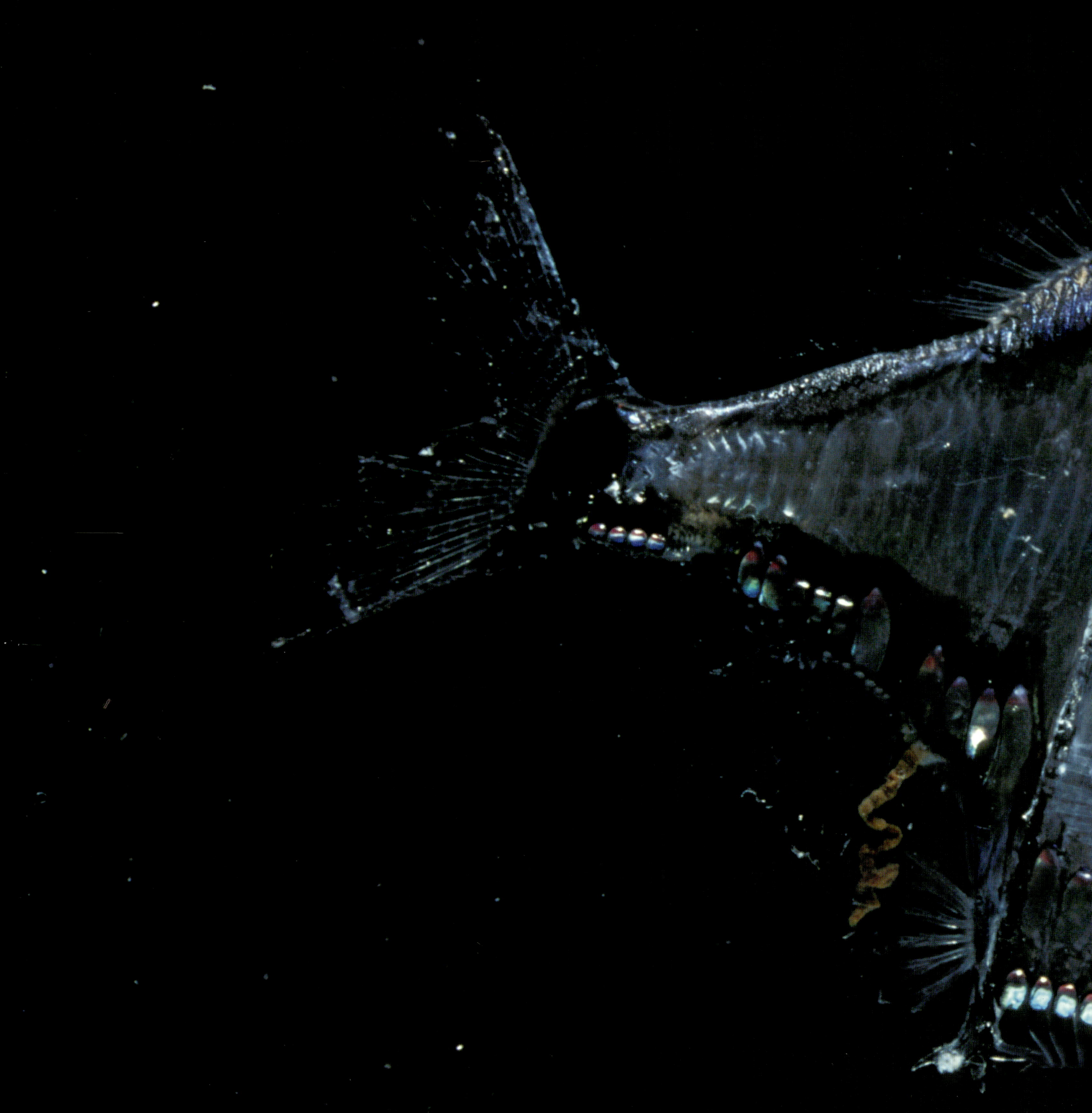

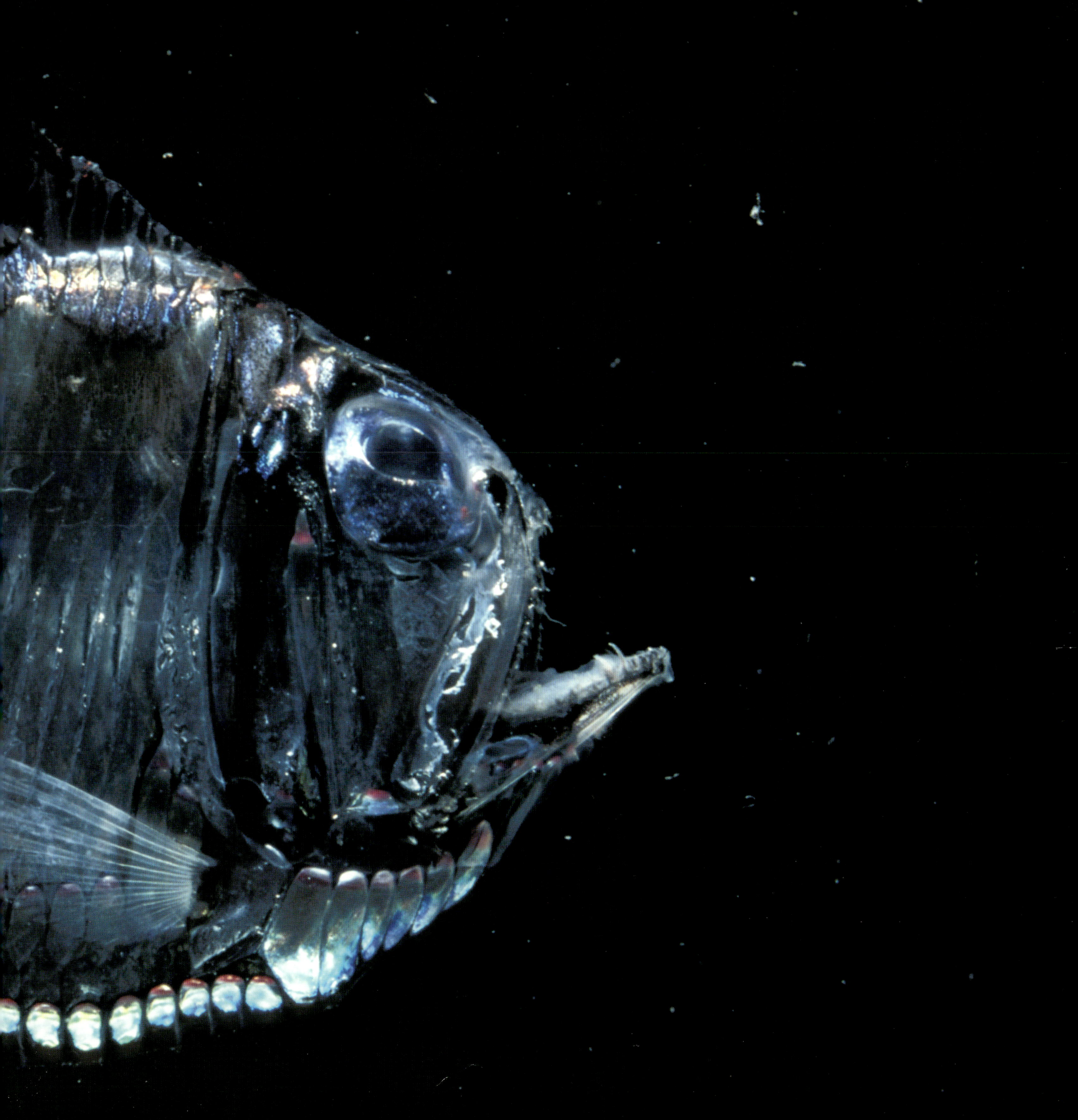

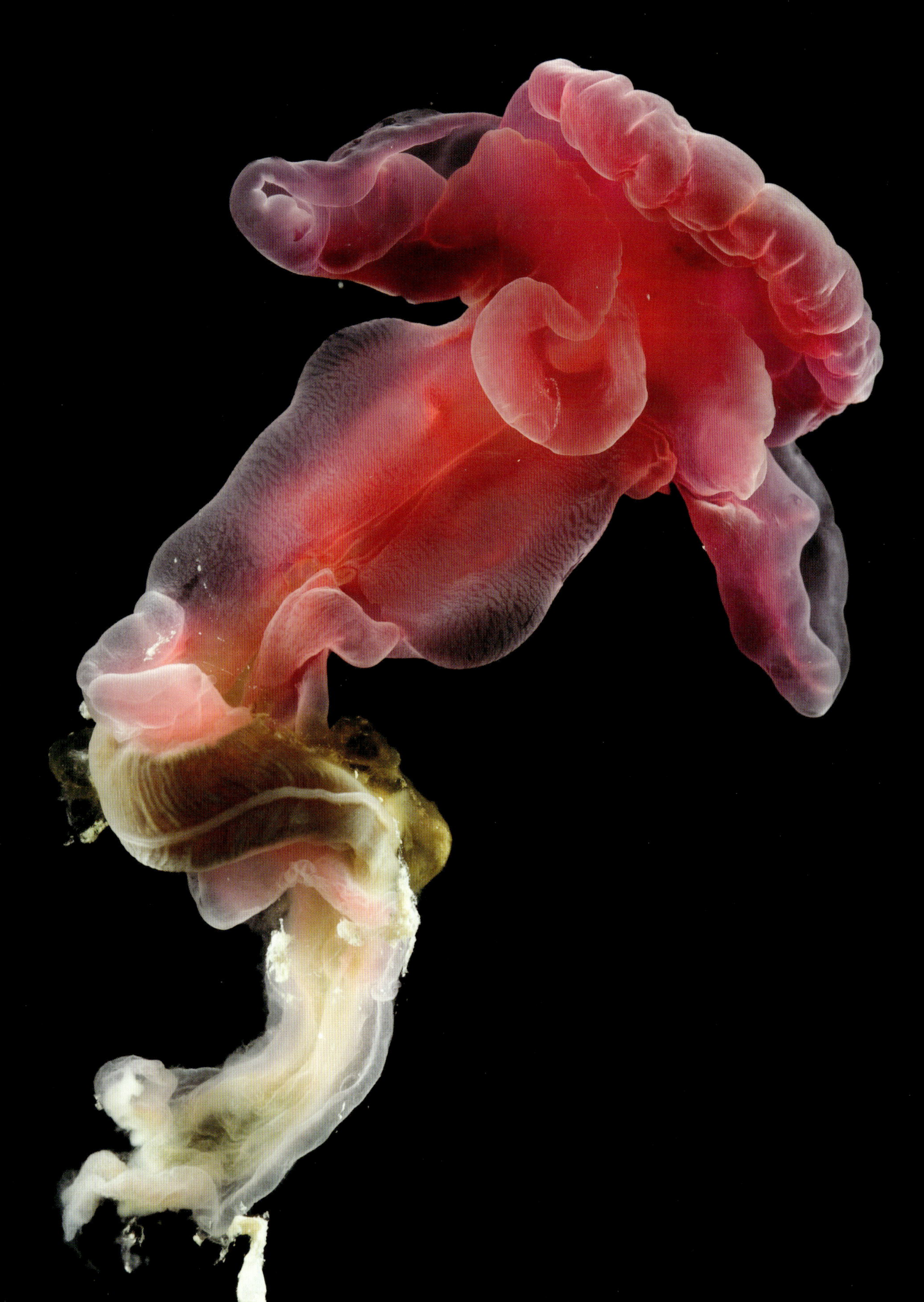

These patches of chemosynthetic life are likely to exist all over the planet, at various depths around the ridges and in the holes and ravines where faults in the crust lead volcanic heat and minerals to interact with water. They create unique ecosystems, with the range of species found in one chemosynthetic environment different to those found in another, due to the varying geological processes supporting their existence. So how many more species might we discover? Are there other alternative sets of rules for thriving ecosystems still to be found? Quite possibly. The oceans are deep in content for us to mine and to muse upon.

We cannot leave the subject of the depths without speculating on one of the most fantastic aspects of life down there—one that plays with our fears, stimulates myths, and continues to pose unresolved questions for marine scientists. This is the topic of deep-sea gigantism. This name is used to describe the phenomenon where animals of the deep are often exaggeratedly larger versions of creatures that also can be found in shallow water. The giant squid, and the even larger colossal squid, are examples of this. These creatures, which may reach up to 46 feet (14 meters)—and perhaps longer, as we have hardly seen them—have only occasionally been found as dead specimens either washed up or caught by trawlers. We do not really know how they live or how many of them might be in the ocean, as they live up to 6,500 feet (2 kilometers) down, perhaps farther. Nobody has ever seen them truly swimming around in their natural environment, but they are clearly fascinating and important creatures in the ecosystem as they affect other life. For example, dissections of sperm whale stomachs indicate that almost 80 percent of their diet by mass may be constituted of colossal squid. The food is not so easily acquired: Sperm whales often have scarring on their backs that indicates they have been attacked by the hook-like tentacles on the squid, perhaps as part of its attempts to defend itself against being eaten. Other examples of gigantism include a fish with the misnomer "the king of herrings," which looks not much like a herring at all and is a giant oarfish of up to 56 feet (17 meters); the seven-arm octopus (so-called because one of its eight arms is coiled and largely hidden, being specialized for use in fertilization), which reaches lengths of 11 feet (3 meters) and possibly more; the Japanese spider crab, with a leg span approaching 13 feet (4 meters); and the giant isopod, which looks like its landlubber cousin the woodlouse, but contrasts dramatically, if not scarily, in size by being the dimensions of two large hands rather than a fingernail. The giant amphipod is another off-putting beast, being somewhat similar to a cockroach but a lot larger. Dr. Alan Jamieson, mentioned for his hadal-zone research, was involved in discovering this species at about 23,000 feet (7 kilometers) down. He commented: "It's a bit like finding a foot-long cockroach. I stopped and thought: 'What on Earth was that?' This amphipod was far bigger than I ever thought possible."

We are still unsure as to what drives gigantism. One theory is that it is due to the colder temperatures in the depths, but this doesn't always hold up. The huge pressures of deep water have also been speculated on as a reason, but this doesn't explain size consistently. The largest jellyfish, for example, the lion's mane jellyfish, is found in cold waters, but near the surface. No easy rules apply to gigantism in general. What we can guess is that we might be in for some more surprising encounters with monsters of the deep as yet undiscovered.

When we relate to life on the land and sea, we tend to identify with and most easily take an interest in things that look or behave a bit like us. That's also why we perhaps try to find smaller and land-based references for the weird giants of the depths. We also anthropomorphize creatures we find down there—give human properties to creatures that are anything but human. It can seem to work (as well as mislead) with our pets, but quickly falls apart with most wildlife. With sea life we are less tempted because things look so different, but that also suggests we are less emotionally connected. Only with the creatures that we can care about a bit, such as whales, turtles, or migrating birds, do we more easily connect. But this is to deceive ourselves over the nature of the ocean. As the Japanese novelist Haruki Murakami puts it:

We get into the habit of thinking, This is the world, but that's not true at all. The real world is in a much darker and deeper place than this, and most of it is occupied by jellyfish and things. We just happen to forget all that. Don't you agree? Two-thirds of the earth's surface is ocean, and all we can see of it with the naked eye is the surface: the skin. We hardly know anything about what's underneath the skin.

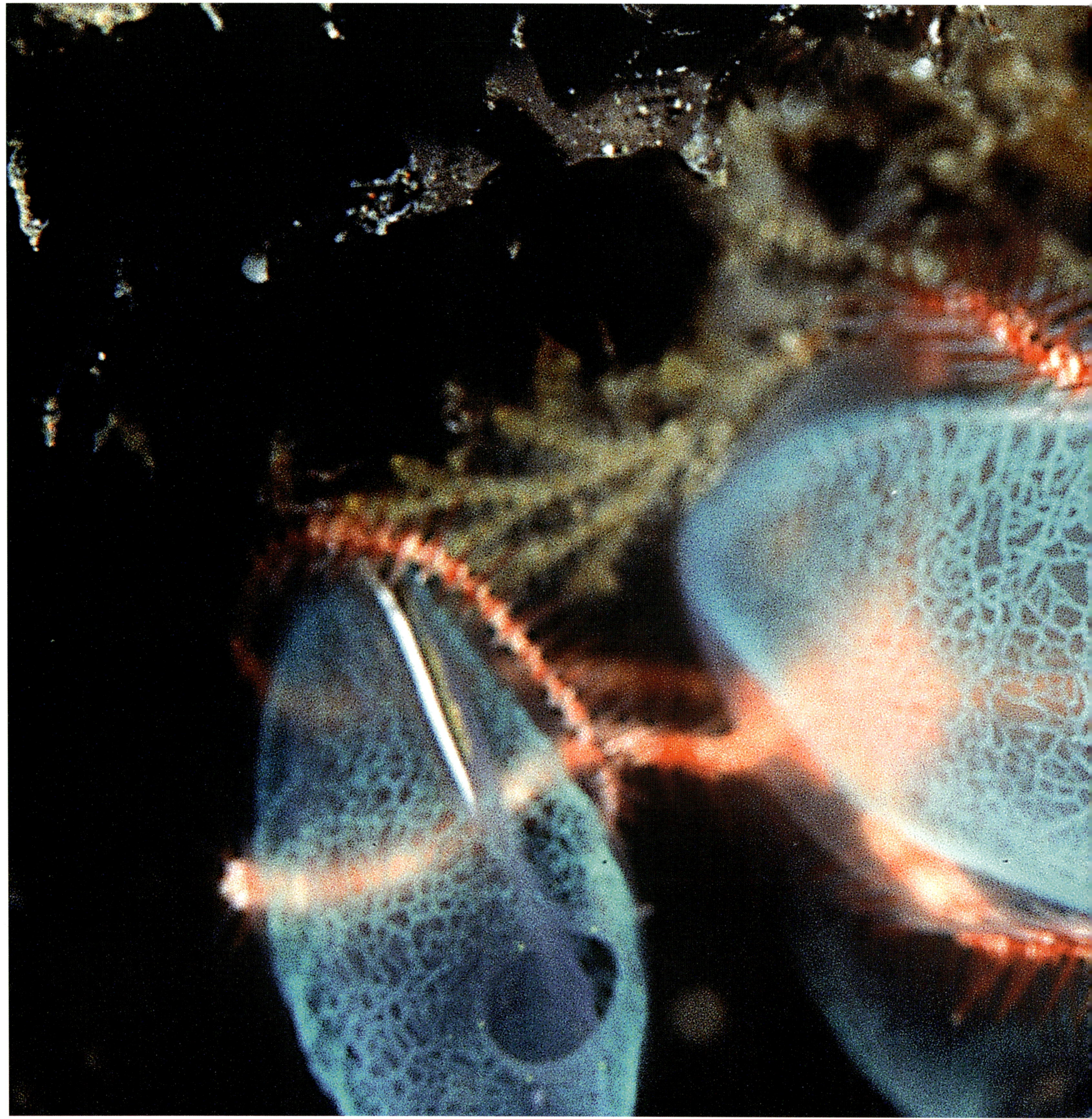

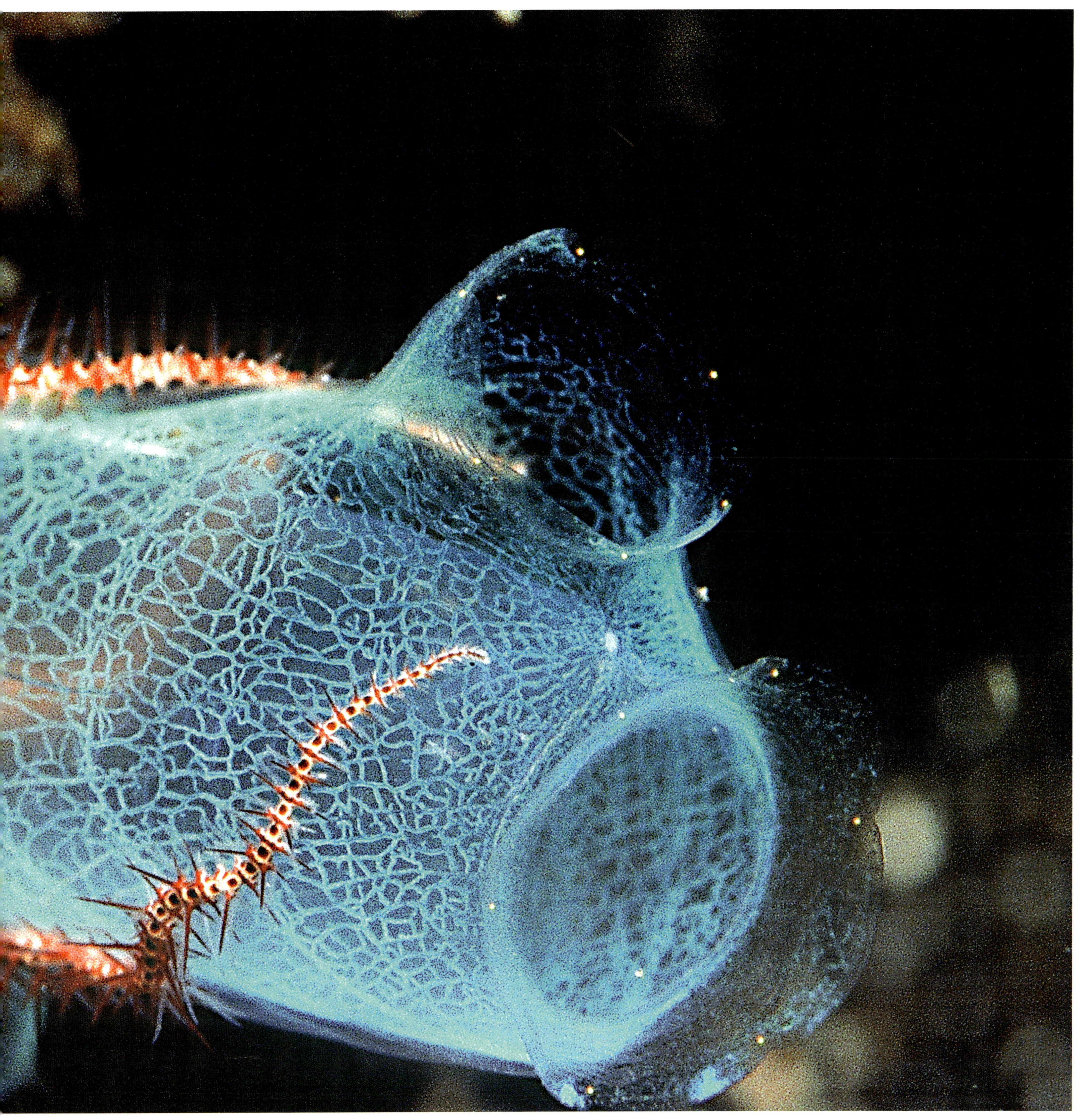

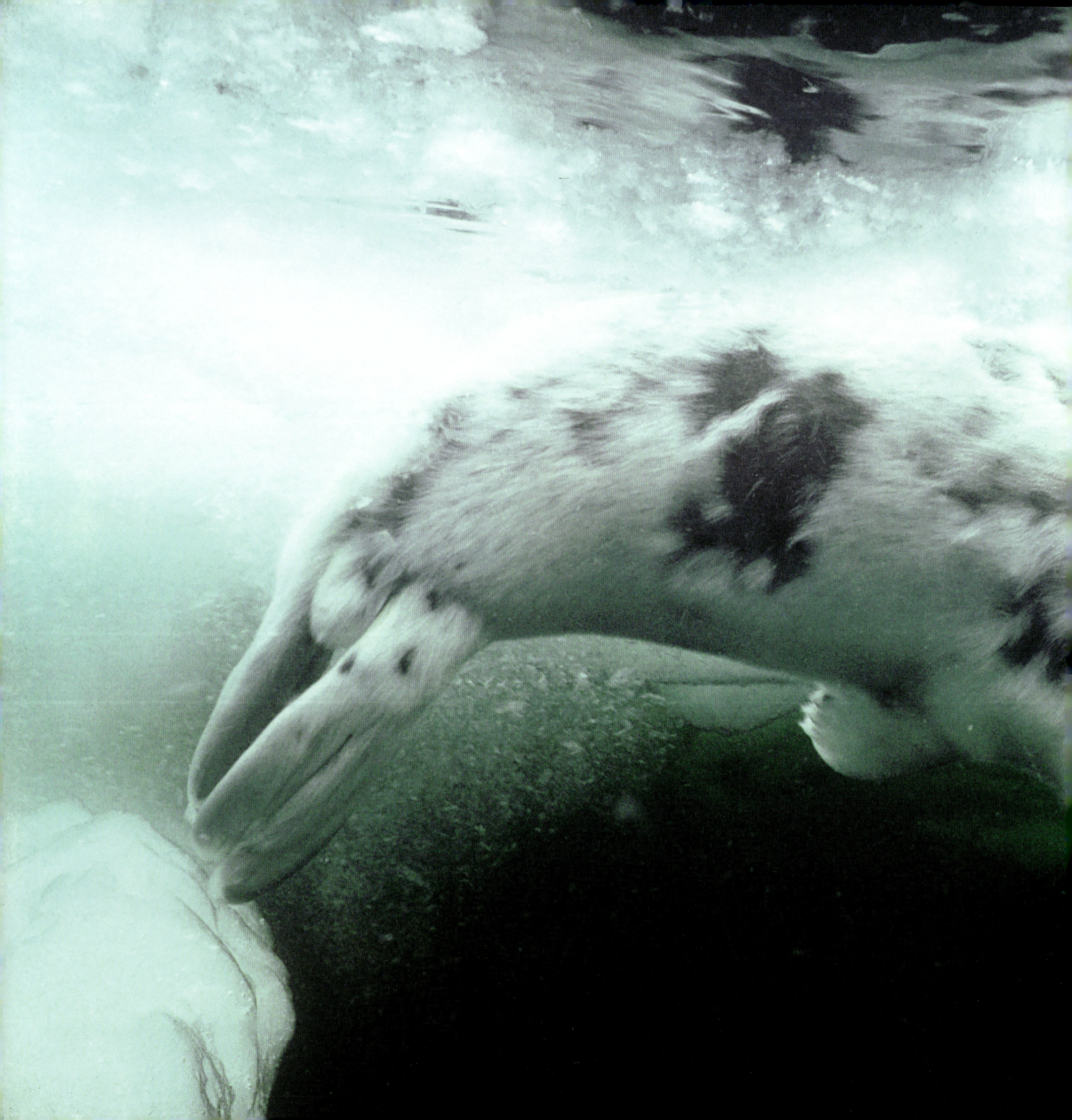

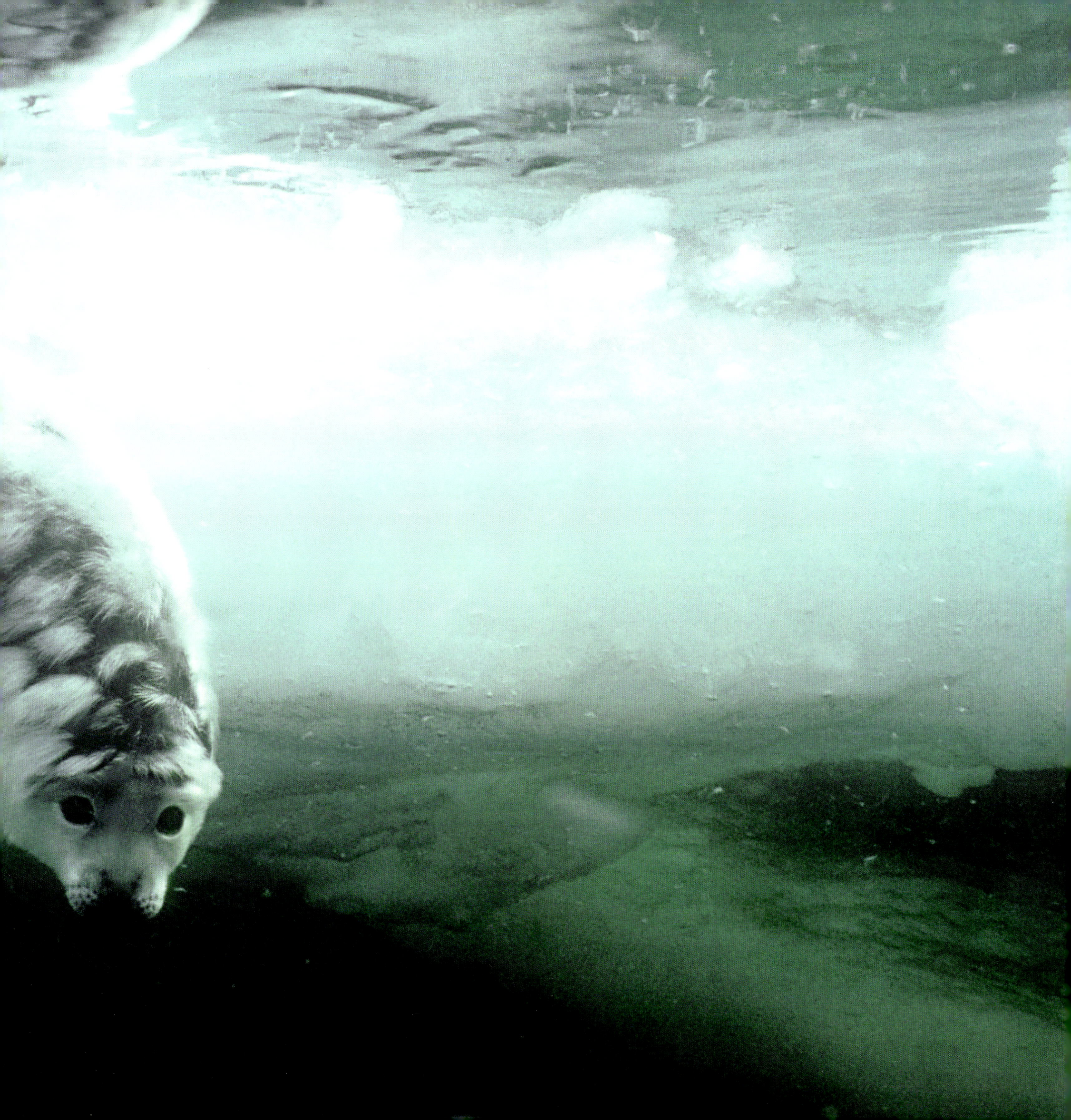

Tame

It is hard to conceive of living in a world where we do not know of the sea and do not use the sea in our life. But, in a very distant way, this would appear to be the case of how we came to live on Earth. We are supposed to have emerged from the forests, descending from an ape-like life in the trees, somewhere in Africa, and then gradually developed into the creatures we are today. We may imagine that at some point our ancestors, out hunting dinner, chased some wild animals farther than ever before, and while heading along a new path, crested an incline to see in the distance a strange, shimmering, silvery blue. When they reached the shore they must have felt like they had reached the edge of the world. And, in a way, they had. Until they started to find ways to travel across it.

Homo sapiens dates back to somewhere between 100,000 and 200,000 BCE, and for a long time the earliest sea travel was thought to have occurred between 40,000 and 60,000 BCE. But no more. It would appear that our ancestor species—both Neanderthals and *Homo erectus*—had mastered some use of marine life as a food source and had even constructed craft for traveling across the sea. There is one theory that the first mariners may have been as long ago as 800,000 to 900,000 years ago, according to evidence of the presence of *Homo erectus* on the Indonesian island of Flores. Tools and other artifacts found on the island indicate that our ancestors reached the location despite a sea crossing of more than 11 miles (18 kilometers) being required. The question has to be asked as to whether that expedition happened by plan or accident: It may have been that these adventurers were washed across on a raft against their will, but travel on water they did and they made a home there. The more usual evidence of a starting point for sea travel is the evidence that early man must have used some kind of boat in order to colonize Australia around 50,000 BCE. These people would be the distant ancestors of today's indigenous Australians while also being the descendants of the first *Homo sapiens* to migrate out of Africa and through Asia. At what point the understanding of boatbuilding began is unknown, but we can assume they did not set off with vague expectations of arriving at Australia without generations of sailing experience.

The first crafts would most likely have been rafts and then possibly dugouts, canoe-like crafts fashioned by simply hollowing out a suitably large tree trunk. As the Norwegian adventurer Thor Heyerdahl demonstrated, early man may have been able to make surprisingly long voyages on a raft. In 1947 Heyerdahl and his crew set off on his raft, *Kon-Tiki*, from Peru: 101 days, and nearly 5,000 miles (8,000 kilometers) later, they arrived in French Polynesia. Despite being uncomfortable about traveling in deep water (he nearly drowned as a child and feared for his life at times on the raft), Heyerdahl showed that primitive craft could achieve remarkable distances. His achievement made him famous around the world, with films and books documenting his achievement. Besides proving that the raft structure was durable for the arduous journey, he also showed that it worked well as a station on which to attract fish, which could be caught and eaten while en route. These fish also provided a source of adequate liquid intake so that early sailors might survive for many days at sea. Heyerdahl also showed that boats made from papyrus could sail great distances. Anthropologists question the direction of Heyerdahl's travel (they think early human migrations mostly expanded eastward, not westward) but the demonstration of just how robust simple rafts and boats can be has been further tested and proven.

Our taming of the seas for travel, fishing, and exploration would seem to have come not so long after our emergence from the forests. While it is risky to float on a primitive boat on uncharted seas, with very little in the way of navigation knowledge, this has to be balanced against the risks of everyday existence on land for our ancestors. Hunting down food that could run fast or fight back was the general nature of land-based hunting. Going out for lunch when on land was always a dangerous business (and can be still). In contrast, the basics of fishing have not changed a lot from then to now: For all the shifts in technology, in crude terms you throw a line or net over the side and hope for the best, having made some effort to understand where and when the fish are likely to be around. In a way, fishing is a form of hunting that half shifts the process closer to farming. While not nurturing a crop, the early fisherman may have preferred the risk-and-reward levels of inshore fishing to hunting dangerous animals that can kill, maim, or easily evade capture. In contrast to the high-risk issues of hunting (resulting in feast or famine), a good shoreline can yield a steady, daily stream of produce. As nets, pots, and multiple hooks developed, alongside simple scavenging for shoreline crustaceans and molluscs, the appeal of a coastal existence most likely grew and is growing still. The oldest fish hooks found to date go back 42,000 years and indicate that our ancestors knew enough to catch tuna, which implies some extent of deep-sea fishing—as tuna swim in deeper waters offshore—rather than hanging out in the shallows. These earliest hooks were found in a cave in East Timor and were made from bone. It may be that the catches were of more juvenile fish, which come closer to shore, but the fishermen would still have had to use boats in relatively deep water to set nets or lines. This discovery is from so long ago that similar sites will be difficult to find because they most likely have become submerged by rising sea levels or otherwise covered by land changes.

So before we ever thought about traveling across the water, we would have picked along the shores and found a rich harvest in the intertidal areas. The processing and consuming of the various creatures and plants of the shore and rocks would have provided the basis for whole communities to develop. And at the same time, the improved traveling capabilities afforded by early boats would mean that what was plentiful in one place, but rare elsewhere, may have been traded with what was plentiful in another place. While a sheltered coastline might be rich in produce—as there are few other environments as generous as a sheltered bay with the right aspect and a sloping seafloor to invite the chain of sea life to make a home—there is only so much you can rely on one spot before the shoals move away or the diet gets tiresome. Some species of sea life are keen on one area for a long time, others may be elsewhere more often, and this creates a natural limitation to trade. Some early hominids may well have grazed quite happily

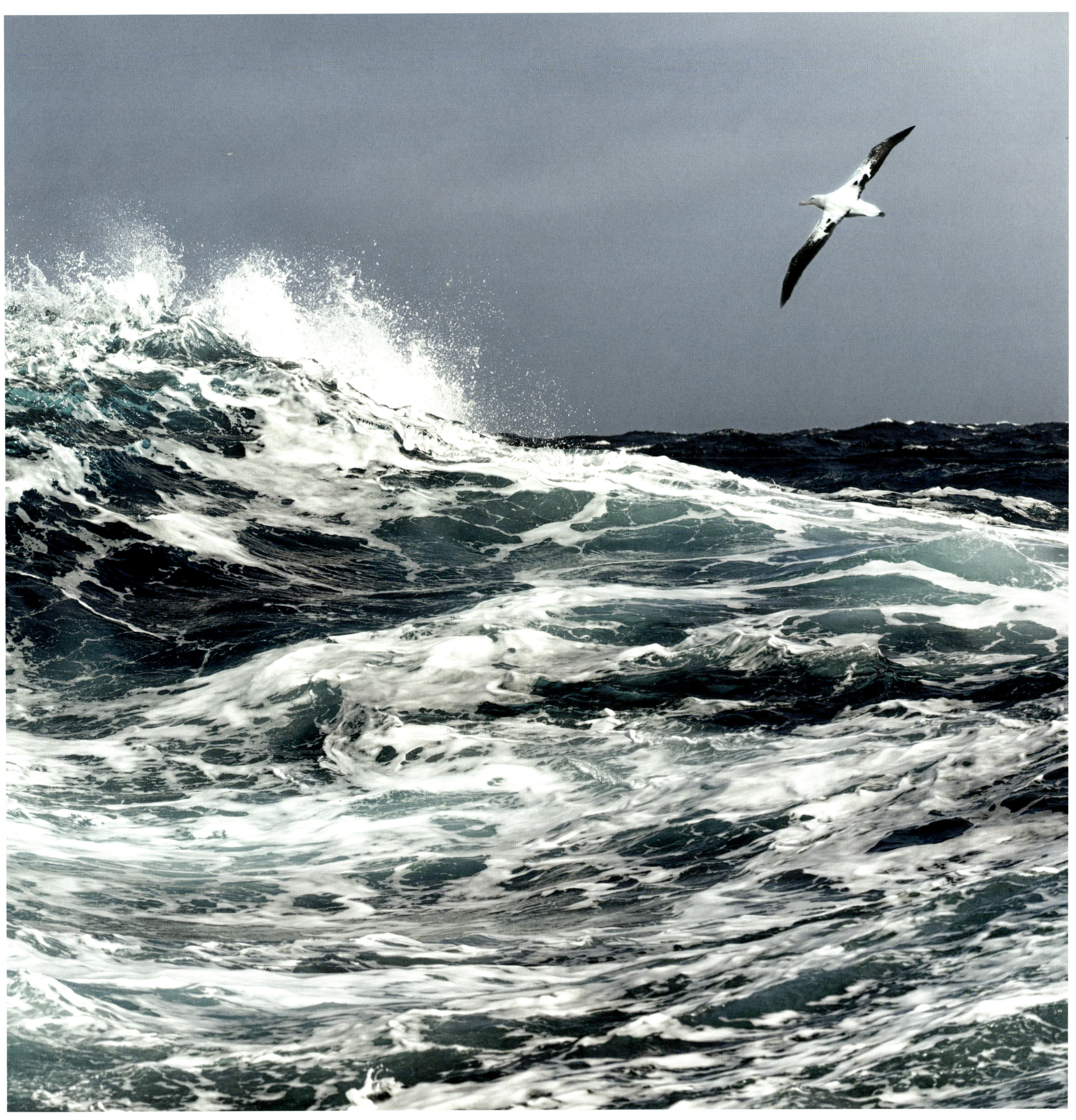

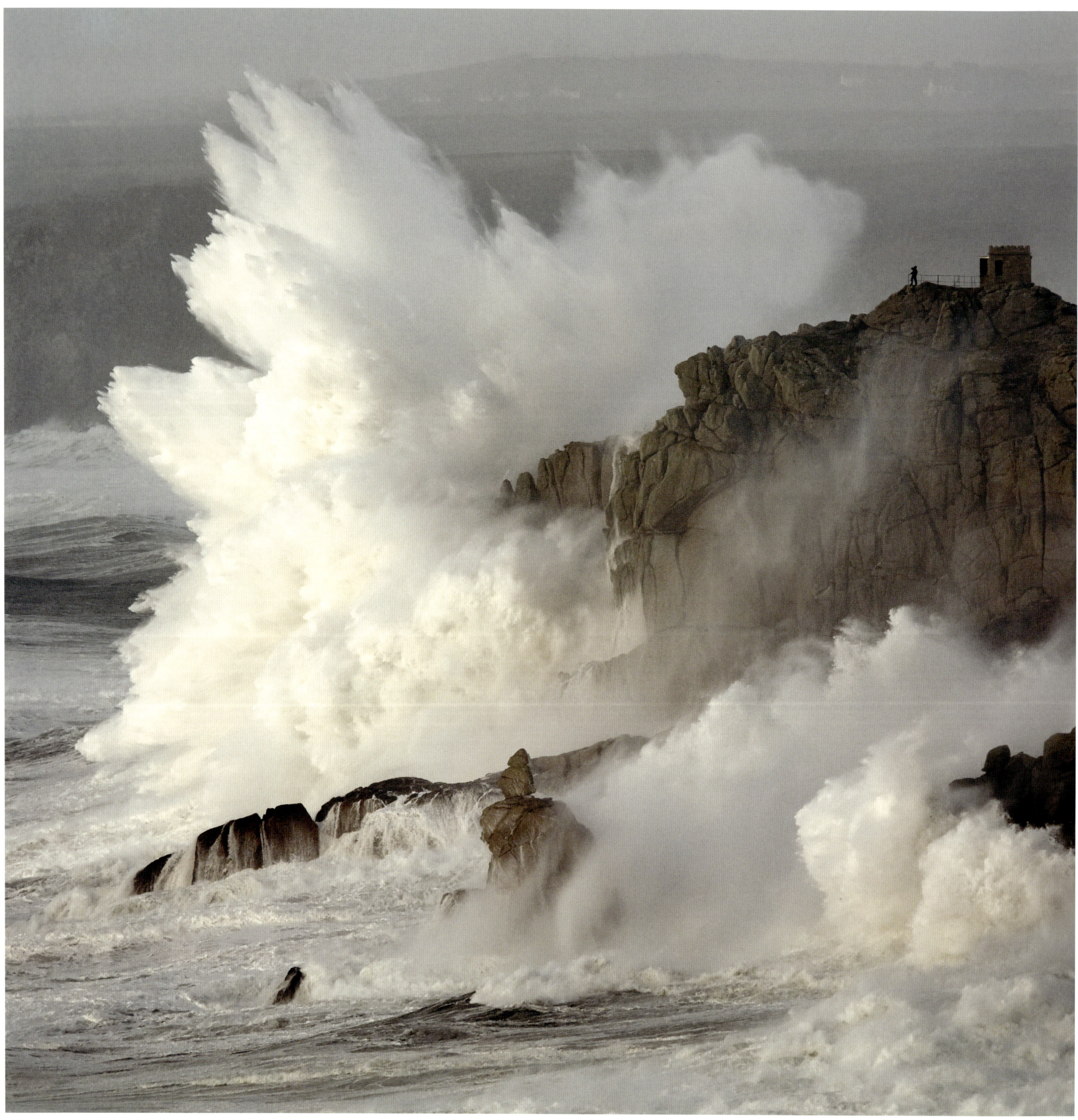

for much of their diet in such locations and had a surplus that they could expand their wealth with. In contrast, the land-based hunter-gatherer lifestyle would have required constant moving on to track the food. The right rich coastal area, with favorable conditions and currents, brings fresh food on every tide. We can but speculate on those early settlements, but given the potential for comfortable food supply, some coastline locations may have seen the earliest permanent settlements of size, although these would have been completely eroded by sea level changes and weather erosion.

We know the ancestors of the Greeks must have been using boats to carry out trade between islands even earlier than we can find clear documentation of it. This seafaring activity is implied by the location of pieces of obsidian—a glass-like volcanic rock that could be used for tools—dating back to around 7000 BCE, found on islands lacking any of the mineral naturally. However, this is evidence that implies the presence of boats rather than depicting them. Around 5000 BCE, more advanced boatbuilding and sailing technology was in use, according to the evidence depicted on pottery found in parts of Mesopotamia including Saudi Arabia, Bahrain, and Qatar. Boat remains of the period were also found in Kuwait, which may have been the leader in boatbuilding at this time. Using the seas as a highway for trade was fundamental to the economies of these desert-backed societies. The sea was where most of all the life and resources lay. It seems the boats may have been made from reeds, with a waterproof coating protecting the hull from absorbing water. The latter was interpreted from the discovery of pieces of barnacle-encrusted bitumen. These chunks may have been an example of early recycling; the material taken off a disused boat to be melted down and reused as a sealant on a new boat or for repairs. The bitumen, of course, is an early sign of the underlying oil wealth, a surface leaking of the deep reserves of crude oil that have now transformed the economies and politics of the Gulf and, indeed, of the world.

The life of the early sea merchants forms the basis for one of the most important and somewhat mysterious early empires: that of the Phoenicians, who are thought to have originally come from the Arabian Peninsula before moving to trade around the Mediterranean. It was a power built simply on the skills and methods of sea-merchant life. It links the ancient Egyptians with the rise of ancient Greece and then Roman civilization. Phoenicia really wasn't so much a place as a loosely connected federation of city-states along the shores, among which trade was brisk. What distinguished the Phoenicians was their innovation and early mastery around marine travel, resources, and trade. This enabled their cities to become wealthy magnets for people from other cultures to move to. An example of their innovativeness would be the dominant business they did in Tyrian purple. This was a clothing dye, one that was highly valued, as it was a striking color that was highly durable in the bright sun, seeming to strengthen rather than fade with use. The origin of this dye is from the sea snails of the family Muricidae. So sought after was this dye, and so associated with this civilization, that it is actually the basis of the civilization's name: Phoenicia

means "the land of purple." They were probably not the originators of the knowledge or the only users. Ancient Mexicans also sourced dye from these snails, while Minoans of 2000 BCE were possibly the earliest users. Pollux, the Roman writer of mythic stories, suggested that the appreciation of the dye originated when the Greek hero Heracles saw his dog's mouth go purple from chewing on the snails. Whatever the origins, the Phoenicians were known to be trading in it as far back as 1500 BCE and it is known that a thousand years later the dye was still immensely valued, being worth its weight in gold at some points in history. If anything, it is surprising it was not even more costly: A gram of the dye would take the best part of ten thousand sea snails to produce, and this may only have been enough to do the hem of a robe. The high value attached to this dye can be seen in its appearance in ceremonial robes and from its association with the clothing of rulers. An indigo-blue dye, derived from a different snail, was also trafficked by the Phoenicians and is perhaps the origin of the strong use of indigo colors in some North African cultures today.

While Phoenicia declined shortly after 400 BCE, being subdued by Alexander the Great among others, the idea of the seafaring merchant nation stayed with us. The Phoenicians traveled around the southern Mediterranean, from Crete and the Turkish coastline to Gibraltar and the Balearic Islands, mapping out an empire of sea-based commerce, some of which came by force. This model of power, built on trade backed by military force, survives until the present day and was central to the evolution of many of the greatest empires of all time, perhaps culminating in the crucial sea power of the British Empire in the nineteenth and early twentieth centuries. We even have an impressive word for rulers whose power is built on the sea: "thalassocracy," a concept that links the early Phoenicians to the Sea Peoples—a loose gang of seafaring traders from around 1200 BCE who greatly annoyed the New Kingdom Egyptians—and comes down through those merchant warriors of the Venetian republic, then the fifteenth- and sixteenth-century explorer empires of the Portuguese and Spanish, who were succeeded by the Dutch. The Dutch in turn lost out to the British Empire, which grew to rule over a fifth of the world's population, with a spread around all the oceans of the world. From the mid-twentieth century this was succeeded by US and Soviet naval powers, and now Chinese merchant sea power is a significant influence in the world, shipping minerals from Australia and Africa, and with investments including the funding of a new canal in Nicaragua to challenge the power of the Panama Canal. Almost all great nations with some coastal areas tend to have sea power as a crucial part of their means of communication and control, as well as being vital to the trade around the empire that enables it to flourish.

There are few lasting powers that did not have strength on the oceans, because it has always been vital to trade, crucial to the acquisition and sharing of resources. Even in medieval times, for example, kippers prepared somewhere in the North Sea were traded around until they cropped up in Constantinople—sea travel and sea products have been a vital part of our civilizations. While we tend to think of the value of the sea as being in the fish and

The sea always filled her with longing,

though for what she was never sure.

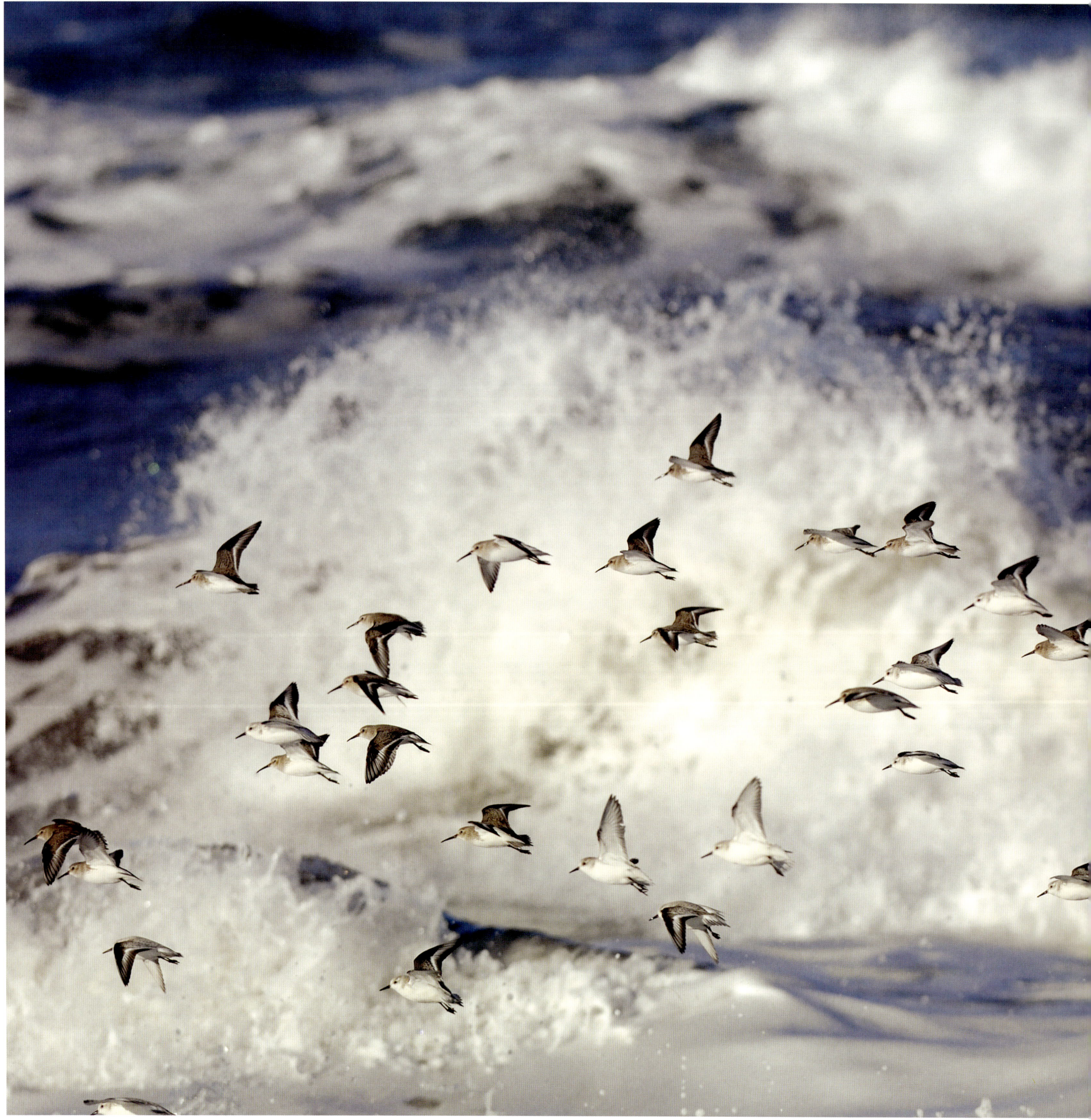

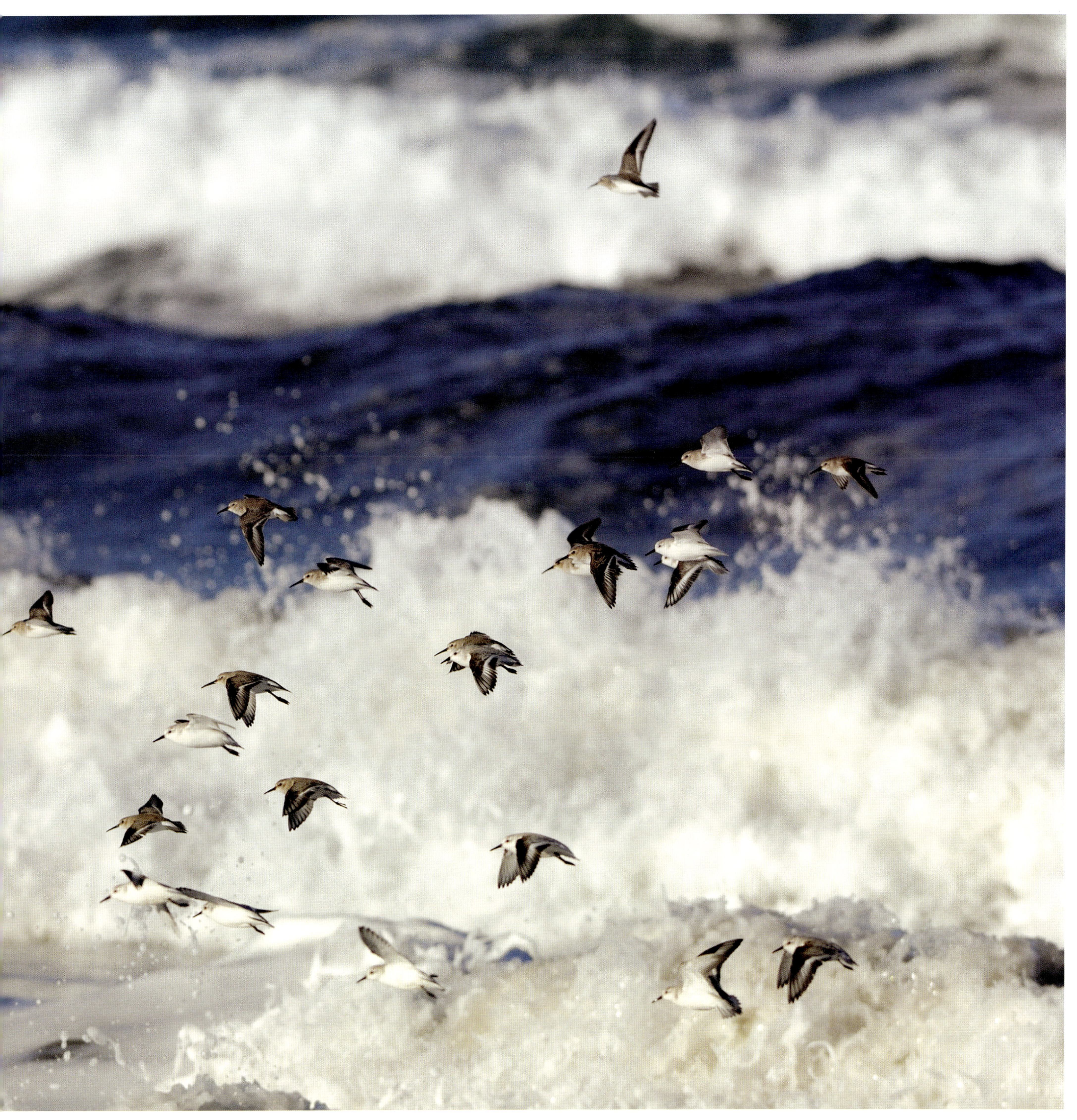

other produce we might take out of it, arguably its greatest value has been in just being the sea: a conduit for travel, with its water, winds, and currents offering a means of travel that permitted us to ship ourselves and our goods from one place to another with relative ease and speed, compared to lugging everything overland (if, indeed, there was a land route available).

The wonders of snail-derived dye are just one of many thousands, if not millions, of resources and products that have been derived from the seas and have built nations and empires. Besides eating anything from seaweed to whales, we have increasingly grown to depend on minerals that come from the sea. In recent years a rising scramble for seafloor has started to rival the nineteenth-century "scramble for Africa" between the colonial powers. Now we see attempts to exploit seas around the Arctic and the Antarctic, and arguments over continental-shelf claims that suggest a new battleground in the making. The Russians symbolically laid a claim on the seafloor at the North Pole by dropping a rust-proof titanium-metal flag from a submarine 13,000 feet (4 kilometers) down. While this is well beyond the 200-nautical-mile (370-kilometer) economic zone extending from their country's borders, the Russians argued for an extension based on the Lomonosov Ridge, which runs out from near Siberia all the way to Canada. Needless to say, the claim is highly disputed.

The United Nations has estimated that today the value of marine-derived production may exceed $7 trillion, and says that by 2025 as much as 75 percent of the world's population may live in coastal areas, deriving some form of economic and cultural influence from the seas. Today, perhaps the greatest value of ocean resources, based on simple economic analysis, might be seen to be in sourcing the vast energy reserves of oil and gas that still lie beneath the oceans. These are often in extremely inaccessible places, such as the Arctic, but when oil prices rise, the market creates conditions for energy companies to explore previously off-limits parts of these polar seas and other deep and inhospitable places. If they make good business sense to explore and recover, environmentalists may not view it this way. Climate-change experts fear that releasing more of the locked-up carbon store under the oceans will greatly accelerate the problems of global warming. Instead, they would rather see exploration of sustainable alternatives that the sea can also offer. The alternative great resource of the oceans lies with its "green power," by finding a way to harness the movement of the tides and the waves. For decades attempts have been made to come up with ways of using the oceans to power our lives. The earliest study of tidal power came out of the United States in the 1920s, and then later work was done in Canada, but nothing was built. The first, and for a long time the only, tidal power barrage—which is essentially a type of dam across an estuary, that harvests tidal energy—was built at La Rance in France in the 1960s. It was a lone pioneer for some years, but in recent decades the field has grown. The biggest tidal plant today is in South Korea, but new ideas continue to be advanced. Among growing interest in the technology around the world, a huge lagoon is in development for the coast of Swansea Bay in Wales, United Kingdom. The Bay of Fundy in Nova Scotia, Canada, which has the

greatest tidal range in the world, is one of the locations to test a new form of tidal power, in-stream turbine technology, which doesn't require a dam, and could be the cheapest and least ecologically damaging form of tidal power, able to be quickly installed in a whole range of locations. Instead of building massive structures to hold back the tide, in-stream technology just captures the water as it rushes past a turbine, without forcing an extra head of pressure. Wave power is another way of harnessing sea energy, through wave-energy-converter technology, but this is less developed, despite the first wave-power product being patented in France in 1799. Since then, many other hopes of harnessing wave energy have been raised, along with investments, only to be lost. There are many research projects currently out in the world's seas, pioneering (or attempting to) ways of using the very apparent, but frustratingly elusive, power of waves. If and when it is cracked, and if it is efficient and also competitive with energy from fossil fuel, fortunes will be made and the world outlook on climate-change threats may change dramatically. At that point, we would have found a way of sustainably, almost infinitely, capturing energy that may power our needs while also helping power actions to wind back some of the damage to our environments. But don't hold your breath; it has been a long wait so far and it may be a while yet before we tame the waves.

In fact, when it comes to who, or what, is taming whom, we might need to realize that it is the seas that have shaped us, rather than us simply exploiting the seas' resources. We are increasingly beholden to the seas, needing to draw on their produce and other resources in greater than ever quantities, with more of us than ever before living close to the coast and dependent on what the oceans deliver. And, of course, all of us depend all the time on the oceans to perform a healthy balancing act on the condition of our atmosphere. For our future health and even survival we will need the oceans to be more productive. Instead of the dubious production of fish farms, which can damage large areas of the sea and its wildlife, we need to find ways to better manage the sea as a wild resource from which we draw a limited amount of its production in a sustainable way. If we don't, it has been estimated that all fish stocks may collapse by the middle of this century, with a third having already collapsed. To do this we may need to evolve what we eat, moving away from difficult-to-sustain or threatened species to those that have greater supply. One of the most efficient, if extreme, dietary moves would be to become a species that eats more of the plankton that the chain of life in the sea itself depends upon, rather than requiring this foodstuff to be processed extensively up the chain through other creatures before we are ready to eat the resulting level of life. Great claims have been made for plankton—including seeing it as a miracle cure for cancer—but to most it would be impossible to conceive of eating the almost invisible plankton as a staple part of our diet, given that our mouths and digestive systems have not been shaped to sieve seawater in the way that fish, whales, and other plankton-eating sea life do. However, as long ago as World War II plans were being put forward to feed people on plankton. British scientists saw this universally present sea life—which is rich in protein and, in theory, highly nutritious for a human diet—as a way of coping with food shortages caused by the conflict. Between 1941 and 1943 they carried out extensive research at various Scottish lochs

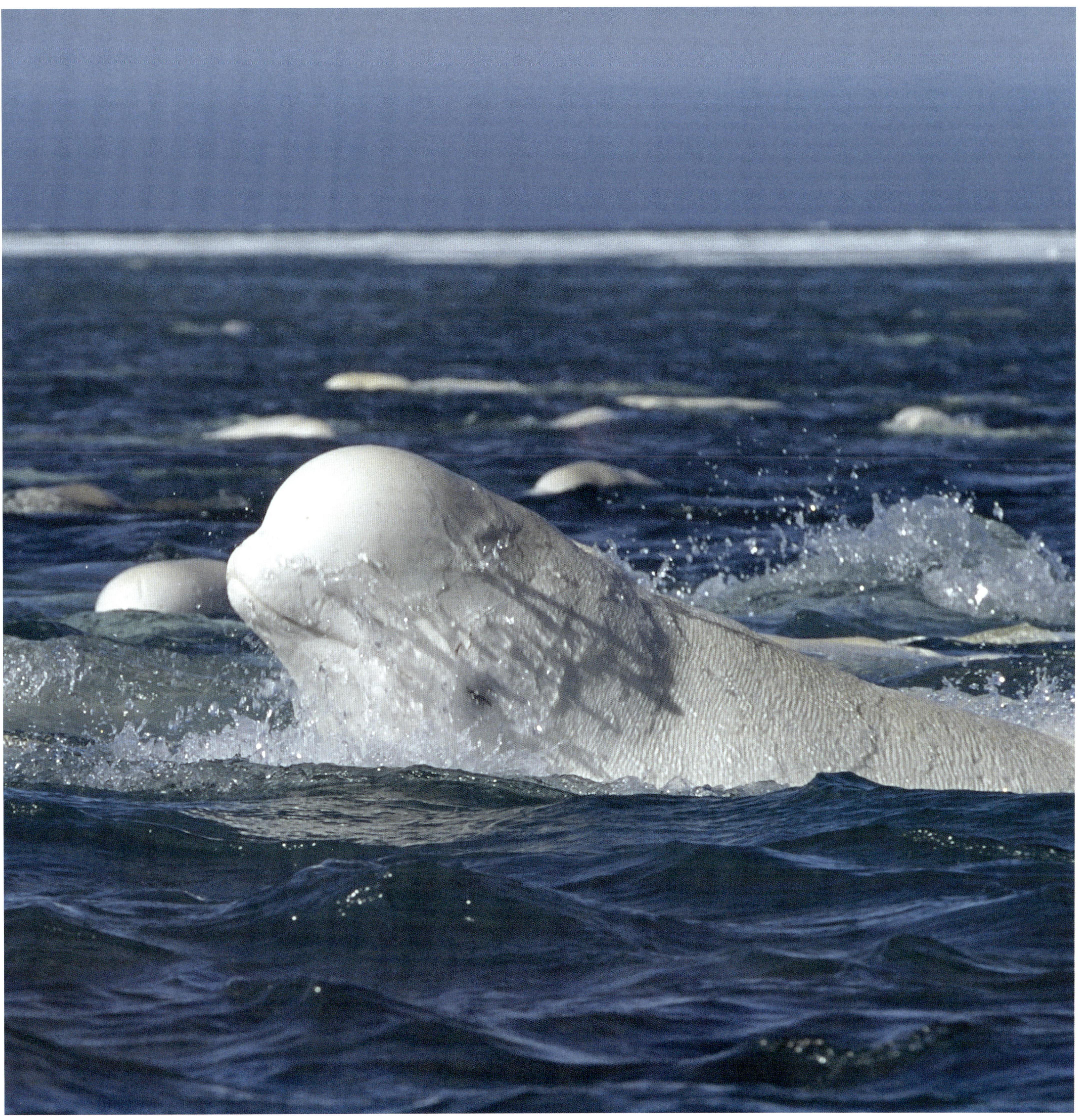

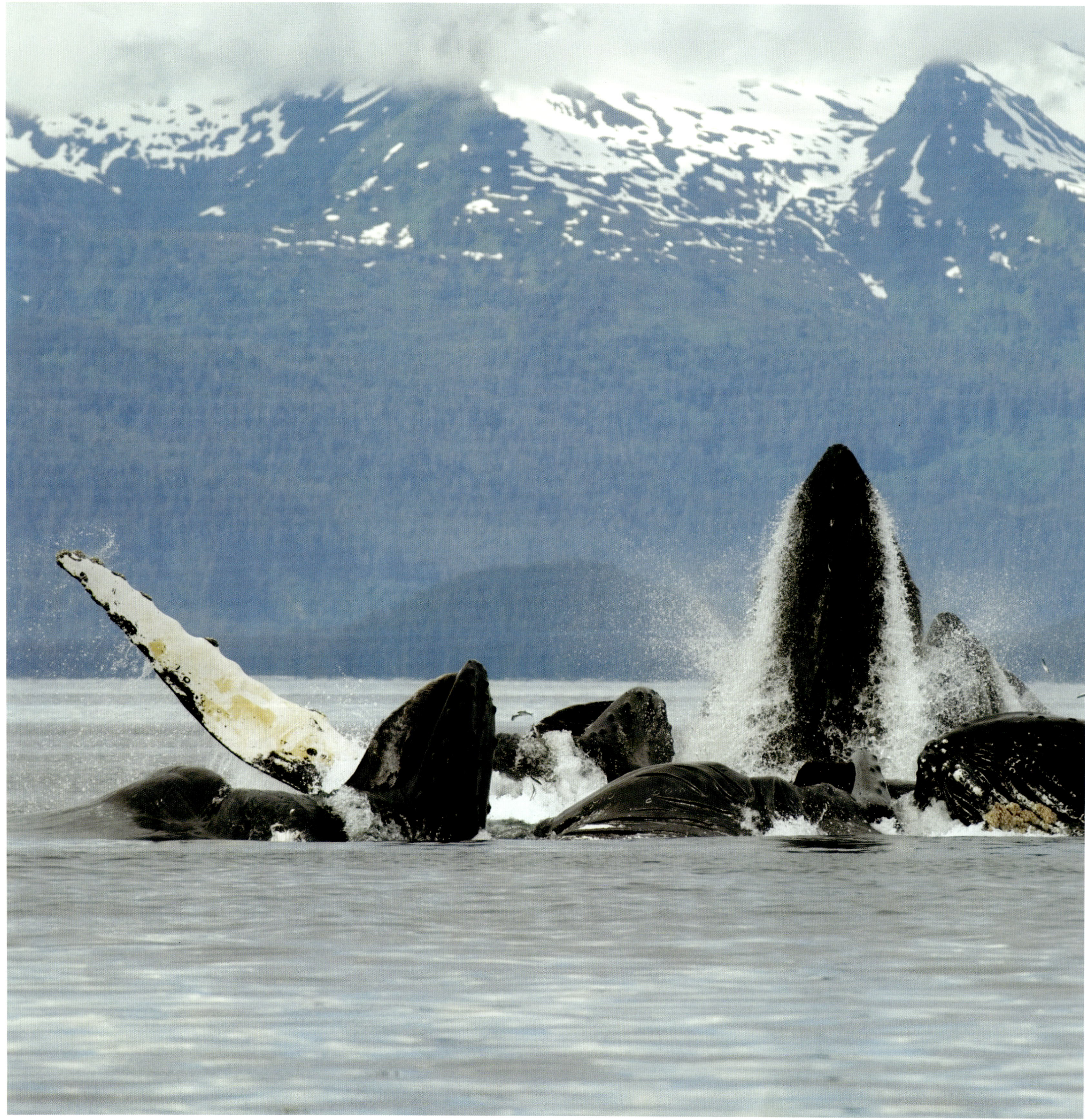

to see if they could extract sufficient quantities. They found ways of trawling out tons of plankton. Calculations showed that in a twelve-hour period nets of 320 square feet (30 square meters) could haul out enough plankton to feed 357 people. The idea was that this would be processed through drying and then mixed in with other food. However, there is no record of how the scientists planned to make this appetizing for public taste. Desperate as the wartime population may have been, they may have struggled to swallow this idea. It would have needed some kind of mashing, as the exoskeletal nature of plankton would have made any meal rather gritty if eaten straight. The scheme was ultimately shelved when the course of the war turned. Plankton is consumed, and commonly, in some Japanese, Korean, and Scandinavian dishes that draw on making edible products from algae. Japan is probably the leader in algae culture. This is mostly noted in the production of the seaweed that gets used in nori, familiar around the world as the dark seaweed sheets for wrapping sushi.

In algae, and its farming, may lie a solution both to our food and fuel needs if J. Craig Venter—the biochemist and entrepreneur who was the first to sequence a human genome, and who led the project to publish the first synthetic human genome sequence—gets his way. He has been investing time and money in looking at how algae farming could generate bioenergy and foodstuffs on a vast scale. To deliver this idea would effectively turn parts of the sea into massive farms. As part of the research for this, Venter's team undertook a massive sampling of genomes of algae drawn from seawater all over the planet over a two-year period. In this work it was said that they discovered 95 percent of all genomes known to science. From this brain-stretching research to any kind of industry that delivers food and fuel is still a long way off. But then it was a long way from a snail on the shoreline to a color that defined an empire. It just takes a little imagination and a lot of hard work to convert the sea into a wonder that changes lives.

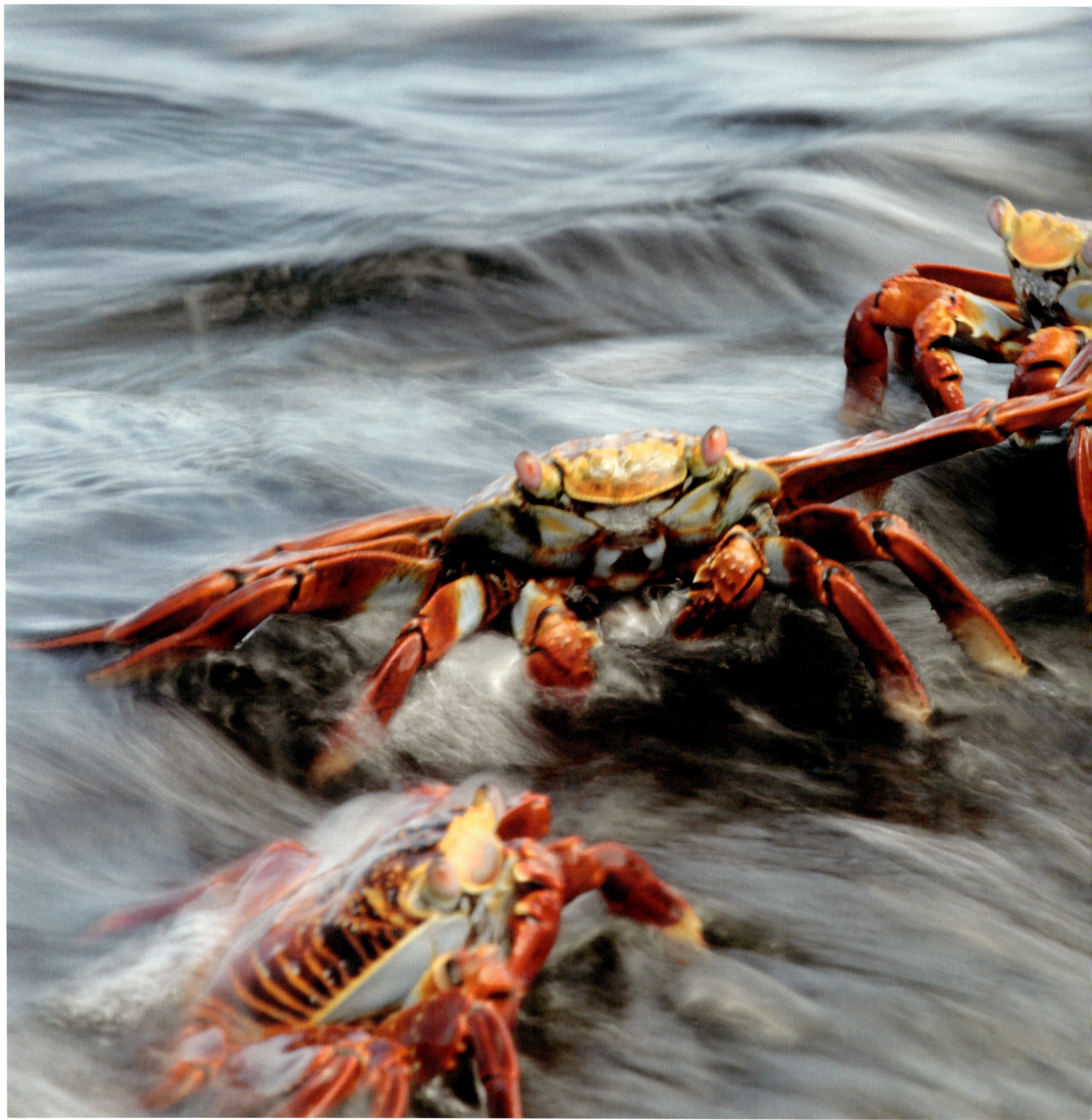

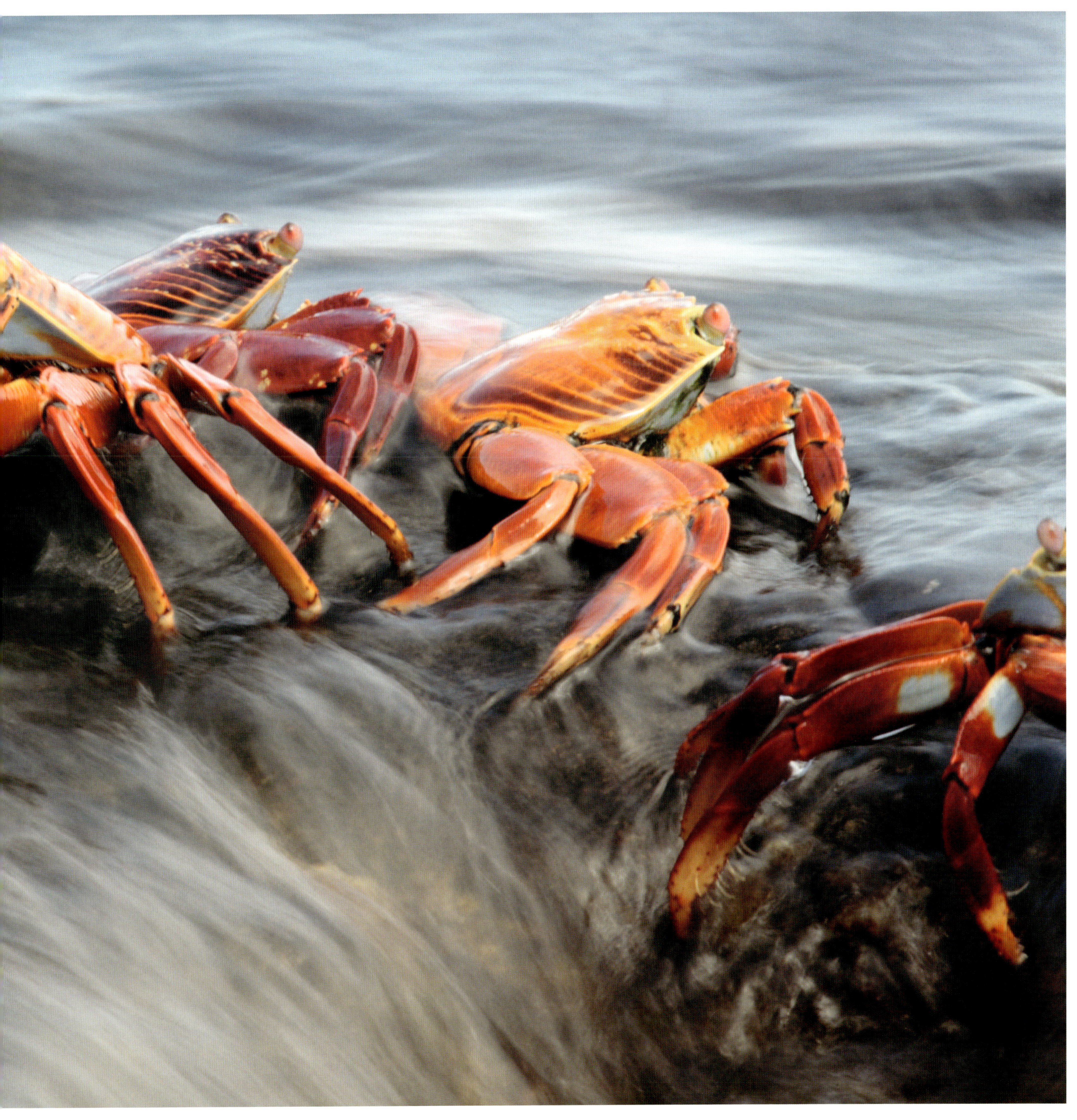

The voice of the sea speaks to the soul.

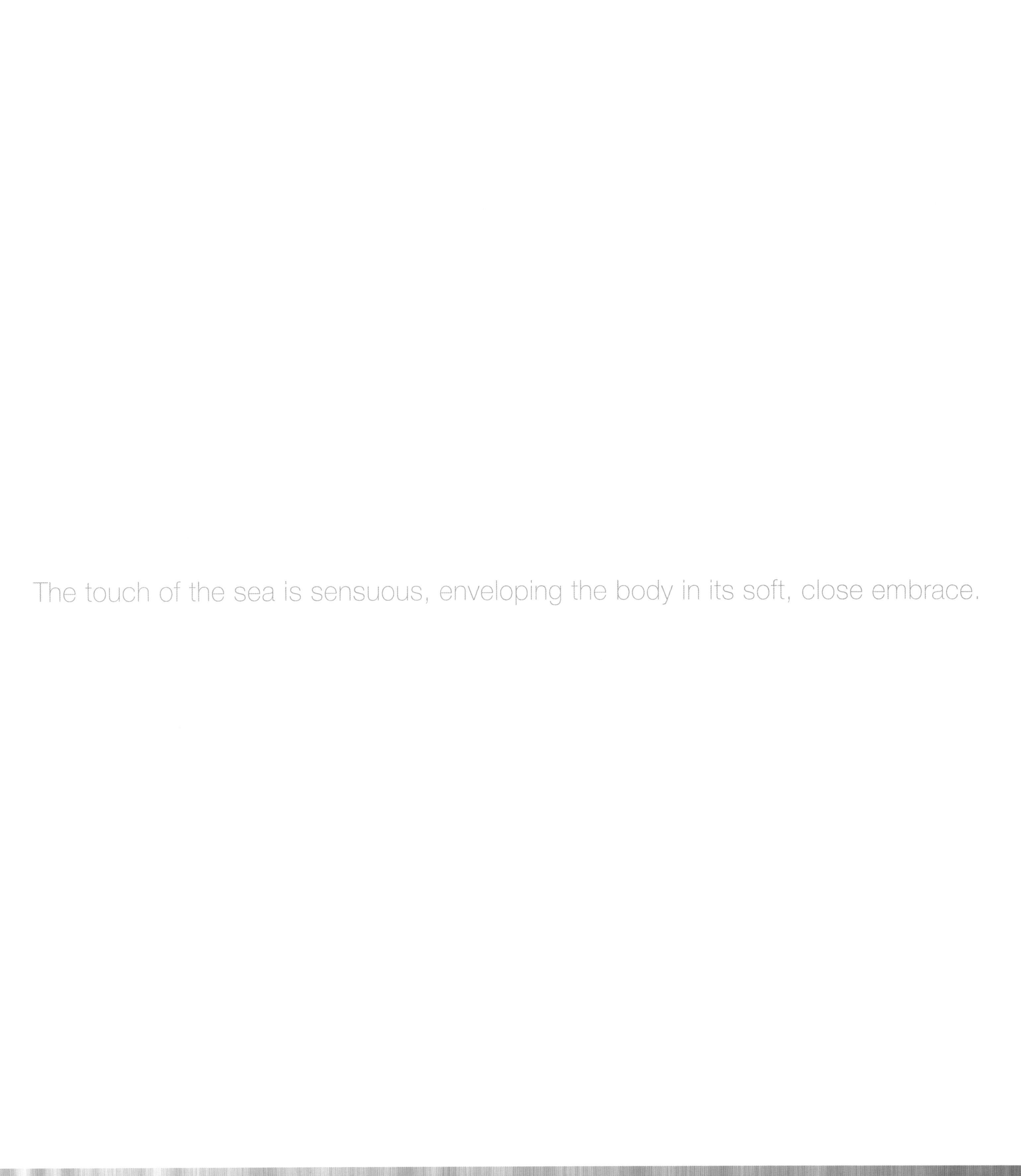

The touch of the sea is sensuous, enveloping the body in its soft, close embrace.

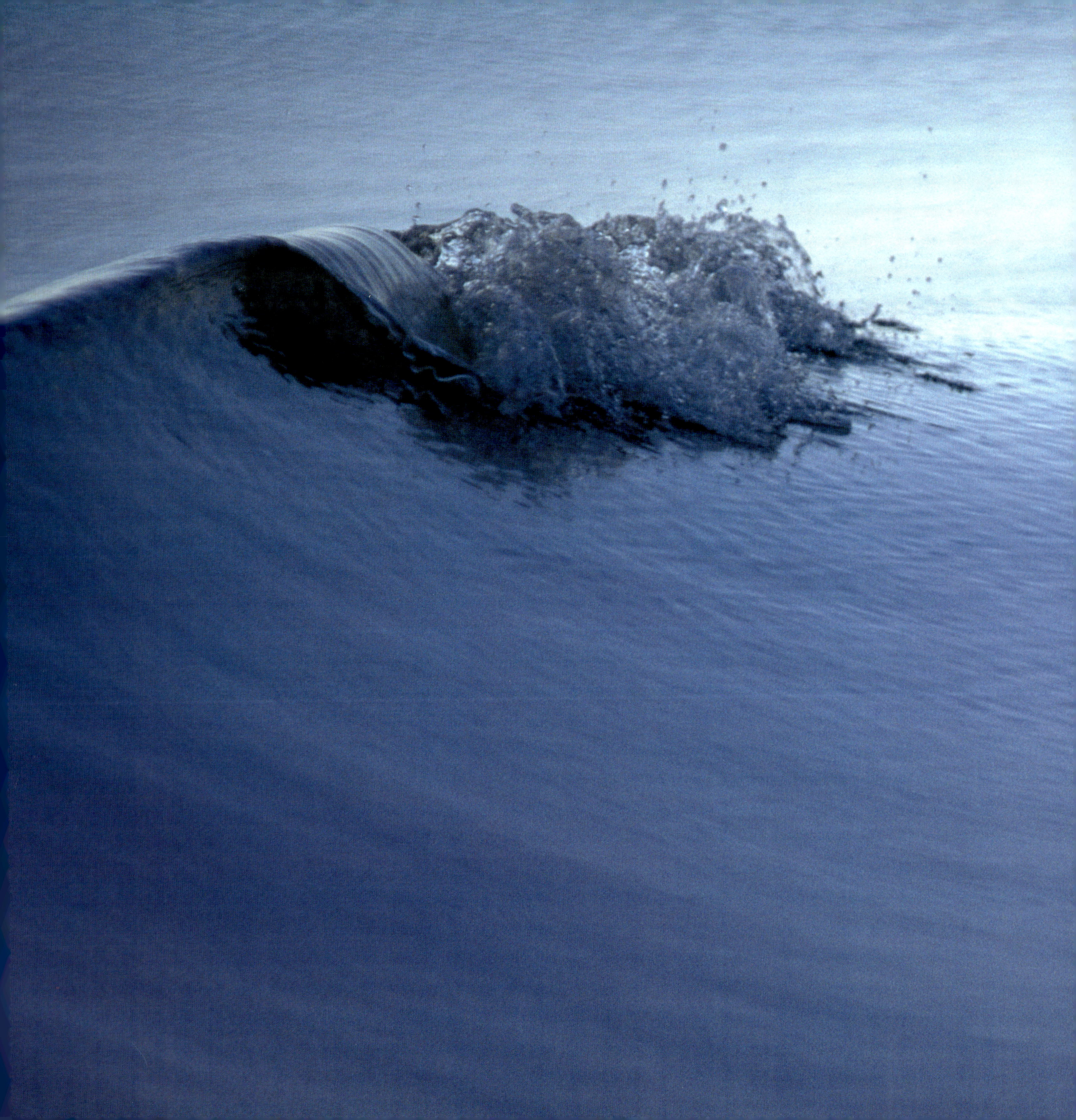

Beauty and the Beast

For this chapter, you might want to sit yourself down and turn on some background music. Play something inspired by the sea. Our subject now is how the ocean inspires our creativity, stimulating artists to feed us with sounds and visions that interpret, reflect, and respond to all things oceanic. For many of us, indeed, the ocean is in our imaginations, as we live inland and have limited experience of it for real. How many of us have looked under the surface of the water? For many the sea is an occasional sight of waves, seabirds, coastal environments, illustrated by half-remembered images from documentaries and stories from books. But even a brilliant film on the undersea world is but another creative act, a filter that the viewer bases a reimagining upon. This is how it has always been, most of us living on received visions. The condition of the ancients—for whom most of the world and the sea was a place that could only be imagined—continues, as we know for certain so little. Artists have been vital in giving us a vision. The oceans are vast in reality and in our ignorance, and we cope by filling in the gaps imaginatively and collectively. This other world becomes a rich subject for invention as well as investigation.

So, for your mood music, the accompaniment, what is it to be? The overture from *The Flying Dutchman* perhaps? The composer of this great opera, Richard Wagner, was inspired to the subject by a rough sea passage from Riga to London, his journey delayed en route for a fortnight by storms. Sea travel was perilous in those mid-nineteenth-century times, and timetables were merely sketches of possibility. Weather forecasts were conjecture. His music may capture a sense of the energy of the wind and waves, but it is also inspired by a wonderful, spooky legend from the seventeenth century. This tells of a ghost ship, the *Flying Dutchman*, that is doomed to sail the oceans endlessly, never able to put into land, making it a messenger of misfortune and a window into the land of the dead. A bit gloomy perhaps, if uplifting in passages. OK, perhaps you want something lighter. Would you rather harness the inner child with the catchy theme tune from the children's television show *SpongeBob SquarePants*? While the character may seem only marginally marine, the music riffs on a great tradition, as the melody is lifted from the sea shanty "Blow the Man Down," a tune that may go back hundreds of years. This is not such an innocent song at all, as it concerns men lusting after women and fighting, very much the kind of thing we might associate with rough, tough sailors of centuries past, rather than a family-friendly TV character. Sea shanties are a rich cultural store, a genre of music that was born and flourished as a tool of sea life. Singing them was once a daily part of the routine on board ship for the working seaman, a way of helping haul ropes to a rhythm, as well as to get through the other repetitive and tiring tasks of life at sea. Inevitably, the shanties both describe and exaggerate the life of the sailor, and they are never dull in their subject matter. The composers of sea shanties are lost in history, with usually not just one name behind a perfect version, as they are collaborative efforts between many. The songs exist in numerous variants.

Collaboration also exists in the lost creative origin of the most ancient and greatest story of a sea journey: *The Odyssey*. Written between 800 and 600 BCE, it is attributed to Homer, the blind poet who is the father of all literature, but the epic poem emerged from an oral storytelling base. It was most likely handed down in a tradition of recital, evolving through versions before becoming fixed in a written form. It has a strong central narrative that gives us one of the most archetypal, influential stories, with the sea journey as a metaphor for all our lives. It tells of the Greek hero Odysseus (later called Ulysses by the Romans) and how he endured a most protracted and adventure-laden sea journey home. He wanted to return to his wife, Penelope, on the island of Ithaca after fighting at Troy, on the Turkish mainland, which involved a long journey around the Greek archipelago and across the Aegean. Odysseus had spent ten years in the siege and sacking of Troy, which featured in Homer's first epic, *The Iliad*. His return, though, took another ten years as a result of his encounters with mythical places and beings. In this journey, Odysseus (whose name means "trouble" in ancient Greek) is a flawed hero, struggling to resist challenges. The sea is the medium that delivers and distracts and threatens him, bringing strange encounters and life-threatening tests. When he finally arrives back in Ithaca he has lost his men and comes in disguise as a beggar. However, his old dog recognizes him (then dies) and he gets his wife back, after killing a bunch of suitors. A happy ending, but at a high price.

Through all the verses of *The Odyssey* the sea is the great giver and taker of life, the canvas on which we celebrate and explore our existence. The ancient Greeks would have seen the waters of the Aegean as crucial to their identity as a nation and in determining the bounds of the known world. The risks of sea travel were huge: It only took a strong wind for sailors to be blown off course, and they struggled immensely to know where they were if they lost sight of land. It would not have seemed so far-fetched to suggest there were wondrous creatures and places just over the horizon. *The Odyssey* laid down a sense of the risks and romance of the sea that persists to this day.

The Iliad and *The Odyssey* were among the first epic stories. By the nineteenth century the novel was the dominant literary genre and its ultimate form was another sea story, *Moby-Dick*, a vast novel about a crazed whaling captain's quest to kill a great white whale (actually a sperm whale of a strange bleached color) that had injured him previously, leading to the loss of a leg. In a very lengthy journey to track and take on the whale, the reader gains great insight into the ways of the sea and of the whaling industry and of what was understood about whales at the time. Songs and customs are extensively referred to and we are immersed in the life of the Nantucket whaler and learn how it must have been to spend many months at sea. Ultimately, the obsessed Captain Ahab finds Moby Dick—or perhaps, in a way, it finds him—and this leads to the destruction of his ship, the *Pequod*, and his death. Only the narrator, Ishmael, survives. Written by Herman Melville and published in 1851, this epic is often seen as the first "great American novel." It draws deeply on Homer in the ways in which it uses the sea as a metaphor. But deep roots and profound thoughts did not count for much, as the book had a poor reception and effectively destroyed Melville's reputation. It was only several decades later, when sales started to take off, that its critical reception changed, and by the early twentieth century it was seen as a masterpiece. It is a monster of a book and is unlikely to be read much outside of English literature courses. Reference to it lives on, though, in many a coffee shop around the world, through the name of the first mate of the ship: Starbuck.

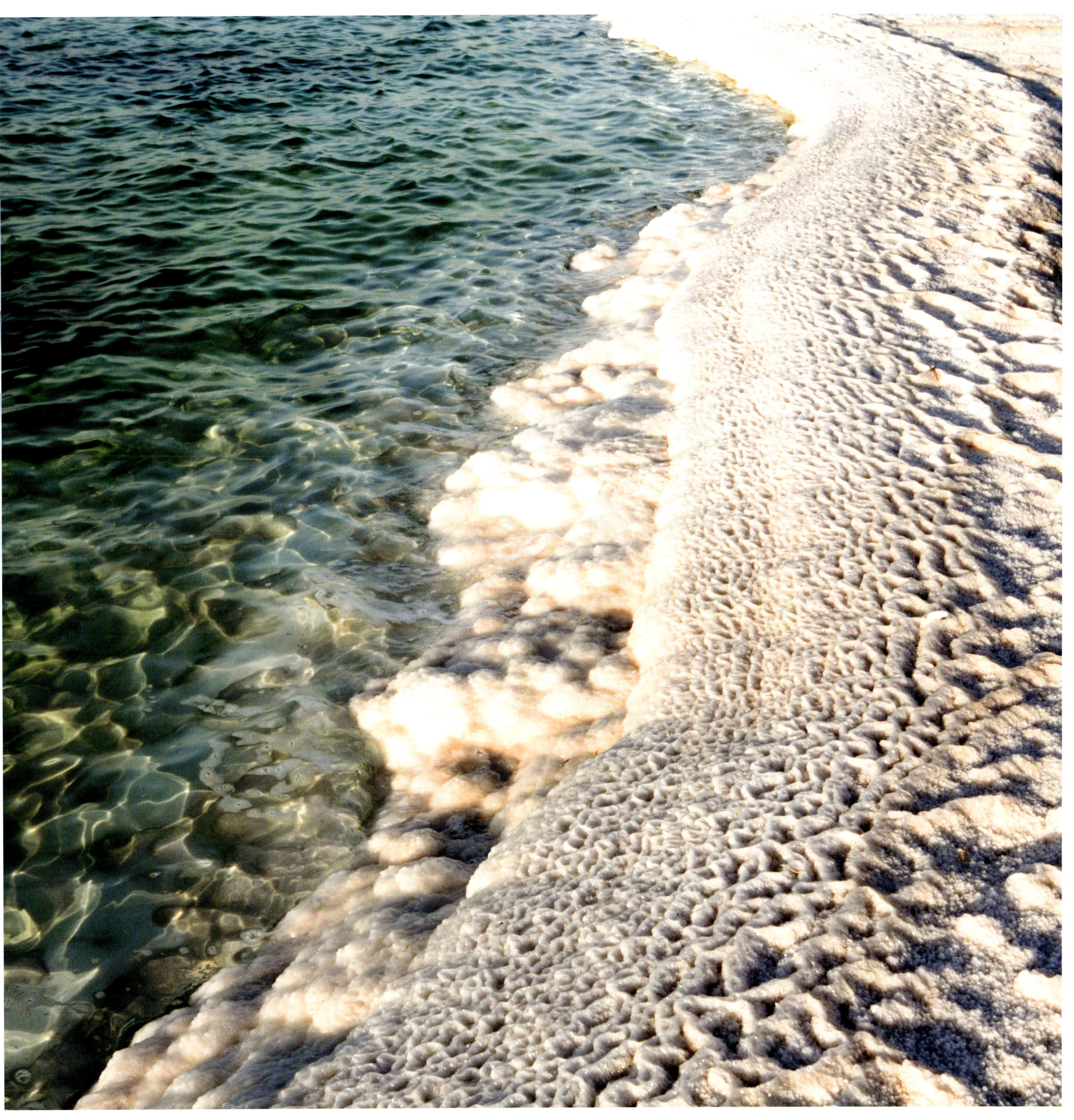

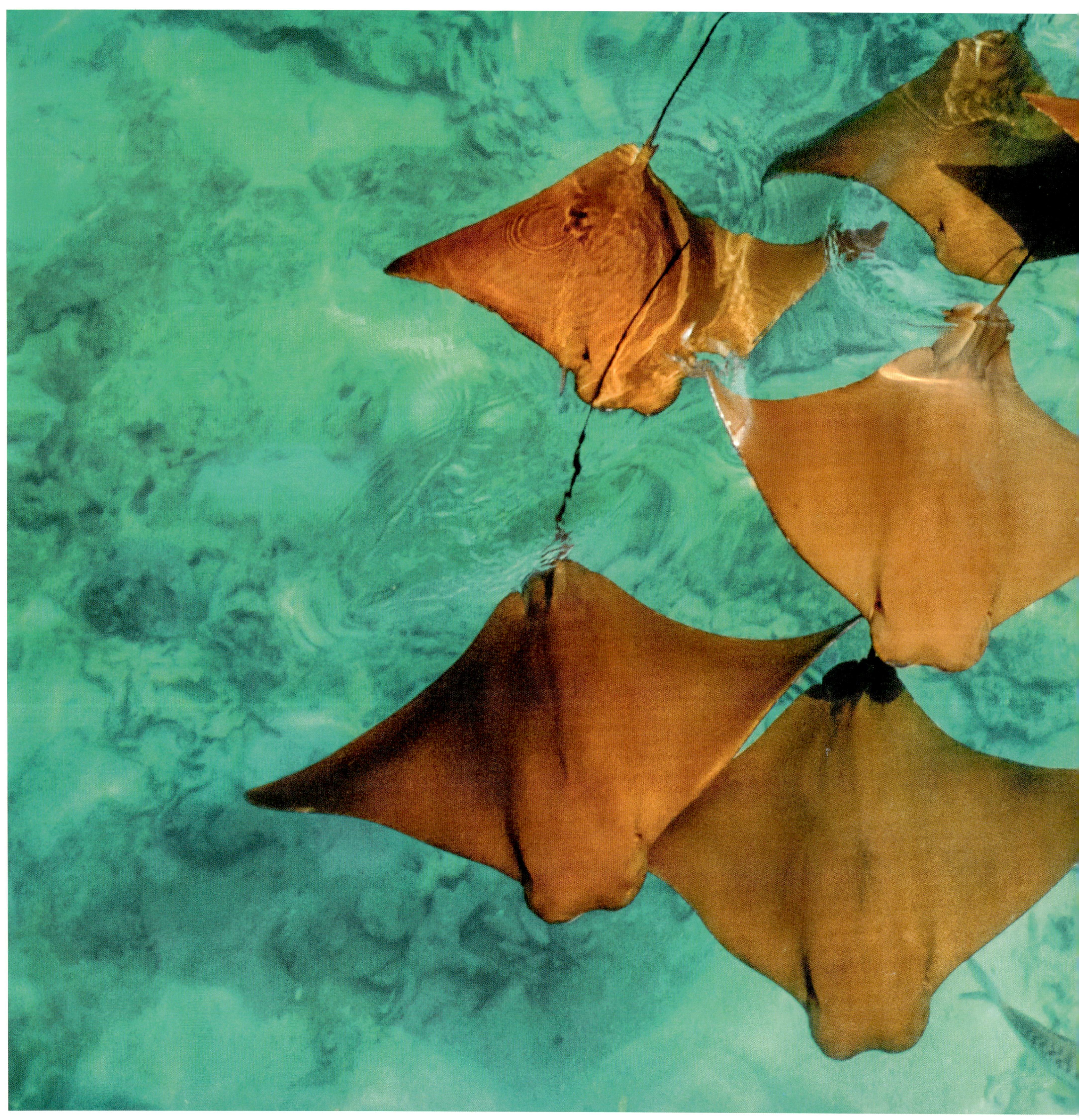

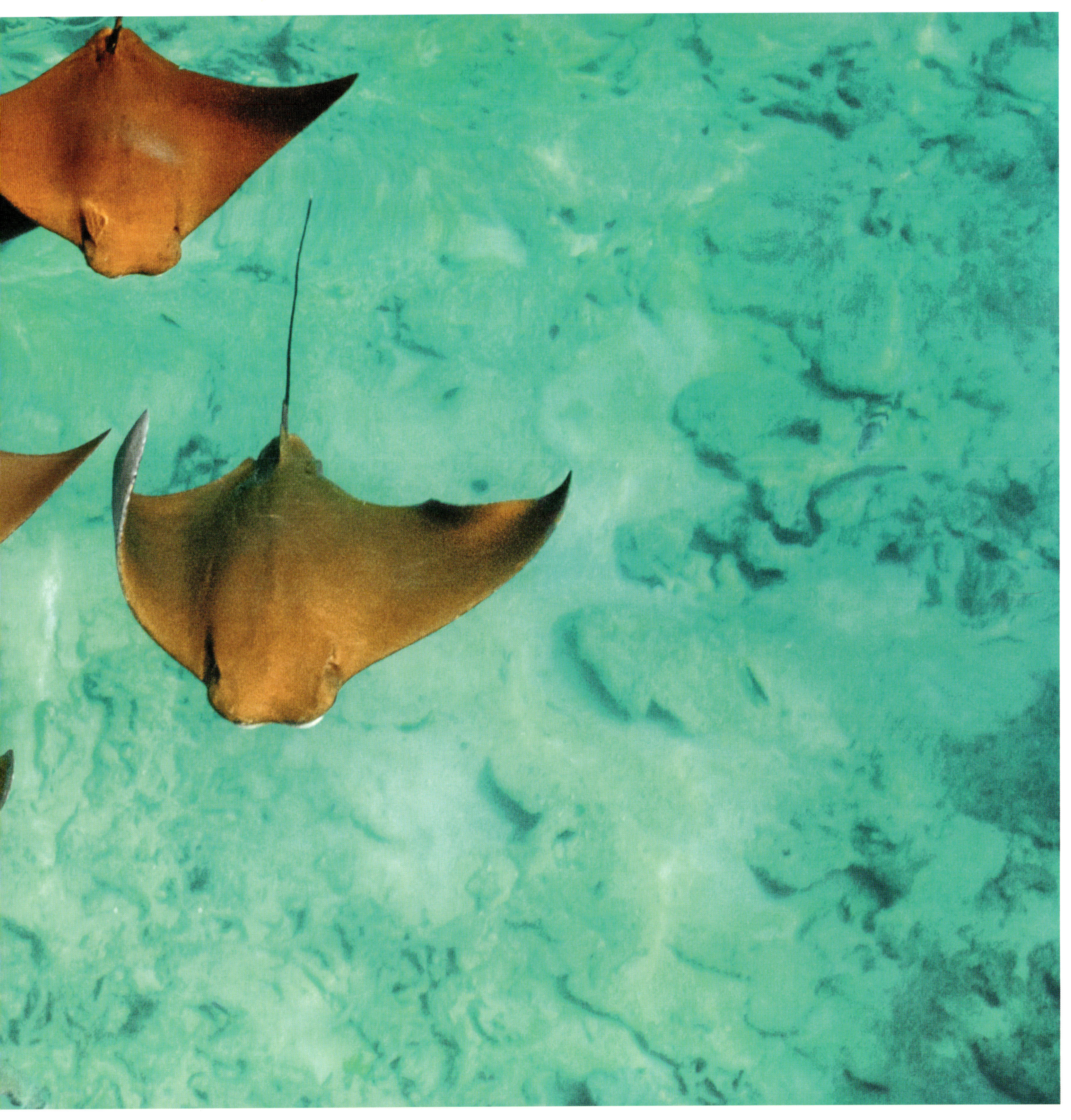

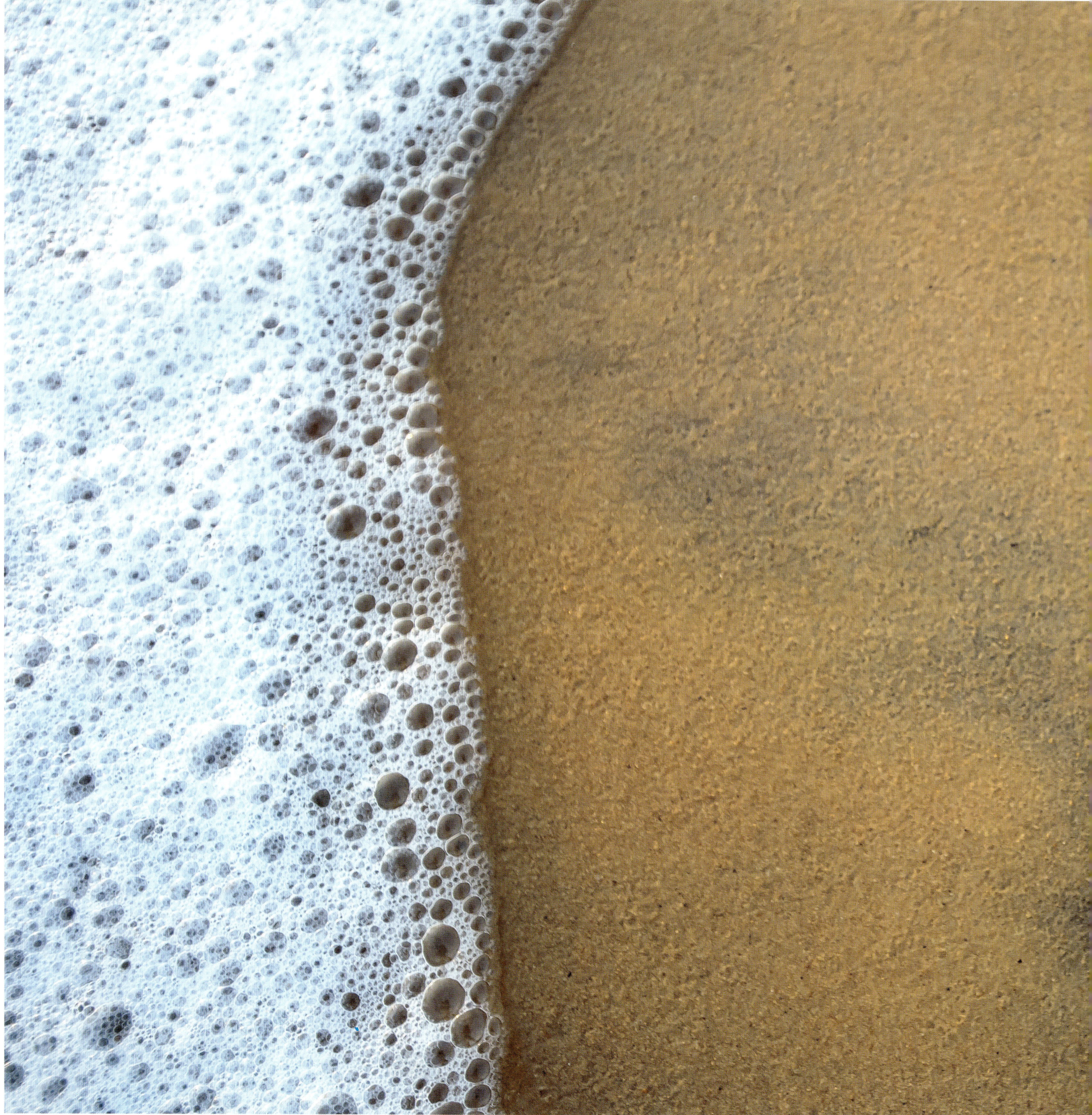

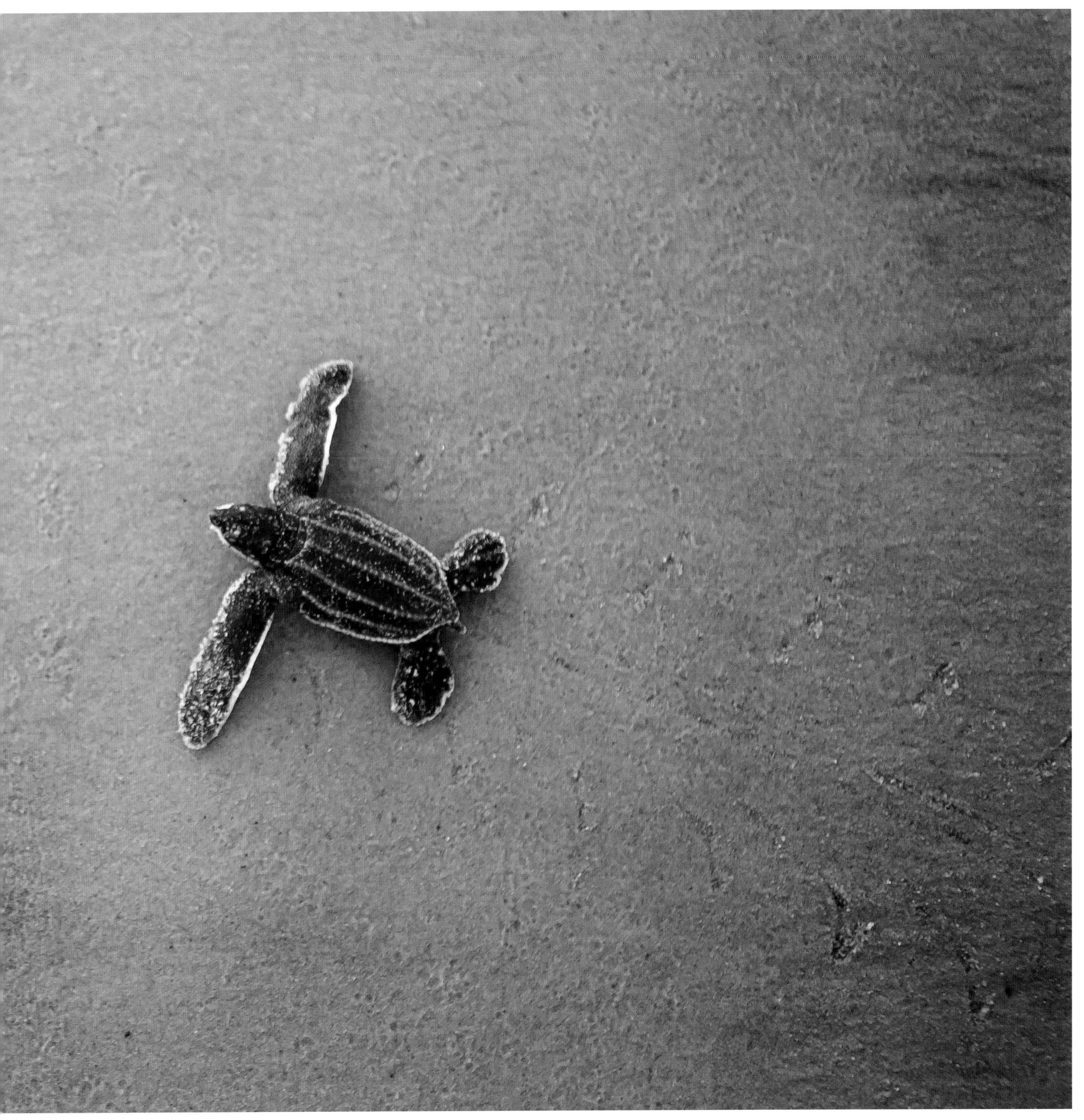

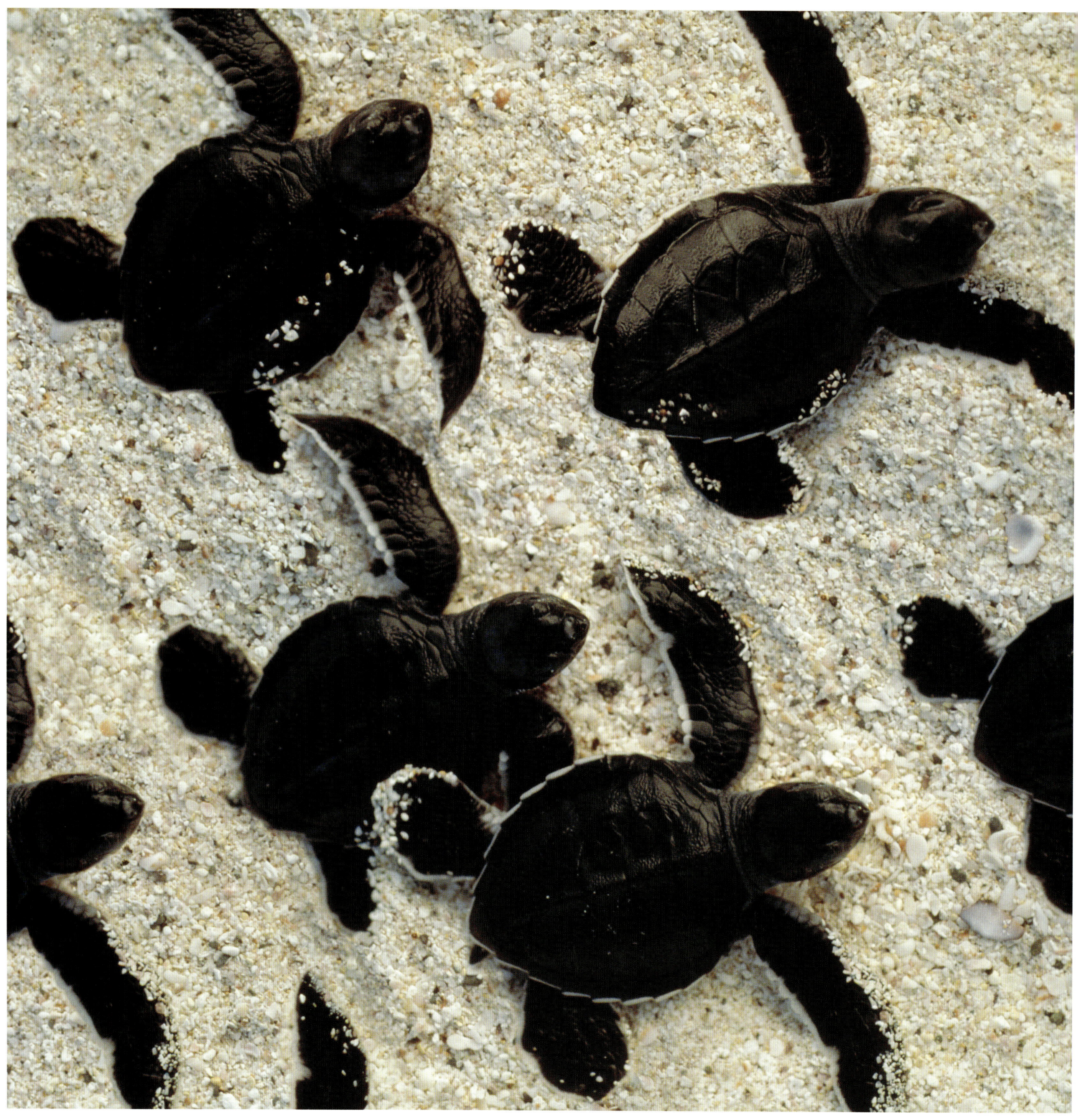

Insult to injury for poor Melville, who only saw the book sell 3,200 copies in his lifetime and never experienced the charms of a skinny cappuccino with chocolate on top.

Moby-Dick is not entirely invention but is based quite closely on the real story of how the whaling ship *Essex* was attacked and sunk by a sperm whale in 1820 in the middle of the Pacific Ocean, thousands of miles from land. As a result, the surviving crew were forced to take to the lifeboats. Ultimately, only eight survived in three boats, which became separated, with the men in two of the boats resorting to cannibalism. In this, the true story was perhaps even grimmer than the fiction. Melville also drew on a story of a whale killed off Chile in the 1830s that had numerous harpoons in it, inspiration for Moby Dick's great strength and endurance. The Chilean whale was nicknamed Mocha Dick and was reputedly white—both references that Melville took in naming and visualizing his own heroic whale, a force of primeval power, of nature versus man.

This approach to depicting the sea as a crushing force against man carries right down to our time, with a movie like Robert Redford's *All Is Lost* in 2013, where he played a single-handed sailor fighting the odds of survival as his yacht becomes increasingly wrecked. The ocean is a place of dangerous beauty that takes us to the edge and beyond, and is ultimately not that welcoming. But then perhaps it is always uplifting in a spiritual sense (there's an ambiguous ending to the film).

There is more to the sea in culture than awe and dread. The world of the arts also shows that the sea captures us in simple, pleasing ways too. Time to switch the soundtrack, perhaps from Wagner to Claude Debussy, whose *La Mer* is one of the most popular classical music works. It is a symphonic piece in three movements that has depth but never strays far from making you feel safe by the sea. It is like a range of sea breezes across the face, never approaching storm force. Even if it moves from a sense of light, rippling waters to choppier conditions, it leads us back with a sense of calm. Ultimately, all is good rather than lost. Debussy wrote most of this away from the sea, being more inclined to draw inspiration from paintings and other arts in their commentary and insights into the sea than to actually learn by looking. However, he is reputed to have finished the work while staying in Eastbourne on the south coast of England. It is tempting to say that he finished it there because Eastbourne and its fairly dreary views of the English Channel are so uninspiring that he found nothing more to add. A side note: the many works inspired by *La Mer* are said to include parts of the score of *Jaws*, Steven Spielberg's 1975 hit movie about a man-eating shark.

And it is sharks that are the spoilers in Ernest Hemingway's elegy to the sea, aging, and manhood, *The Old Man and The Sea*. They are the destructive angel in sea nature, eating the giant marlin that the old man of the title has caught, leaving only skeletal remains for him to show for the record catch that he so nearly had. This tale of a fisherman risking everything to prove himself, perhaps one last time, in harmony with the challenge of nature, focuses simply on the catching of a great fish. It is somehow both peaceful and violent, capturing the ambiguity of the ocean's power. It is also expressive of our dilemma: We live (rather like the shark) off the death of other life, and however much we respect it, we need to destroy it.

But back to Debussy. He could have spent many a day happily viewing seascapes if that was his inspiration, as painters have used the subject extensively in all cultures across time. Whether depicting ships setting sail, or mythical or actual creatures of the sea, or just attempting a form of documentation of weather and scenes, paintings of marine environments go back almost to the origin of art. The challenge of depicting in one image a world where the elements are endlessly, rapidly changing makes an intriguing test for the artist. Caspar David Friedrich, the great Romantic artist of the early nineteenth century, occasionally touched on seascapes as subjects for his meditative images, projecting outwardly our inner life and values into inspiring and portentous scenes. *The Wreck of The Hope*, also called *The Polar Sea* or *The Sea of Ice*, painted between 1823 and 1824, is one of his most mysterious and yet pseudo-documentary of images, imagining a world of broken shards of sea ice that look highly realistic, and yet the artist never traveled to Arctic waters and he had no photography to refer to. The painting is dominated by the ice, in which we find a ship being crushed. Almost at the same time, on the other side of the world, Katsushika Hokusai was in Japan creating his masterpiece, *The Great Wave off Kanagawa*. Here, a wave of tsunami proportions rears up majestically, dwarfing a distant Mount Fuji and sucking into its grasp the boats below. Once again the beauty and meditative qualities of the sea turn destructive. Around the same period, this balance could also be seen in the work of J. M. W. Turner, who began a journey to near-abstraction in wrestling with the light on water, waves, clouds. His images of placid waters in the Thames estuary, ships floating in the sunset, are popular and almost "easy viewing," and yet they are fraught with a sense of the temporary, that all things are changing. The sunset turns to fire, the ships dissolve in the light. The sea is there, too, in almost endless variations, within the works of impressionist and post-impressionist artists, where the brilliance of waves and big marine skies fueled the artists' search to show that, in a painting, all is light. The paintings are easy on the eye to our way of seeing now, but they were highly disruptive of the rules of depiction arrived at in previous eras. Turner's seas were revolutionary.

Today every tourist seaside town is likely to present us with a gallery of local artists' output, which extends, in its modest way, the tradition of marine art. Whether the paintings are good or bad or entirely forgettable, we can perhaps rest our critical faculties and just enjoy the process. Painting or drawing a subject that is so restless is an interesting, meditative activity that can have great value in just the doing. To use a popular psychology reference of these times, you can practice mindfulness by just studying the image of the sea. To give an artistic response to the endlessly changing sea and sky is to comment on life. Leonardo da Vinci saw no boundaries between the arts, science, and daily observation, and he was one of the first to realize that there was once no boundary between what was land and what is sea, and vice versa. He commented in his notebooks: "Why are the bones of great fishes, and oysters and corals and various other shells and sea-snails, found on the high tops of mountains that border the sea, in the same way in which they are found in the depths of the sea?" That sense of the mutability of all things perhaps sits within the smile of the *Mona Lisa*.

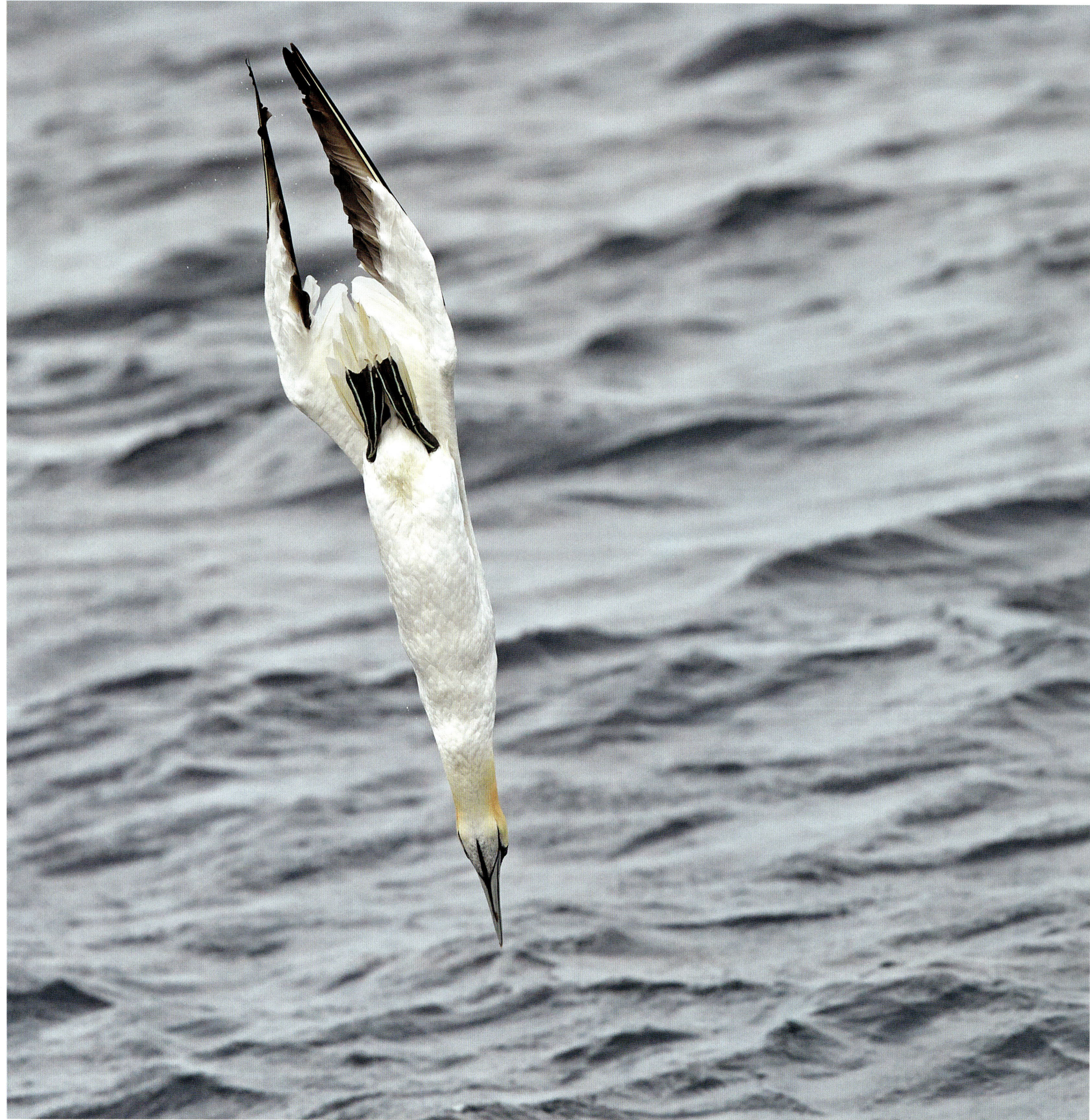

As air is a sacred gas, so is water a sacred liquid
that links us to all the oceans of the world and ties us back
in time to the very birthplace of all life.

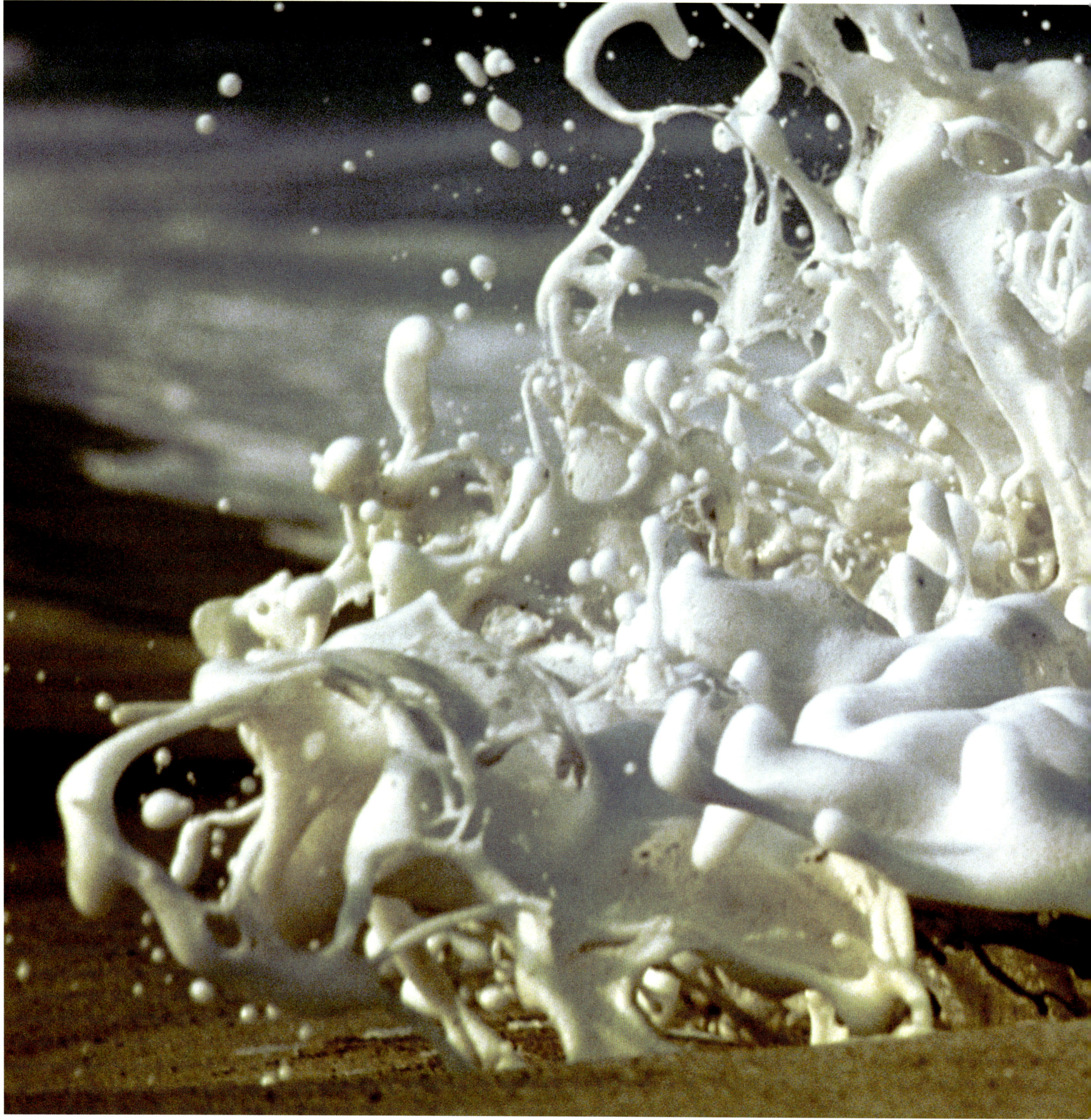

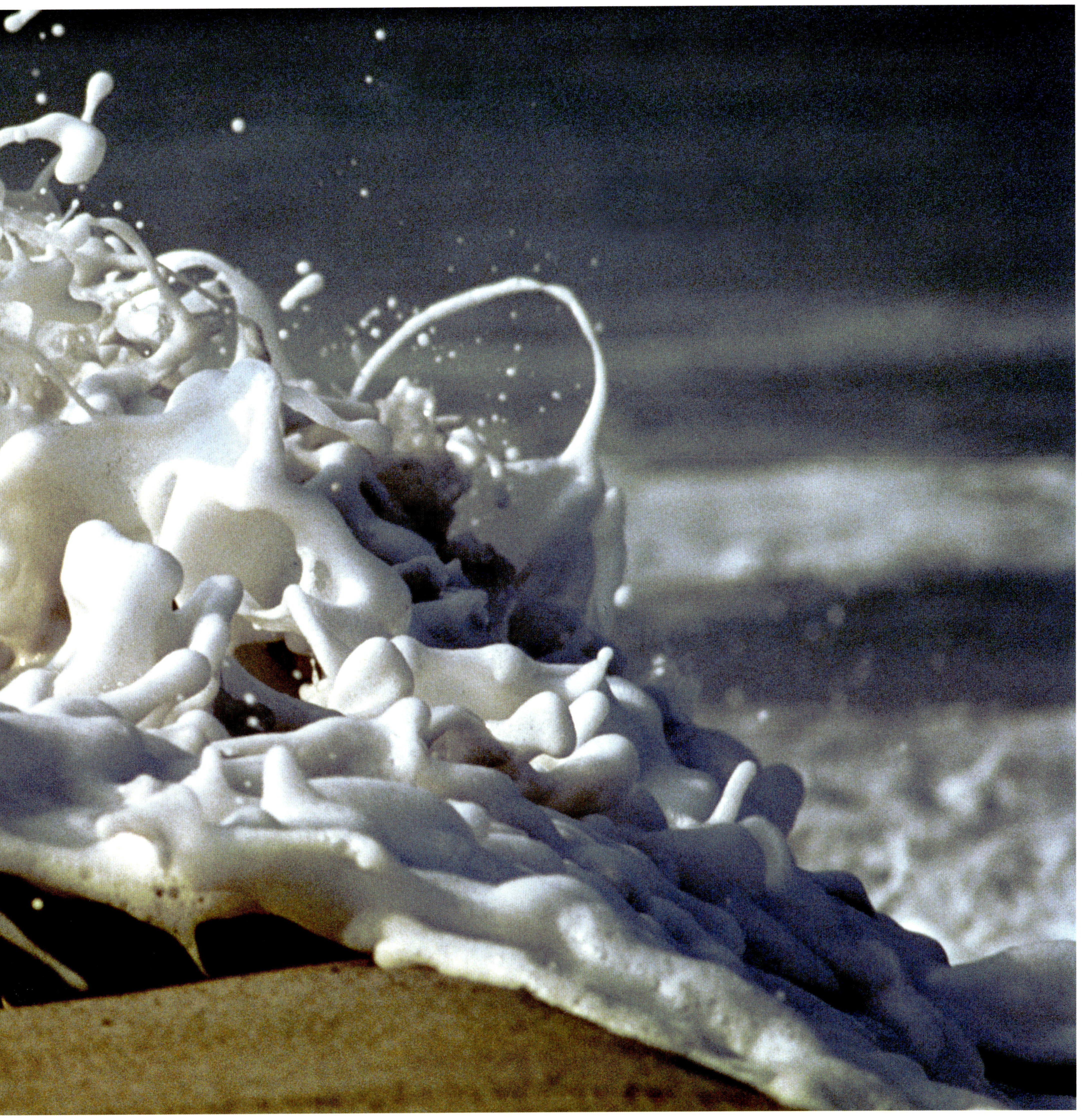

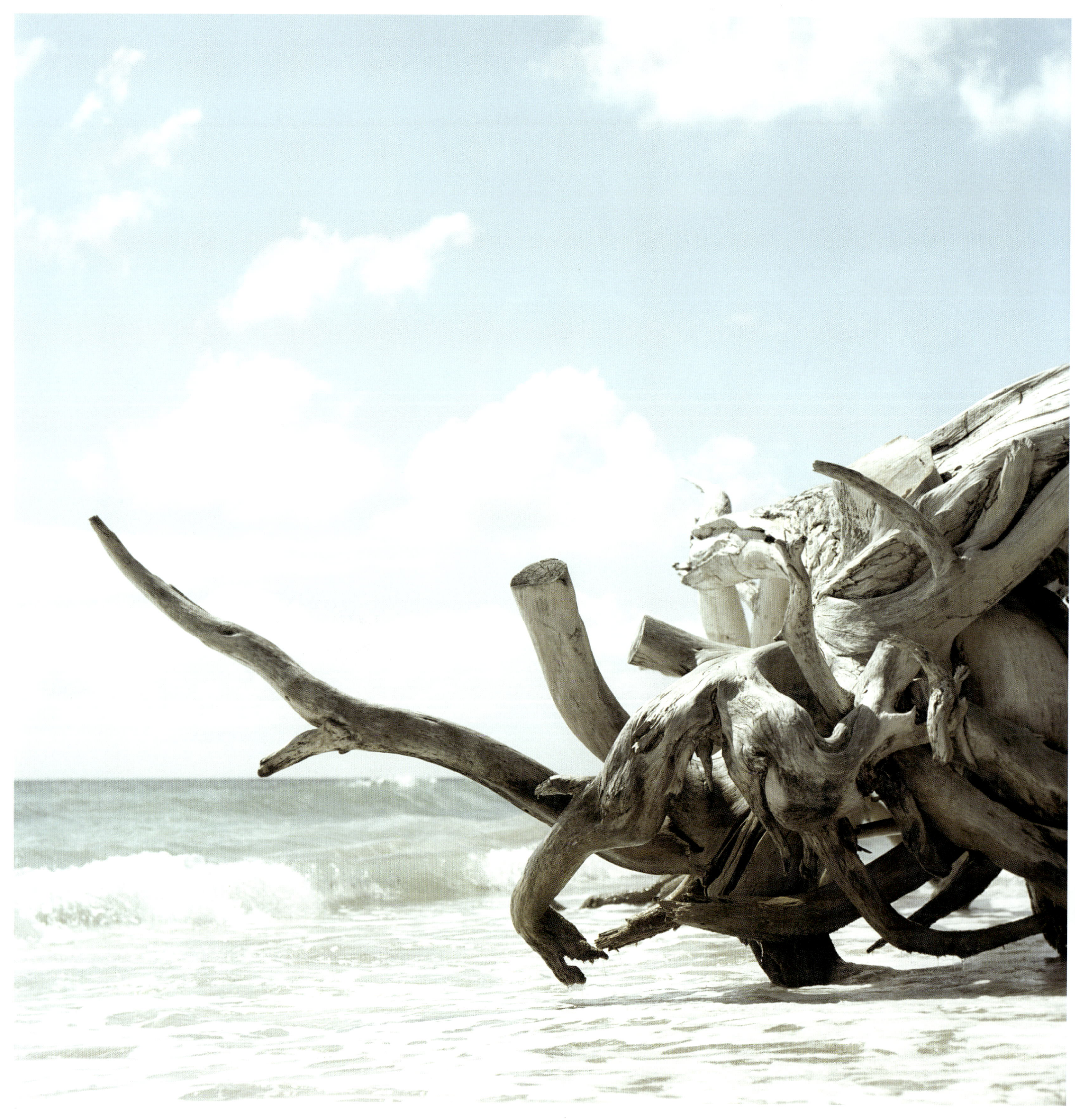

In every outthrust headland,
in every curving beach,
in every grain of sand
there is a story of the earth.

Beyond the Horizon

There are six famous, haunting lines early in William Shakespeare's *The Tempest*. They charm and dismay in equal measure:

Full fathom five thy father lies;
Of his bones are coral made;
Those are pearls that were his eyes;
Nothing of him that doth fade
But doth suffer a sea-change
Into something rich and strange.

This tragicomic meditation on mortality, power, and creation is set on a mysterious island where, with magic and fantasy as everyday elements, the plot wraps around issues of life and love. It makes our human existence seem but fleeting against the immensity of nature and the sea and the universe. It was probably the last play Shakespeare wrote on his own, and is a sign-off of sorts, a wave good-bye by the great playwright and poet, putting his own work and profession into small relief within the great creation and mystery of all things. All done with seemingly effortless lightness and skill, of course.

In that imagery of pearls that were eyes, coral that was bone, and "rich and strange" change as the essential nature of things, Shakespeare presents both a holistic vision of the world and yet one in which we never ultimately know the answers. Change happens, but there is no assurance of a healthy cycle of rebirth. In this, as in so many areas, he was a thoroughly modern thinker. As we look at the future of our planet, and of our engagement with the oceans, we know that things will change and we know that we don't know how it works and will work. What we do know, with some degree of certainty, is that we are unlikely to get on top of what drives "sea-change." In other words, we know that we don't know much and may never know enough. If this sounds confusing, that is because it is. We have oceans out there with life that works to principles that we are yet to discover. We will continue to get more clues about how the oceans operate, but they are bigger than us, they are more complex, they won't be "resolved" easily or perhaps ever. We have to handle with care and yet we seem unable to stop ourselves from wreaking destruction. But we must prevent that damage as much as we can, and even try and reverse it.

The future of this planet will involve us more intelligently and extensively engaging with the seas. As has been said, often in various ways, this planet is less "the earth" and more "the ocean." But how this future involves us *Homo sapiens* is an area of major debate. The signs are that we need to behave differently if this world is to cope. The planet doesn't need us, but we need it, while the indications grow that many species would be better off without us. OK, yes, the dogs would miss us for a while, but they would get over it. We live in an era of geological time that is now called the Anthropocene because it is our species that is driving the shape of all life—changes that are mostly bad news for other species that disappear rapidly whenever we care to count. What this means for the largely unknown world of the ocean, on which all us landlubbers depend to create the air we breathe and more, is perhaps one of the largest mysteries, questions, and threats to which we have to respond.

Before we screw things up totally, become extinct, and hand responsibility to deep-sea crustaceans and chemosynthetic life, along with a few ants and beetles on the desiccated land, perhaps we can improve our game, correct our course. It is not too late, but we must move fast.

Many years ago our politicians set down fine words on how we might manage the oceans in the document known as the United Nations Convention on the Law of the Sea. As of writing, 166 countries and the European Union have signed up to terms that largely date back to the 1970s. It is supposed to set out the rights and responsibilities we all have in using the oceans, but there is an emphasis on how to divide up the spoils and exploit the resources rather than on holistic management. And given that one non-signatory is the United States, which continues to feel its economic and security interests are not fully protected by the terms, we can see that "the freedom of the seas"—the principle that the seas belong to everyone—which underpins the convention, is still some way from being achieved. International laws and treaties also exist relating to fish, mammals, birds, and other conservation and usage issues, but we remain fragmented in our approaches to the oceans and patchy in our consistent application. Whales are still hunted, sharks are butchered in the thousands just for the fins, and worldwide fishing quotas are evaded or ignored. Meanwhile, the rules are sometimes insensitive: Many fish are trawled up only to be thrown back for being the "wrong" type of fish, outside the quota. Hardly a victory for conservation.

It is often the small things that matter most in conservation and get overlooked until too late in the day. For example, it is hard to get passionate about saving a species of seagrass. But if we can't retain and restore the meadows of Johnson's seagrass, then the endangered green sea turtle, whose diet, it is thought, may largely depend on it, will suffer. It is not easy to make that the stuff of law, or raise widespread public concern. Multiple small-scale actions are required to protect and enforce the restoration of a species. And can this happen for many thousands of species in thousands of places?

The Great Pacific Garbage Patch, the vast area of polluting plastics and other waste that is caught in the North Pacific Gyre, is big enough to catch global attention. But it is also an example of a problem that belongs to everybody but nobody: It is outside the protected zones of the United Nations convention, beyond any one government's care. Who really wants to pay for sorting the problem? Various nongovernmental conservation bodies and innovative individuals are bringing the problem to the world's attention, along with possible solutions, but the cause and the cleanup will take concerted, consistent actions. Meanwhile, more and more micro-plastic particles are finding their way into the food chain that sea life, and all life, depends on. Like a shark scenting blood, these issues will come back to bite us, even if the problem only appears to be the scale of a nibbling sand fly at present.

While we wait for international law to catch up with conservation pressures, we need to work out what we, as a global society, or societies, think should happen to the seas. How do we move from messing up the oceans to nurturing them? They must be sustainable so that we can feed and fuel ourselves in the future. Who writes the business plan for that?

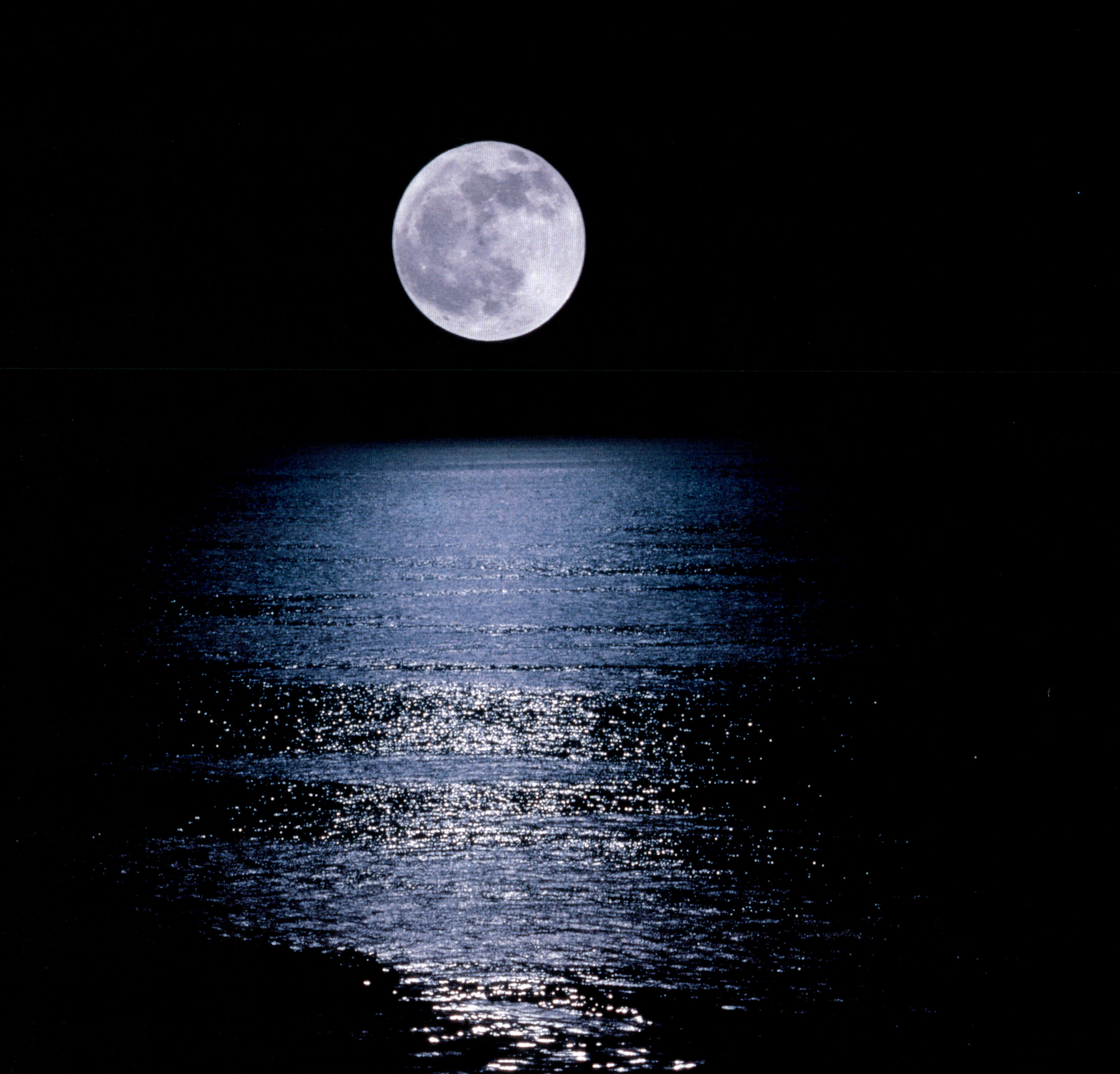

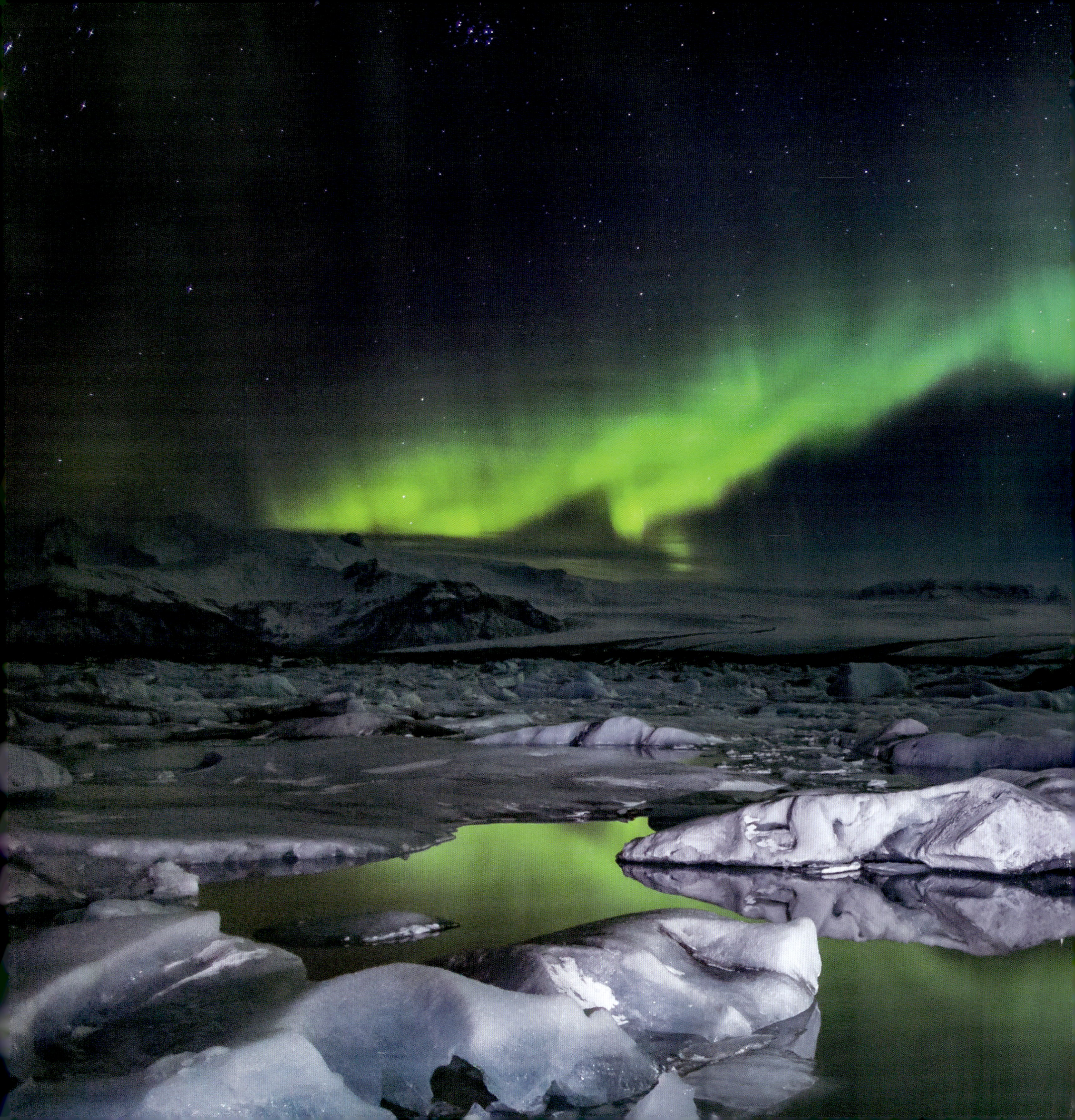

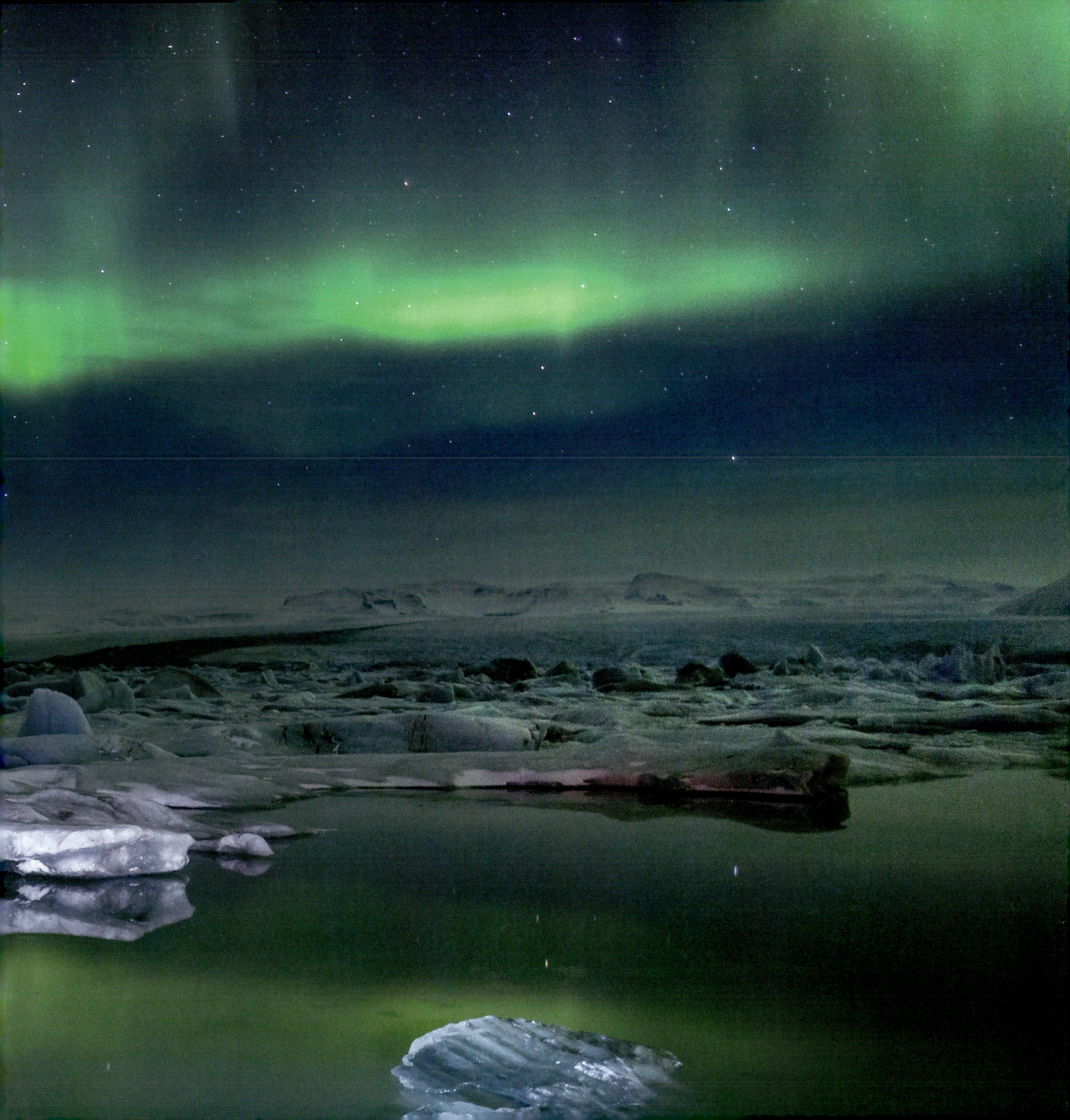

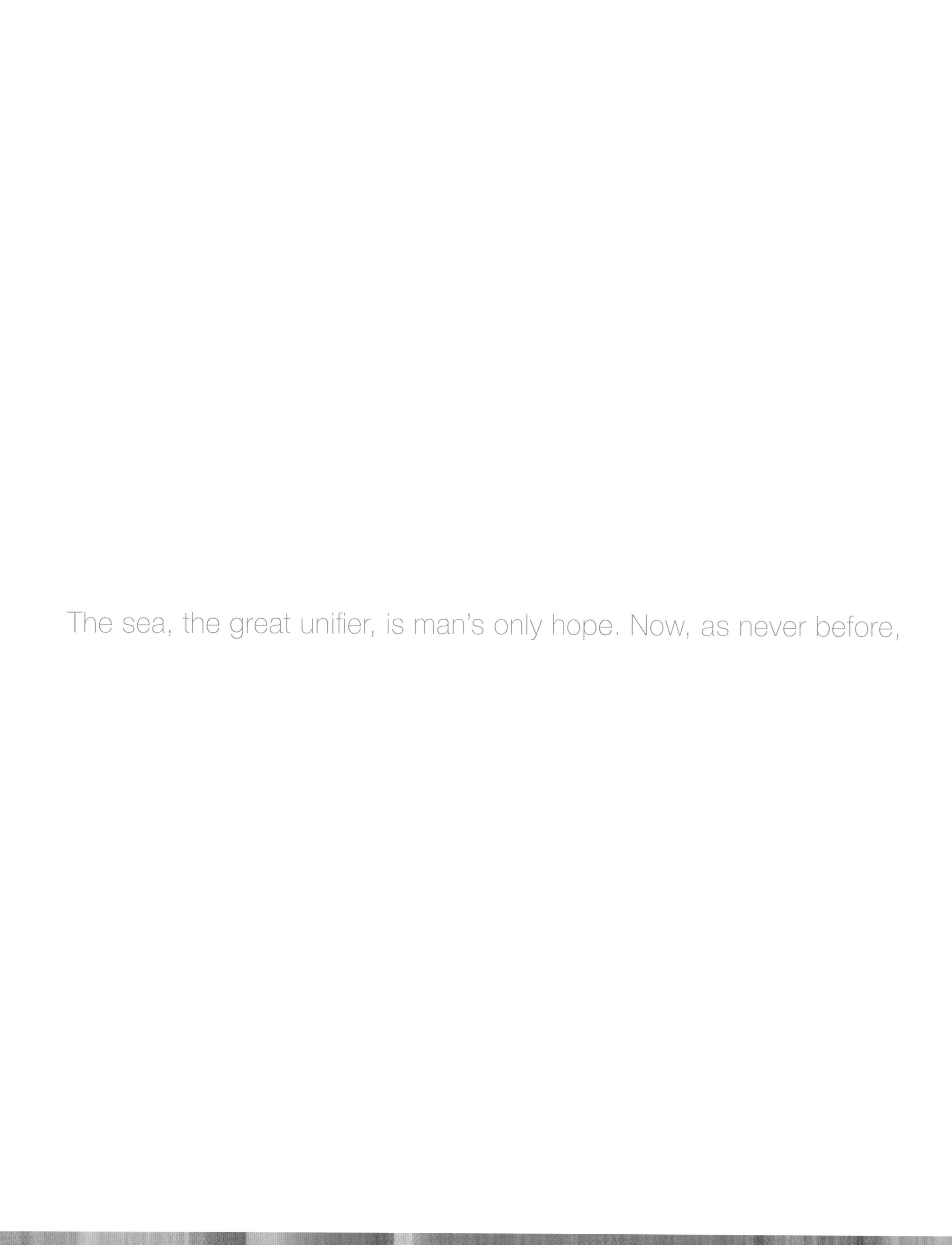
The sea, the great unifier, is man's only hope. Now, as never before,

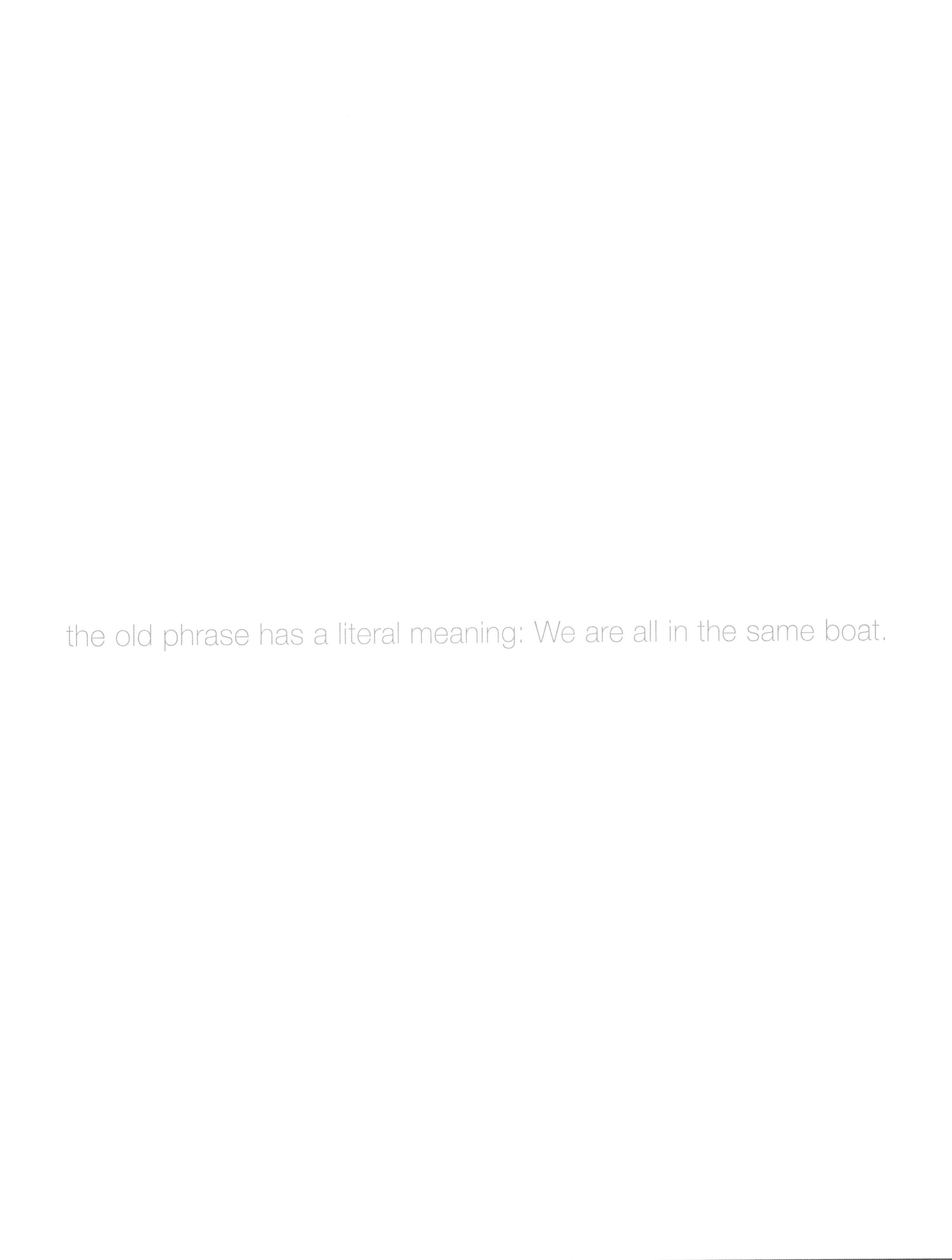

the old phrase has a literal meaning: We are all in the same boat.

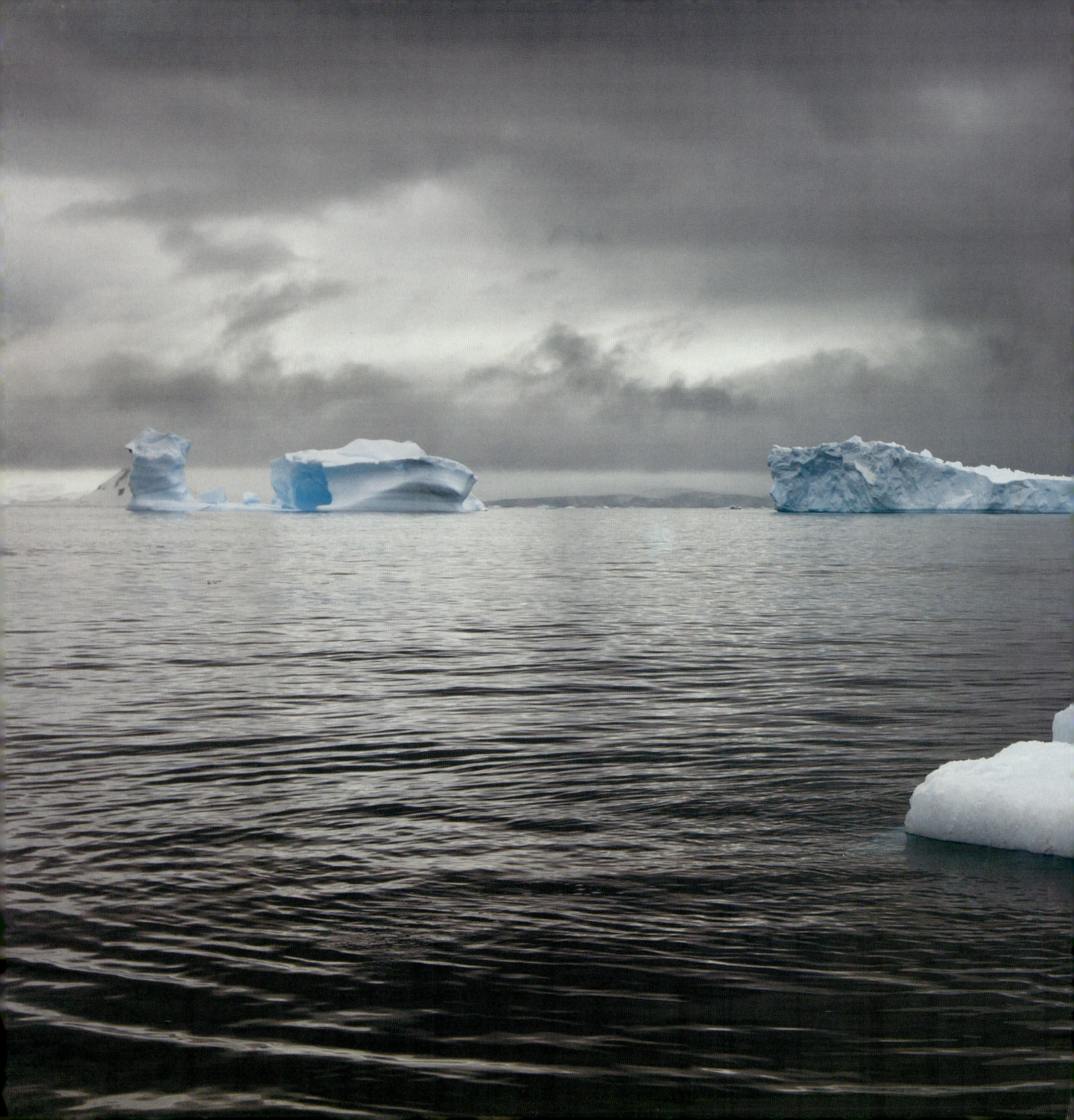

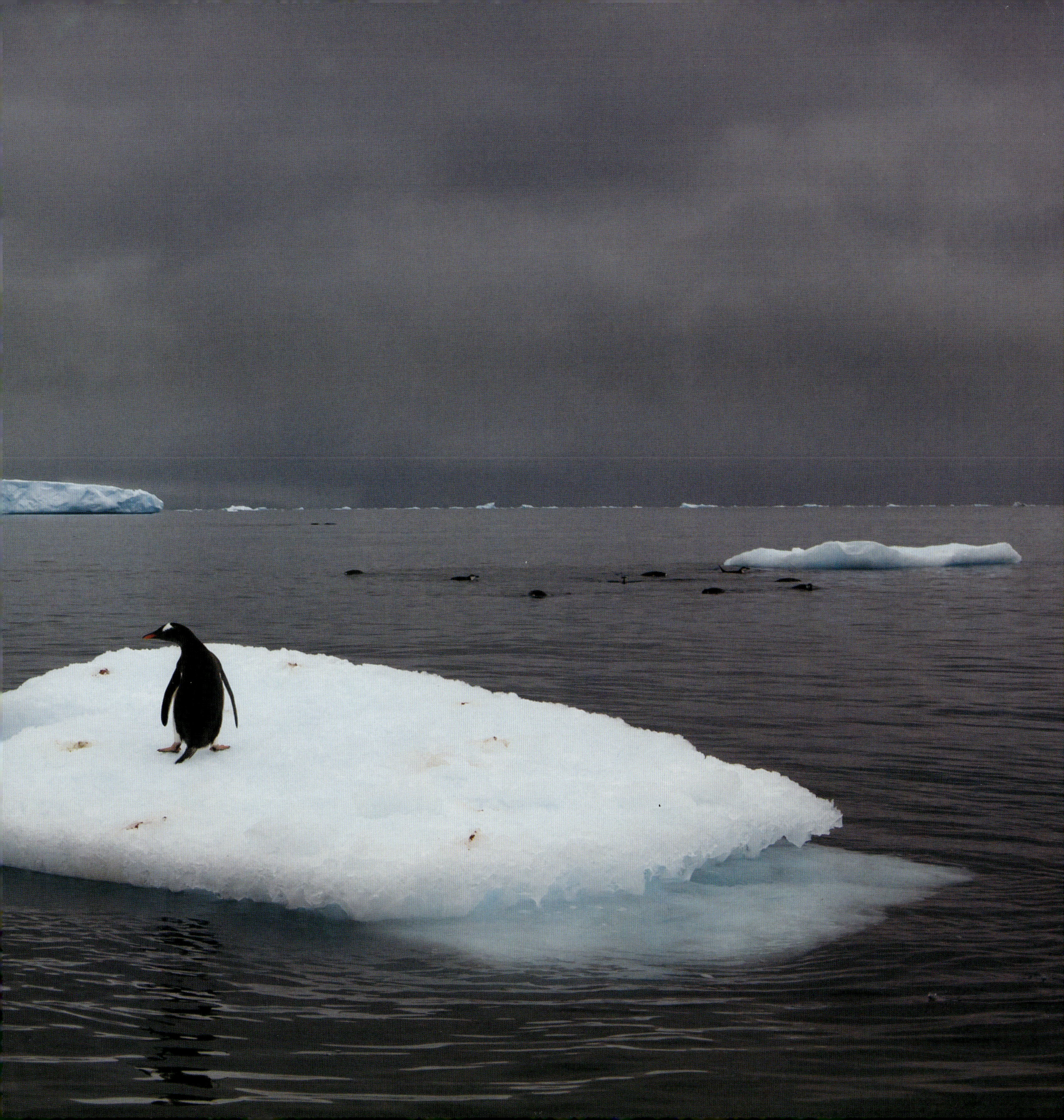

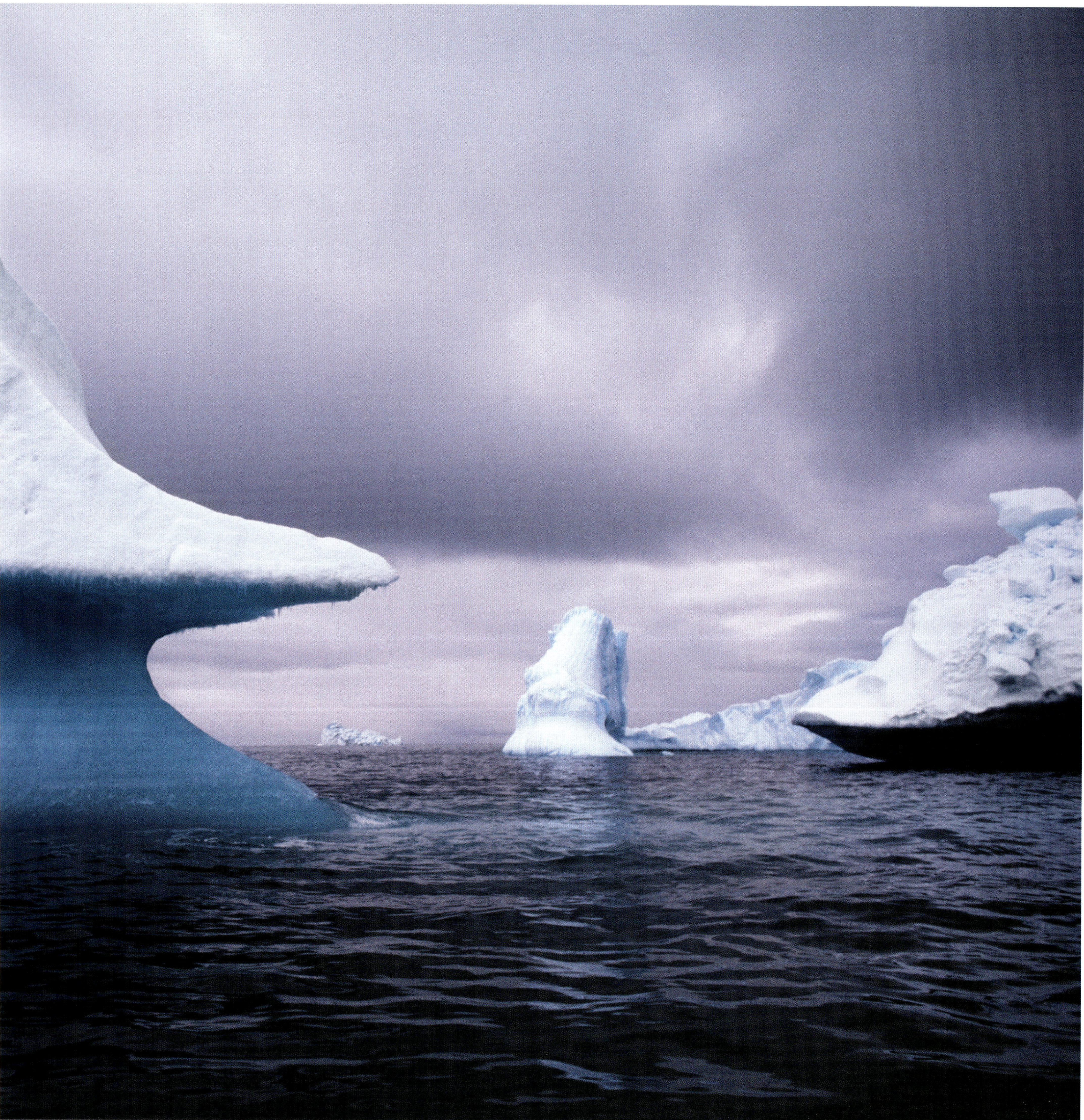

Education, lots of it, would seem a key part of this. We all need to learn to love the seas, and this begins with more understanding. It needs to be in our nature to want to understand how to care for the ocean and we must have an unquenchable drive to explore them further. Why do we invest in space travel and getting to distant icy worlds when we have a world down in the depths that is still so unknown and rich in possibilities? Instead of viewing the seas as a place to pump up fossil fuels from underneath, or haul out nets of flesh from the waters above, we need to find a new harmony with the seas, with new insights into the lasting wealth that is there for all of us.

It may take a James Cameron, with his passion for exploring the deepest trenches, or a billionaire or two who are not beholden to any country, or just a genius with an idea, to turn a switch on for us to see the possibilities. Could it be, for example, that the floating cities envisaged by the Seasteading Institute may offer a better future? Founded by software engineer and theorist of political economy Patri Friedman and funded by billionaire Peter Thiel, who made much of his money with PayPal, this project proposes creating communities of permanent homes on and under the ocean. Hmm. Sounds a bit Jules Verne, science fiction from the past. But it is real in its intent, and a heavily funded project. It is supported by other serious backers and scientists, many of whom go under the neologism of "aquapreneurs." The institute argues for "eight great moral imperatives." These include restoring the oceans alongside cleaning the atmosphere, feeding the hungry, caring for the sick, enriching the poor, living in balance with nature, using sustainable power, and stopping fighting. Hard to argue with any of that, but also hard to think we can get there any time soon. The institute proposes harnessing power from the temperature differential between surface and deep water (rather like ground-source heat pumps) and also proposes new farming technologies to use the sea for harvesting plant products, with a move away from overfishing. "Algae stores more carbon than the world's forests and can be sustainably harvested to transform carbon dioxide into food and fuel," says the institute. Can we imagine a future where algae and other fast-growing, abundant life-forms feed us? We will not see these ideas appearing today or tomorrow, but none of the science is that crazy. It is the will to apply and test this thinking at scale that we need to find first, so I am not laughing at the Seasteading Institute. Power to them, I say. A key part of the institute's thinking is also at a supranational, geopolitical level: It notes that much of the ocean is beyond the 200-nautical-mile (370-kilometer) economic zones controlled by sea-bordering nations, and is a free zone of sorts, out there for innovators to experiment in. The great ocean that covers most of the planet is a world waiting to be embraced with a new improved vision that we must rapidly engineer and build, argue the institute's evangelists. These are big ideas that would stretch the bank balance of even several billionaires if serious large floating communities were to be prototyped. But I would love to see them try.

Right now, however, we can hope to see more commitment to protecting the most valuable places of the oceans. What has happened with the Pitcairn Islands Marine Reserve (see "Wild," p. 105), the world's largest area of protected water, must spread rapidly elsewhere to other important territories at sea. At present there are just over five thousand such reserves around the world, but most are small and the total area encompassed is only 2 percent of the ocean. We need to increase this by at least tenfold as a next goal. Just the thinking and communication of such an intent would raise our consciousness about the value of the ocean. At present, the United Nations has a target of only 10 percent of the oceans being protected, while many countries already exceed that in their ambitions for fragments of habitat. A joined-up global ambition is needed next because having patches of protected water doesn't really work: The seas are all connected, meaning pollution or other devastation in one area quickly affects another. We need protective efforts to be made on a larger scale. We might think that this is easier to comprehend and do in the richer countries rather than the poorer ones, but we would generally be wrong. It is the big, industrial and postindustrial, consuming countries that struggle with curtailing their actions. Meanwhile, those nations that are less developed, perhaps living by and from the sea, can all too clearly see the need to live sustainably. It is those who use more resources who drive the pollution. Excessive consumption fueled by unequal wealth is a problem for the conservation of our planet. It might be fun to eat as much as we like and flaunt our material goods, but such behavior consumes the hopes of our descendants.

If we had to sum up all the issues and debates relating to conservation, climate change, resource use and exploitation, and so on, we might arrive at the following question: How do we meet the growing needs of our species while balancing them with retaining and fostering the health of a global ecosystem? In this, the oceans are the only big engine we can still fire up for survival. The land can't meet our needs, and we are destroying it as we overexploit it.

There is no easy equation to balance these conflicting pressures. We have to deal with many interlinked ecosystems, ranging from big systems like the Gulf Stream or the Great Barrier Reef down to what happens with a particular kind of phytoplankton—its food, relations, and predators—a phytoplankton that just might turn out to be the building block on which everything else is resting. Such systems are not of greater or lesser importance, but are always interdependent.

It's that interdependence that we are still struggling to appreciate. Although we might worry about the food miles behind our fruit or the carbon footprint of our travel, we feel incapable of taking effective action. But we should not abandon hope. Collective actions and innovations are in our nature just as much as the destructive potential we display. For example, in recent years the hole in the thinning ozone layer has started to shrink, and this would seem due to the ban on CFC gases that were causing the damage. Human collaborative actions can deliver change. If you can believe in the power of democracy to represent each of us on an agreed path, then we can agree on the power of small collective actions to drive environmental change. The sea can be saved and restored, and can feed an era of discovery and sustainable wealth. The love that we feel for our environment must prompt us to save and develop the oceans for our future. Some fine lines from *The Tempest* come to mind as a fitting conclusion to summarize our limits and our needs:

This is as strange a maze as e'er men trod
And there is in this business more than nature
Was ever conduct of: some oracle
Must rectify our knowledge.

For the sake of the seas and ourselves, let us find and act on the oracles.

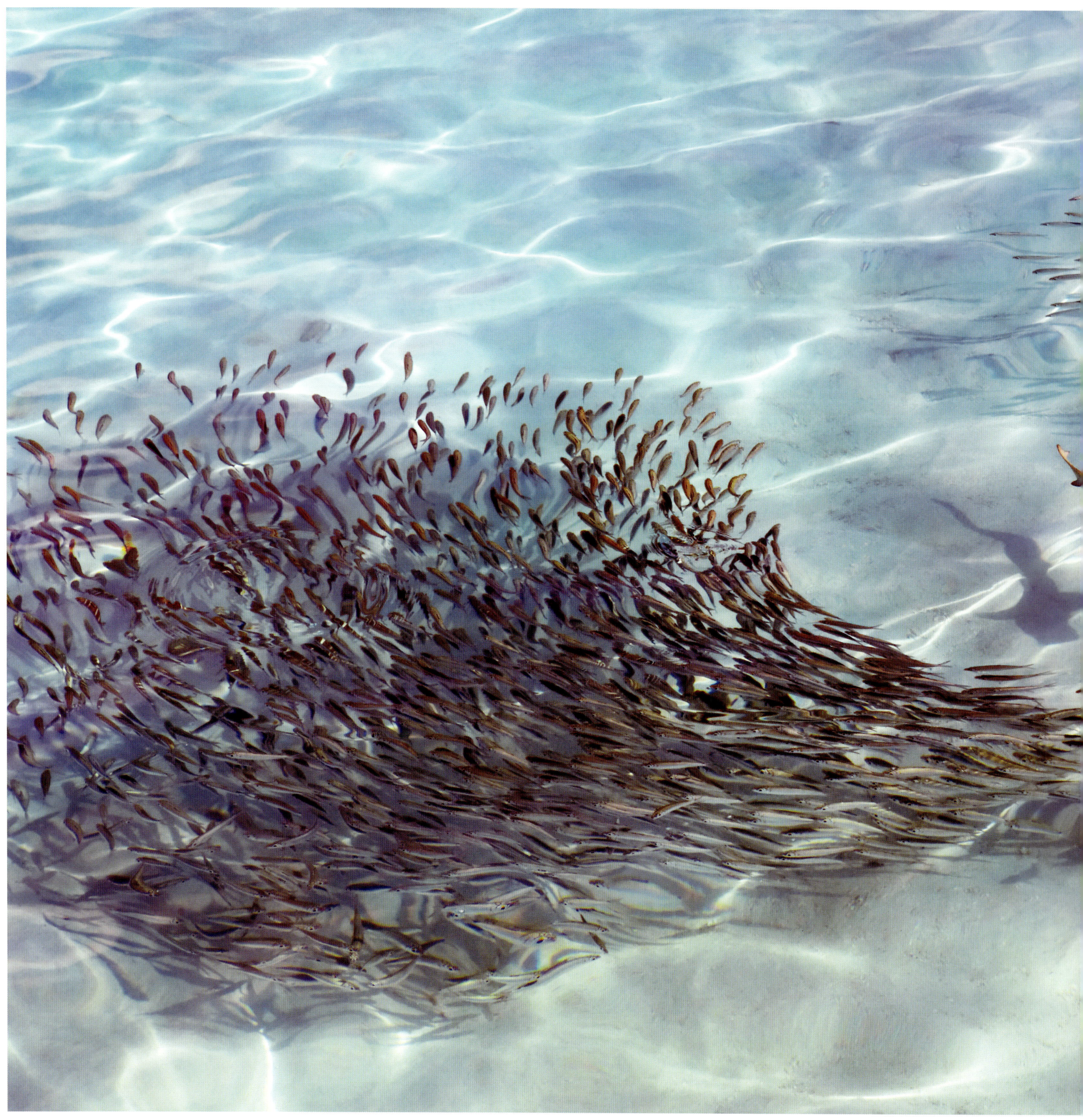

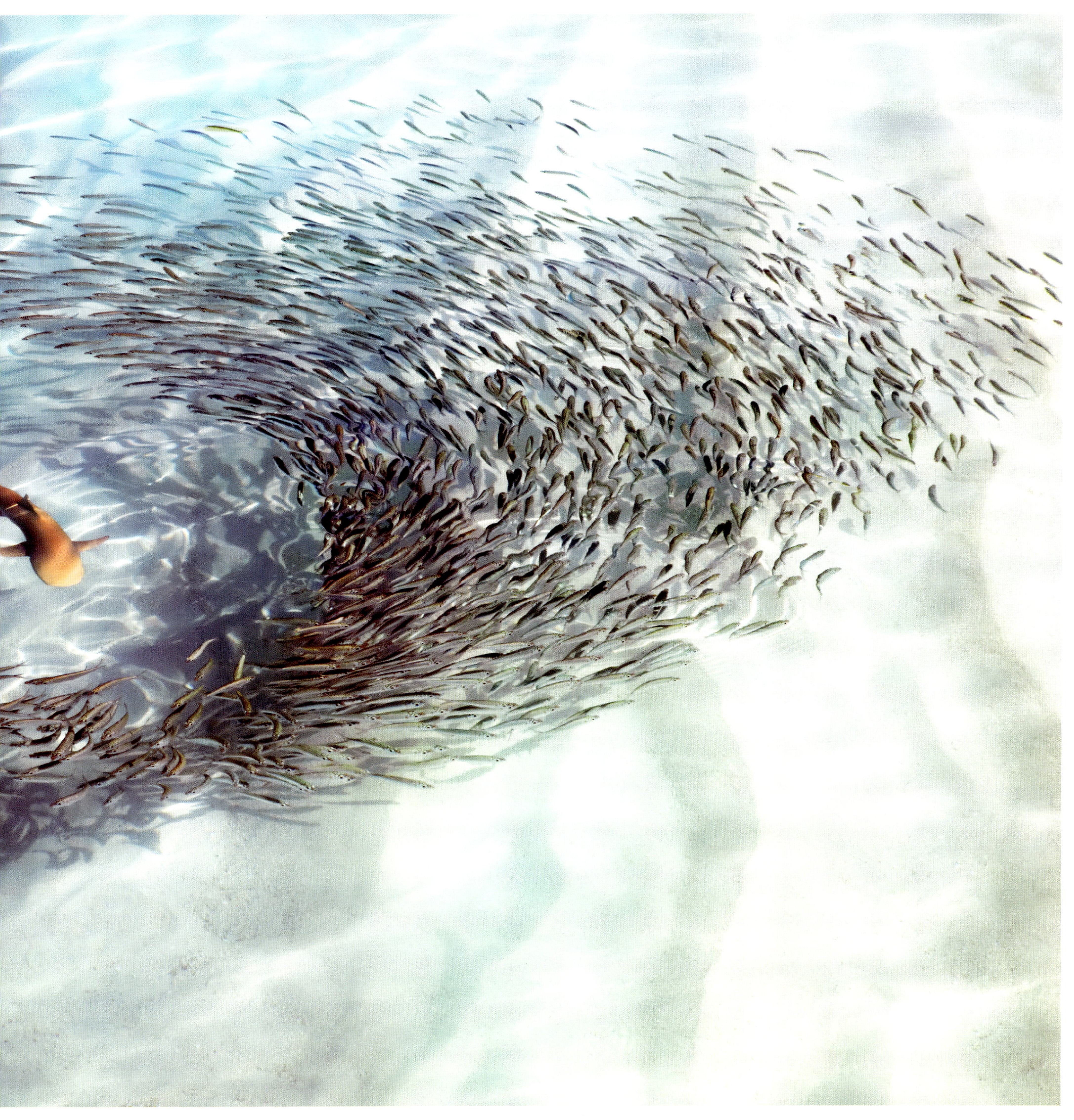

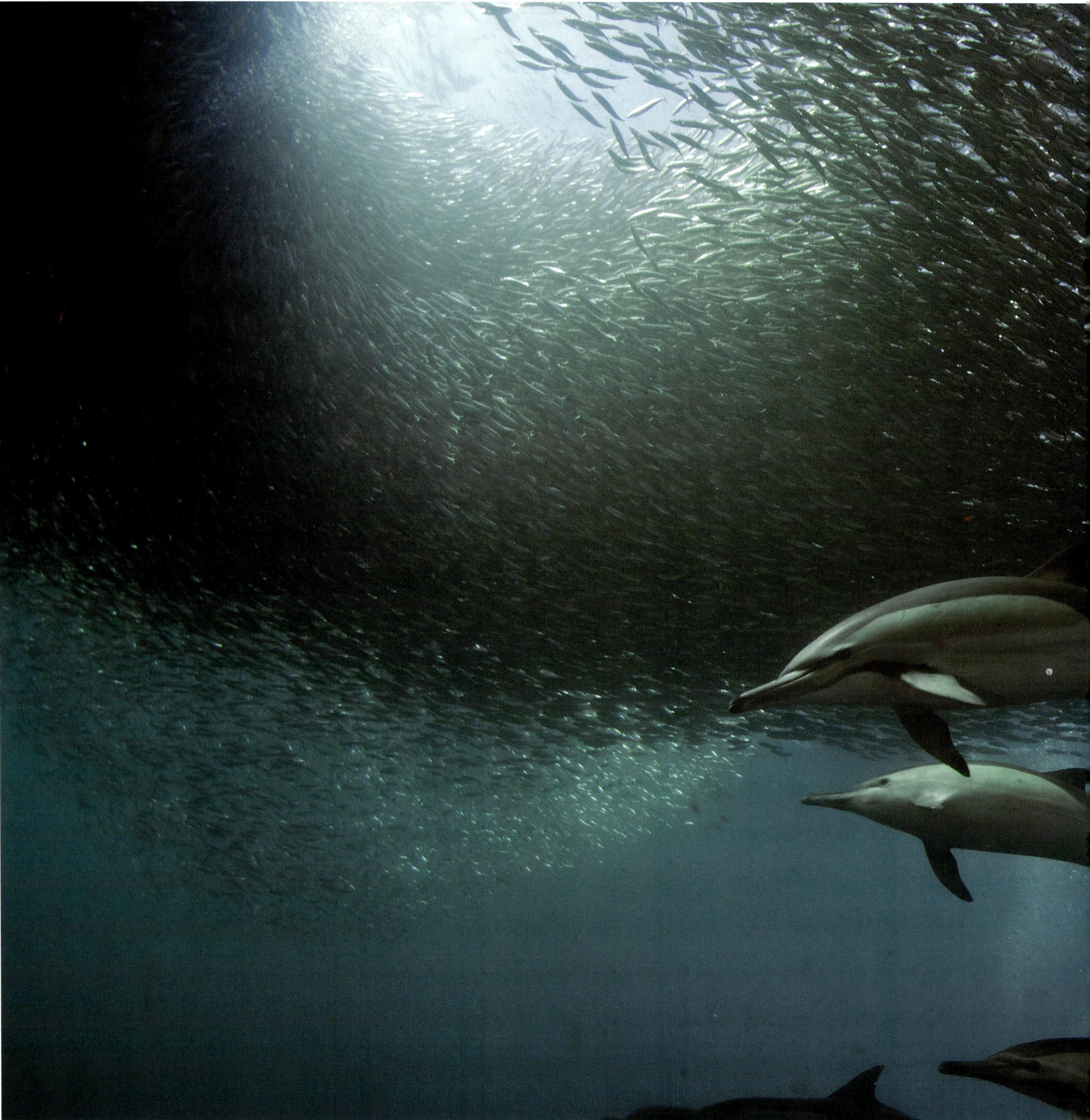

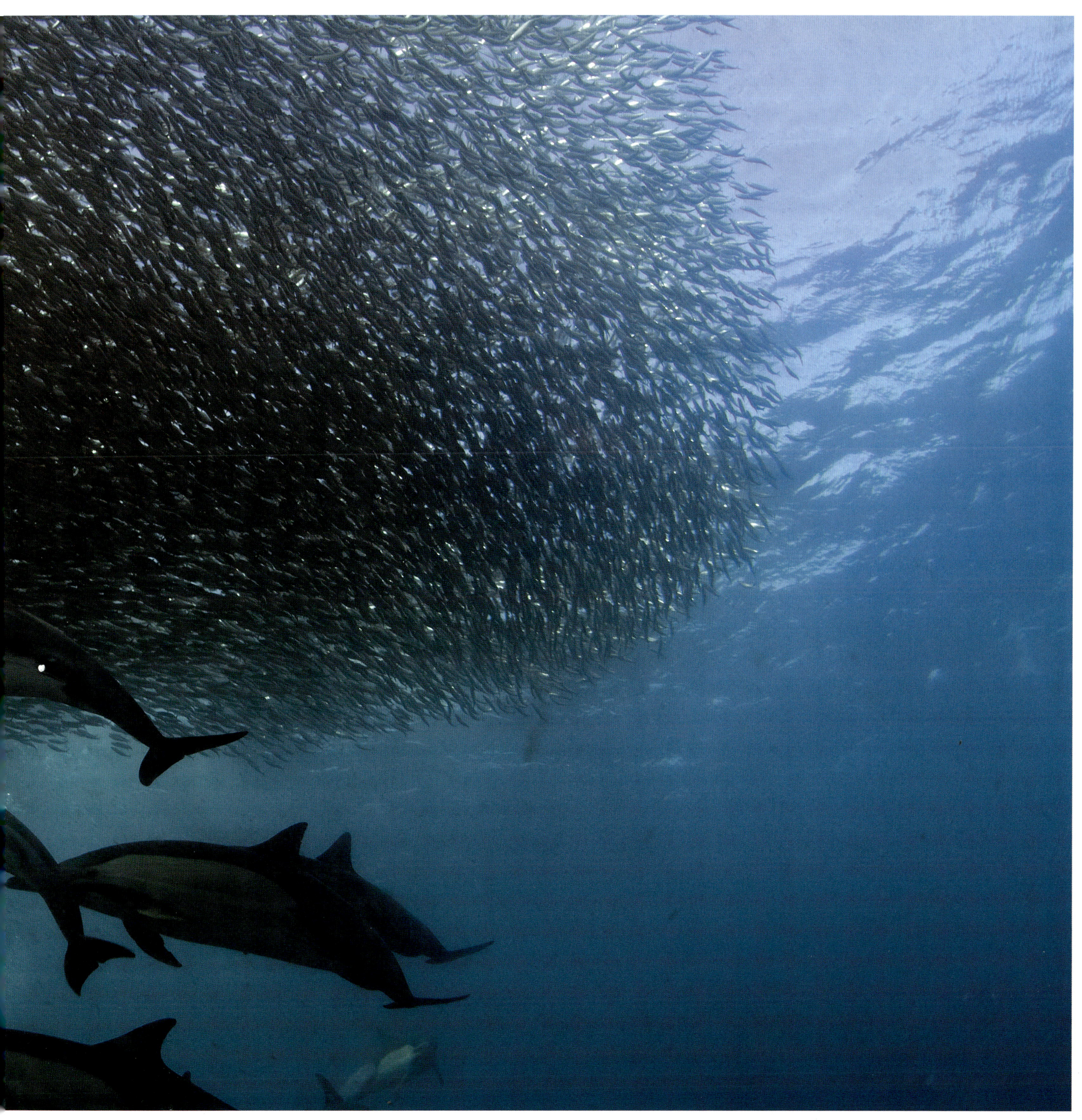

We know that when we protect our oceans

we're protecting our future.

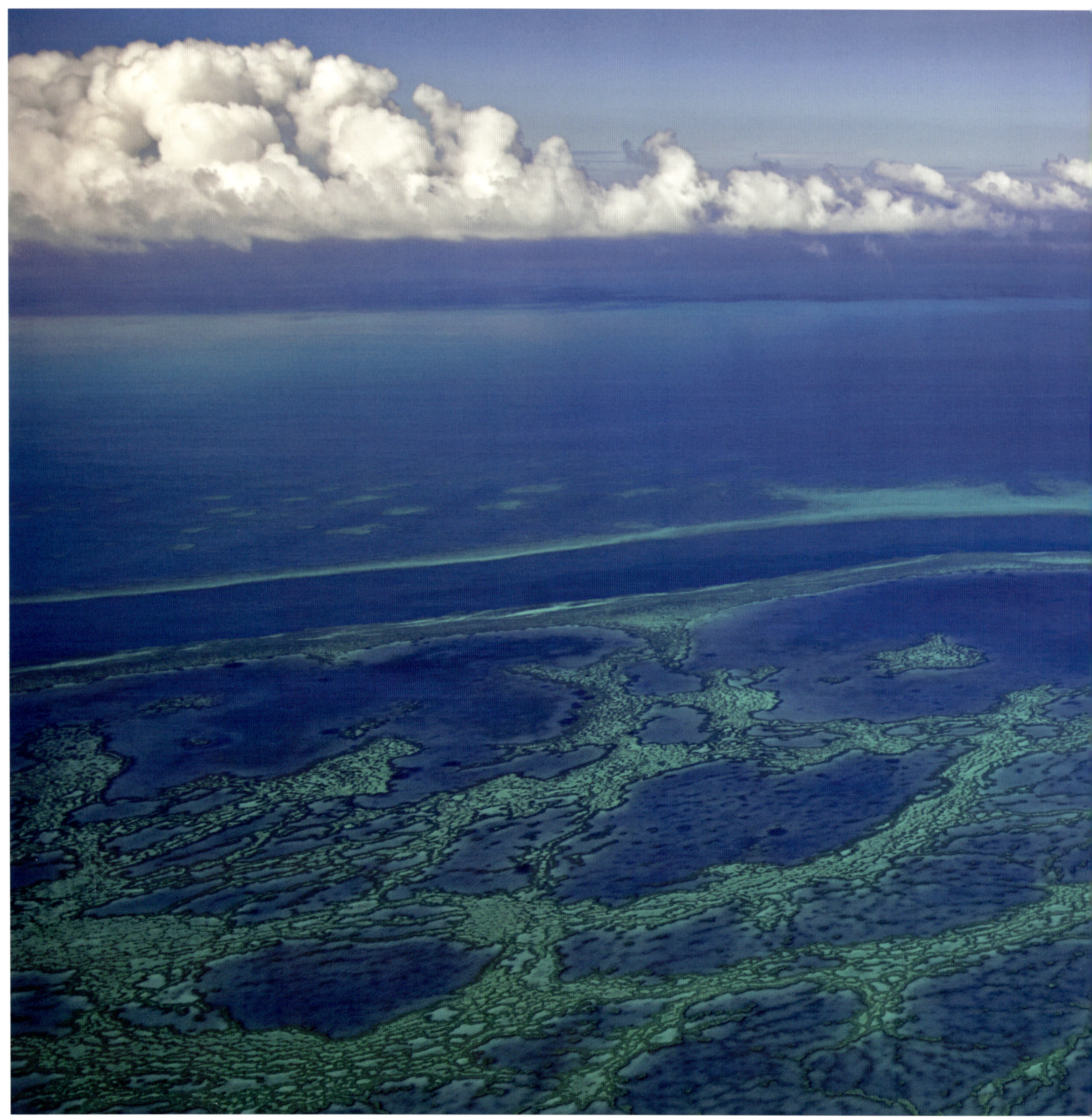

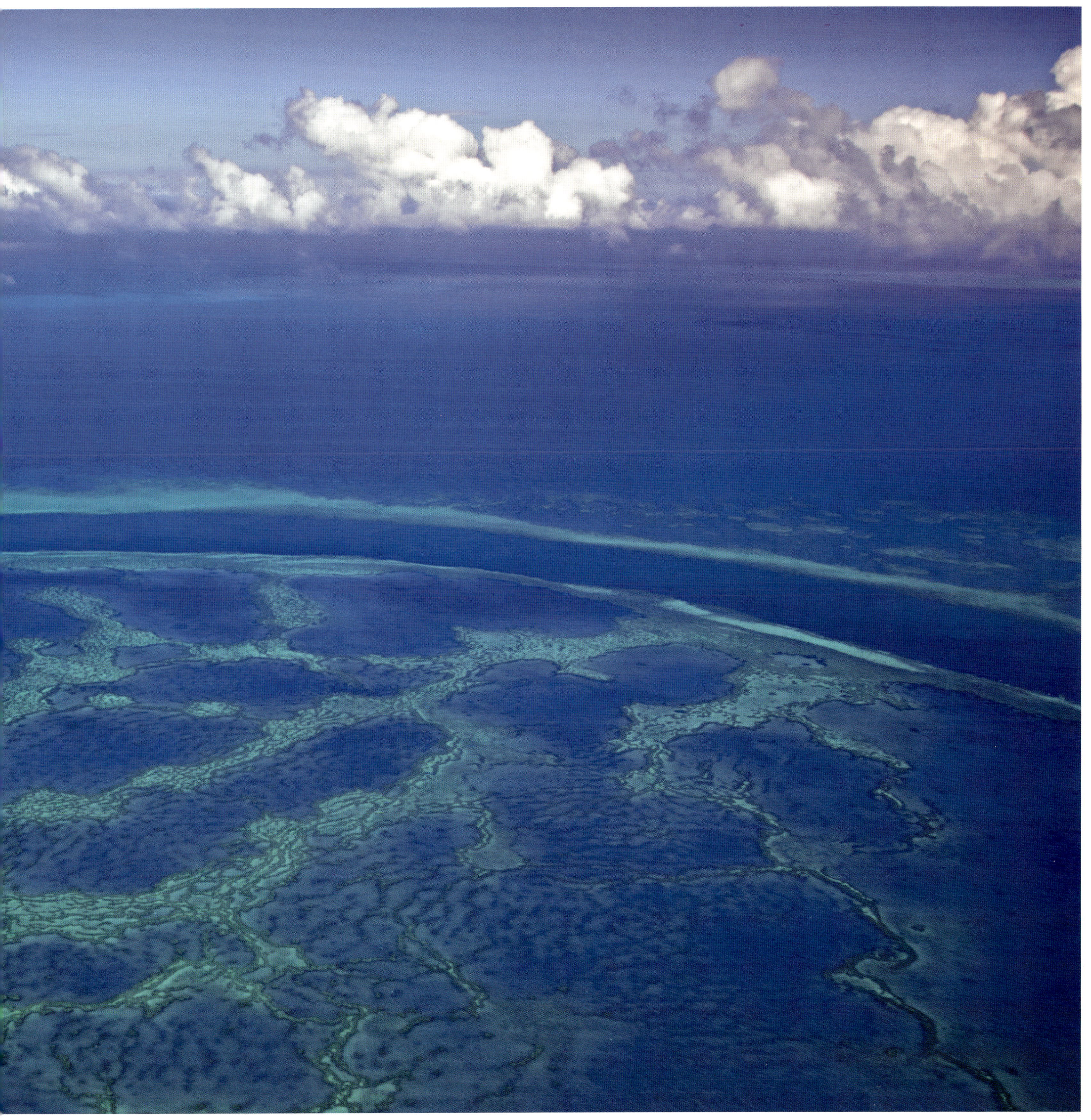

The Quotes

The ocean lends itself to many modes of writing. From the dryly factual documents of scientific bodies, in which might lurk a truly astounding fact or two, to some of the greatest poetry, novels, and plays, there are many ways of singing the song of the sea. The disparate voices we sampled across the pages of this book are a jumping-off point to discover other books, films and documentaries, and living voices. And for want of somewhere else to put them, I'm going to begin the explanation of the various quotes with two more. What lies behind all the evocative quotes is summarized sweetly in this line from the American poet E. E. Cummings (1894–1962): "It's always ourselves we find in the sea." And then, I feel I must counterbalance the good feelings engendered by some of these quotes with this sobering alternative from the great novelist Joseph Conrad (1857–1924): "The sea has never been friendly to man. At most it has been the accomplice of human restlessness."

pp. 12–13

How inappropriate to call this planet Earth when it is quite clearly Ocean.

In capturing how our planet is mostly covered in water, the science-fiction writer Arthur C. Clarke (1917–2008) penned a classic phrase of which its precise wording and origin have been somewhat obscured. It originated in an article in *Nature* magazine, but multiple minor variants exist across usages spanning decades of reference. Clarke, born British, moved to Sri Lanka in 1956, a country he went to in part to indulge his interest in scuba diving.

pp. 20–21

There's nothing wrong with looking at the surface of the ocean itself, except that when you finally see what goes on underwater you realize that you've been missing the whole point of the ocean. Staying on the surface all the time is like going to the circus and staring at the outside of the tent.

This must count as one of the more serious reflections by Dave Barry (1947–). The Pulitzer Prize-winning American author and columnist is known for his humorous reflections on the irrational nature of life, for which jokes are a defense mechanism against the daily problems of exuberant absurdity. One might speculate that inside the ocean's "tent," he expects to find some clowns. Well, there are clownfish.

p. 35

Limitless and immortal, the waters are the beginning and end of all things on Earth.

This comes from the thoughts and studies of Indologist and historian Heinrich Zimmer (1890–1943), as written in his *Myths and Symbols in Indian Art and Civilization* (1971). Zimmer interpreted Hindu and Buddhist thought and helped it be understood alongside the Western tradition.

p. 45

When one has been long at sea, the smell of land reaches far out to greet one. And the same is true when one has been long inland.

Nobel Prize-winner John Steinbeck (1902–1968) is known for classic novels such as *Of Mice and Men* and *The Grapes of Wrath*, but this quote comes from *Travels with Charley: In Search of America*, the nonfiction work he produced after a road trip in 1960 with his dog (Charley). He wrote of the Pacific as being his "home ocean," reaching out to him, drawing him back.

pp. 50–51

The sea is everything. It covers seven-tenths of the terrestrial globe. Its breath is pure and healthy. It is an immense desert, where man is never lonely, for he feels life stirring on all sides.

By Jules Verne (1828–1905), French novelist, poet, and playwright best known for his pioneering science fiction. It comes from *Twenty Thousand Leagues Under the Sea* (1870), which tells the story of Captain Nemo and his submarine (a highly experimental form of machine at the time), *Nautilus*. The tale is told from the viewpoint of Professor Pierre Aronnax after he, his servant Conseil, and Canadian whaler Ned Land encounter Nemo's craft, which they had initially hunted as a mysterious sea monster. On the *Nautilus*, the four journey around the world, beneath the sea. The ship's name references one of the most ancient creatures of the deep, prized for its shell.

p. 55

There's nothing more beautiful than the way the ocean refuses to stop kissing the shoreline, no matter how many times it's swept away.

From the poem "B" by the American writer Sarah Kay (1988–). The poem became well known after Kay performed it at a TED conference in Long Beach, California, in 2011. Since the age of fourteen, Kay has been performing spoken-word poetry to wide audiences in the United States. In 2004, she founded, and has since co-directed, Project VOICE, a group dedicated to furthering the appreciation of the spoken word as an inspirational tool.

pp. 66–67

"Master, I marvel how the fishes live in the sea."

"Why, as men do a-land; the great ones eat up the little ones."

Fine words from the pen of William Shakespeare (1564–1616). Or are they? The quote comes from *Pericles, Prince of Tyre*, a late play by Shakespeare that is thought to have been written only in part by the Bard and also displays signs of an inferior writer (most likely the otherwise little-known George Wilkins) fleshing out the storyline. These fine lines are surely from the genius himself. Pericles, the troubled hero, is a prince from the great seafaring empire of the Phoenicians, and a shipwreck is a key driver of the plot.

p. 79

My soul is full of longing for the secret of the sea, and the heart of the great ocean sends a thrilling pulse through me.

From the poem "The Secret of the Sea" by Henry Wadsworth Longfellow (1807–1882), which appeared in his anthology *The Seaside and the Fireside*, published in 1850. This seminal nineteenth-century American writer had a great love of the sea, and it often appears as a metaphor in his poetry. Living in Portland, Maine, in his early years, he witnessed dramatic sea events such as the fight between British and American ships in 1813. The rhythms of Longfellow's heavily metric verse have been compared to the lapping of waves.

p. 85

No aquarium, no tank in a marine land, however spacious it may be, can begin to duplicate the conditions of the sea.

These words are attributed to Jacques-Yves Cousteau (1910–1997), the oceanographer, filmmaker, conservationist, and aqualung pioneer (the forerunner to scuba diving). He was talking about the depressing experience of seeing a dolphin he had attempted to keep in captivity die after banging its head against the glass of its tank. Cousteau was a great communicator, popularizing the idea of underwater exploration as a great adventure through more than 120 documentaries and 50 books.

pp. 94–95

The sea was the cradle of primordial life, from which the roots of our own existence sprouted. Billions of years of evolutionary development brought forth an enchanting variety of forms, colors, lifestyles, and patterns of behavior.

From *Life in the Sea: As it Is, How it Came to Be, How it Could Become: The Magic of Diversity, its Origins, and Implications*, published in 2001, by the neurologist and psychiatrist Werner Grüter (1929–2014). Grüter combined his enthusiasm for marine biology with a deep concern for how to communicate scientific ideas in powerful ways. His diving and underwater-photography skills helped him make a happy marriage of the two interests.

p. 101

The sea is as near as we come to another world.

This line is from "North Sea Off Carnoustie," one of the works of the American-British poet Anne Stevenson (1933–). It was used on the aquatic attraction The Deep, in the North Sea city of Hull, to set the scene on the whole experience of viewing underwater life.

pp. 106–107

Coral reefs represent some of the world's most spectacular beauty spots, but they are also the foundation of marine life: Without them, many of the sea's most exquisite species will not survive.

A comment attributed to Sheherazade Goldsmith (1974–). English environmentalist, writer, and jeweler, she is a popularizer of green issues.

p. 115

The sea has this contradictory quality, that the more you see of it, the more it overwhelms the eye and disappears in its own brightness.

This observation comes from within an article written by the award-winning British poet Alice Oswald (1966–) for the *New Statesman* magazine. The thought was part of a meditation on the state of mind of refugees prepared to stowaway dangerously on ships to cross the sea in their attempt to get into Britain.

pp. 118–119

Health to the ocean means health for us.

Plain talking from Sylvia Earle (1935–), an American marine biologist who was in 1998 named *Time* magazine's first Hero for the Planet. Earle's long career includes being, at various times, chief scientist at the National Oceanic and Atmospheric Administration (NOAA), leader of the Sustainable Seas Expedition to explore the United States' thirteen marine sanctuaries, and holder of a deep-diving record for women.

p. 123

O, wonder! How many goodly creatures are there here!

Another marine musing from William Shakespeare, this time from *The Tempest*, his last great play, in which a magical island far out at sea provides the environment for drama, humor, and revelation. The quote is perhaps best known more fully as the reference for the title of Aldous Huxley's dystopian novel *Brave New World*:

How many goodly creatures are there here!
How beauteous mankind is! O brave new world,
That has such people in't!

p. 127

Vast movements of ocean and air currents bring dramatic change throughout the year. And in a few special places, these seasonal changes create some of the greatest wildlife spectacles on Earth.

These are just a few of the many insightful words that Sir David Attenborough (1926–) has delivered over the course of a long career that makes him one of the (if not *the*) preeminent natural history broadcasters. He created a global reputation with the work he produced out of the BBC Natural History Unit. Several species are named after him.

p. 145

We know more about the surface of the moon and about Mars than we do about the deep sea floor.

This remark is from the Canadian biological oceanographer Paul Snelgrove (1962–) who worked on the Census of Marine Life, an international project that ran from 2000 to 2010 and aggregated work from 2,700 scientists as they attempted to count marine species. In the process of 540 expeditions, more than 6,000 new species were discovered.

pp. 150–151

Consider the subtleness of the sea; how its most dreaded creatures glide under water, unapparent for the most part, and treacherously hidden beneath the loveliest tints of azure.

In the depths of the sea, Herman Melville (1819–1891) saw not only a largely unknown world waiting to be discovered, but a grand metaphor for life. This quote comes from his epic novel *Moby-Dick or, The Whale*, a work that has come to be seen as the great American novel. In it he took a tale of whaling and built it out into a grand exploration of the human struggle for meaning in our fragile existence. It is full of wonderful insights into sea life, captured during an age of rapid scientific and geographic advances. Give it a go.

pp. 164–165

The sea always filled her with longing, though for what she was never sure.

From *Inkheart*, a novel for young adults by the award-winning German children's fiction writer Cornelia Funke (1958–). Her books have sold more than 20 million copies.

pp. 174–175

The voice of the sea speaks to the soul. The touch of the sea is sensuous, enveloping the body in its soft, close embrace.

From the novel *The Awakening* by American writer Kate Chopin (1850–1904). Set in New Orleans and on the Louisiana Gulf coast at the end of the nineteenth century, the sea features in a sensual but also dark way. It's a beautifully written book and we won't spoil the story here.

p. 187

As air is a sacred gas, so is water a sacred liquid that links us to all the oceans of the world and ties us back in time to the very birthplace of all life.

This quote is from *The Sacred Balance*, a 1997 book by David Suzuki (1936–), which reflects on humanity's impact on the planet and (as the subtitle puts it) argues for us "rediscovering our place in nature" and living within sustainable resources. Suzuki is an academic, broadcaster, and environmental activist who was the highest-ranked person still alive in the 2004 CBC television series *The Greatest Canadian*.

p. 191

In every outthrust headland, in every curving beach, in every grain of sand there is a story of the earth.

Quotable lines come thick and fast in Rachel Carson's writing about the sea, including this quotation from an article she wrote for a special issue of *Holiday* magazine devoted to "Nature's America" in 1958. Six years earlier she had won the National Book Award for *The Sea Around Us*, the second book in her trilogy on how oceans work. On receiving the award she said, "If there is poetry in my book about the sea, it is not because I deliberately put it there, but because no one could write truthfully about the sea and leave out the poetry." Carson (1907–1964) is even better known for her 1962 work *Silent Spring*, which documented the terrible damage that the insecticide DDT was inflicting on nature.

pp. 196–197

The sea, the great unifier, is man's only hope. Now, as never before, the old phrase has a literal meaning: We are all in the same boat.

Another quote from Jacques-Yves Cousteau, this from *National Geographic*, in 1981. He is arguably the individual who did the most to kick-start worldwide awareness of the wonders of the oceans and the need for their conservation.

pp. 206–207

We know that when we protect our oceans, we're protecting our future.

American President Bill Clinton (1946–) made these remarks on signing the Oceans Act at Martha's Vineyard on August 7, 2000. The act helped establish the United States Commission on Ocean Policy, a national advisory board dedicated to studying and promoting balance between environmental and human interests around the use of the seas.

The Photographs

Many of us have taken an image of the sea. It's the backdrop to so many holiday snaps, along with more artistic personal shots of a fine or stormy day by the coast or on a boat. Great as it is that we so often take visual notes of the sea, I would like to put down a marker here for the different challenges that sit behind the photographs that appear in these pages.

This is the first era where we are nearly all photographers, should we care to be. At all times we are armed with a camera that is attached to our phone, ready to share almost instantly with whomever we want to connect with around the world. And we do share some amazing images; some of the most remarkable shots of our day emerge from the rich content of ordinary life and non-professional talent. However, taking great shots of the sea, particularly underwater, is not for the lucky amateur. It requires not only special equipment, but special talent too. It is not only that the photographer needs to know how to work with a different element—as light behaves very differently in water, as does the electricity required to power any flash—but also the movement of water makes the whole world of photography even more uncertain. Skill and application has to remove that uncertainty as much as possible. Preparation is that much more protracted and deliberate, not only for the image content—knowing what to look for, where to go, and when—but for the equipment's and for the photographer's own survival. The tripod and the long exposure are not so often called for in these pages. The photographer works in a world where time is more immediately limited by the boundaries of life and death, and the standard backdrop is so often a shade of blue, which makes for a creative challenge altogether different than that on land.

It's a world where there has been secrecy at times. In the past, the precise location of certain shots and the techniques pioneered by some photographers have, on occasion, been surrounded by more than a little discretion, if not obfuscation. But such behavior has largely gone. Now it is more often a world of incredible generosity and openness. Many professionals (including several represented in these pages) go to great lengths to share their learning through workshops, and see their work, in part, as pushing back the boundaries of knowledge, both by documenting what is down there in the ocean and also by inventing new methods of making these images more effectively.

Countless days, weeks, years, and, I fear, lives have been spent in the pursuit of these remarkable images. It is one of the more dangerous ways of taking a picture, one of the few areas in which the production of a photograph can still involve true exploration rather than simply applying expertise to locating the camera in the right place at the right time. The sea is never quite the same environment twice, while the great majority of it is still unvisited, unknown, in any detail. It is a place of mystery, beauty, treasures, and real risks. So before giving the image information, I would like to thank the extremely talented photographers represented in these pages, and also the many whose work informed and influenced me immeasurably over the years. Thank you all for the inspiration and help in making this book.

Inside front and back cover
Pacific flatiron herring (*Harengula thrissina*) bait ball
Sea of Cortez, Baja California peninsula, Mexico
Photographer: Franco Banfi

Opposite p. 1
Surface of the sea
Photographer: Datacraft Co Ltd

pp. 2–3
School of yellowtail razor surgeonfish (*Prionurus laticlavius*)
Galapagos Islands, Ecuador
Photographer: Michele Westmorland

p. 4
A solid wall of fish almost completely blocks out sunlight
Photographer: Nature, Underwater and Art Photos—www.narchuk.com—Andrey Narchuk

pp. 6–7
Bryde's whale (*Balaenoptera edeni*) feeding on sardines (Clupeidae) during a sardine run
Durban, KwaZulu-Natal, South Africa
Photographer: Michael Aw

pp. 10–11
Guineafowl pufferfish (*Arothron meleagris*)
Photographer: James R. D. Scott

pp. 14–15
Sand patterns
Photographer: WIN-Initiative

p. 16
Current patterns and breaking waves
De Hoop Marine Protected Area
Western Cape Province, South Africa
Photographer: Peter Chadwick

p. 19
Schools of floating, pulsing Mastigias and moon jellyfish (*Aurelia aurita*) in a plankton-rich inland lake.
Photographer: Michele Westmorland

pp. 22–23
School of fish swimming in the ocean
Photographer: Alexa Rae Smahl

pp. 24–25
Manta ray (*Manta birostris*)
Maldives
Photographer: Franco Banfi

pp. 26–27
Northern sea stars (*Asterias vulgaris*) feed on a bed of mussels
Gloucester, Massachusetts, USA
Photographer: Jeff Rotman

p. 29
Lava flow from the active volcano Kilauea enters the sea
Hawaii Volcanoes National Park, Hawaii, USA
Photographer: Doug Perrine

pp. 30–31
Lava flow from Kilauea solidifies upon falling into the sea
Hawaii Volcanoes National Park, Hawaii, USA
Photographer: Art Wolfe

p. 32
Kannesteinen Rock
Vågsøy, Sogn og Fjordane, Norway
Photographer: Orsolya Haarberg

p. 34
Underwater shallow sandbar with ripples
Grand Cayman, Cayman Islands
Photographer: Alex Mustard

pp. 36–37
Aerial image showing sandbanks and islands in the Bahama Archipelago
Bahamas
Photographer: Juan Carlos Munoz

p. 38
Day octopus (*Octopus cyanea*) illuminated by bright sunburst
Hawaii
Photographer: Ed Robinson

pp. 40–41
Great Barrier Reef, Queensland, Australia
Photographer: Art Wolfe

p. 43
Manta ray (Manta)
Similan Islands, Phang Nga, Thailand
Photographer: Nature, Underwater and Art Photos—www.narchuk.com—Andrey Narchuk

p. 138
Christmas tree worms (*Spirobranchus giganteus*)
Fiji
Photographer: Pete Oxford

p. 141
Umbrella squid (*Histioteuthis bonnellii*) changing
color
Mid-Atlantic Ridge, Atlantic Ocean
Photographer: David Shale

p. 142
Close-up of the head of an adult amphipod pram
bug (Phronima)
Atlantic Ocean
Photographer: Solvin Zankl

p. 143
Close-up of light reflecting off the cilia of a comb
jelly
Great Barrier Reef, Queensland, Australia
Photographer: Jurgen Freund

p. 144
Volcanic gases bubbling up from small vents on
the ocean floor
Sangeang, Indonesia
Photographer: Georgette Douwma

pp. 146–147
Broadclub cuttlefish (*Sepia latimanus*)
Great Barrier Reef, Queensland, Australia
Photographer: Jurgen Freund

p. 149
Black-smoker hydrothermal vents on the bottom
of the ocean
Photographer: Science Photo Library

pp. 152–153
Deep-sea hatchetfish (Sternoptychidae)
Atlantic Ocean
Photographer: David Shale

p. 154
A recently discovered species of acorn worm
named *Yoda purpurata*
North Atlantic Ocean
Photographer: David Shale

pp. 156–157
Tunicate with a brittle star wrapped around it
Photographer: Christy Gavitt

pp. 158–159
Two-week-old harp seal pup (*Phoca
groenlandica*)
Gulf of Saint Lawrence, Canada
Photographer: Doug Allan

p. 161
Preservation Inlet, Fiordland, New Zealand
Photographer: Downunderphotos

p. 162
Massive waves breaking on a headland
Cornwall, UK
Photographer: David Clapp

pp. 166–167
Flock of sanderlings (*Calidris alba*) flying over
a rough sea
Berwickshire, Scotland, UK
Photographer: Laurie Campbell

p. 169
Pod of beluga whales (*Delphinapterus leucas*)
Canadian Arctic, Canada
Photographer: Doc White

p. 170
Humpback whales (*Megaptera novaeangliae*)
feeding
Photographer: Kevin Schafer

pp. 172–173
Sally Lightfoot crabs (*Grapsus grapsus*) feeding
on wave-swept rocks
Fernandina, Galapagos Islands, Ecuador
Photographer: William Gray

pp. 176–177
Close-up of sea wave
Photographer: Jochem D Wijnands

p. 179
Salt formation caused by the evaporation of the
water on the shore
Dead Sea, Israel
Photographer: Science Photo Library

pp. 180–181
Cownose rays (*Rhinoptera bonasus*)
Bocas del Toro, Panama
Photographer: Art Wolfe

pp. 182–183
Leatherback sea turtle (*Dermochelys coriacea*)
hatchling heading towards the sea
Cayenne, French Guiana
Photographer: Graham Eaton

p. 184
Green sea turtle (*Chelonia mydas*) hatchlings
move towards the ocean
Galapagos Islands, Ecuador
Photographer: Art Wolfe

p. 186
Gannet (*Morus bassanus*) diving
Shetland, Scotland, UK
Photographer: Markus Varesvuo

pp. 188–189
Foam splashing on the beach
Photographer: Panoramic Images

p. 190
Tree trunk in the sea
Photographer: Stephanie Cabrera

p. 193
Moon reflecting on the sea at night
Photographer: Ken Biggs

pp. 194–195
Aurora Borealis lights up the glacial river lagoon
Jökulsárlón on the edge of Vatnajökull National
Park
Vatnajökull Ice Cap, Iceland
Photographer: Ragnar Th. Sigurdsson

pp. 198–199
Gentoo penguin (*Pygoscelis papua*) on an
iceberg
Antarctica
Photographer: Art Wolfe

p. 200
Icebergs
Weddell Sea
Photographer: Art Wolfe

pp. 202–203
Blacktip reef shark (*Carcharhinus melanopterus*)
and school of fish
Baa Atoll, Maldives
Photographer: Frank Krahmer

pp. 204–205
Short-beaked common dolphins (*Delphinus
delphis*) feeding on sardines (*Sardinops sagax*)
East London, South Africa
Photographer: Chris and Monique Fallows

pp. 208–209
Great Barrier Reef, Queensland, Australia
Photographer: Art Wolfe

The Footage

Go beneath the surface with author Lewis
Blackwell as he documents his inspiration
behind the making of the book with bonus video
footage compiled by underwater filmmaker Steve
Hathaway.

Visit www.thelifeandloveofthesea.com or scan
the QR code below.

In order of appearance:
Large crashing wave; Diver, Poor Knights Islands,
New Zealand; Kelp forest, Aldermen Islands,
New Zealand; Octopus walking across the
ocean floor, Fiordland, New Zealand; Blue
maomao, Poor Knights Islands, New Zealand;
Crown-of-thorns starfish, Kermadec Islands,
New Zealand; Divers, Poor Knights Islands,
New Zealand; Tropical fish in staghorn and table
coral, Vatu-i-Ra Passage, Fiji; Jellyfish (*Pelagia
noctiluca*); Schooling bannerfish, Koro Sea, Fiji;
Lionfish and soft corals, Kermadec Islands,
New Zealand; Smalltooth emperor fish in soft
coral reef, Andaman Sea, Indian Ocean; Spotted
black groper chasing striped boarfish, Kermadec
Islands, New Zealand; Schooling fish, Poor
Knights Islands, New Zealand; Soft corals,
Vatu-i-Ra Passage, Fiji; "Hitchhiker" fish swarm
around a jellyfish, Andaman Sea, Indian Ocean;
Northern scorpionfish and blue maomao, Poor
Knights Islands, New Zealand; Short-tail stingray,
Poor Knights Islands, New Zealand; Long-tail
stingray, Hen and Chickens Islands, New Zealand;
Orca, Northland, New Zealand; Southern fur
seal, Poor Knights Islands, New Zealand;
Long-finned pilot whales, Northland, New Zealand;
Mosaic moray in soft coral, Kermadec Islands,
New Zealand; Common dolphins, Northland,
New Zealand; Lemonpeel angelfish and orange-
fin anemone fish in Merten's carpet anemone,
Koro Sea, Fiji; Blue moki, Poor Knights Islands,
New Zealand; Pacific double-saddle butterflyfish
feeding on hard coral, Vatu-i-Ra Passage,
Fiji; Galapagos sharks, Kermadec Islands,
New Zealand; Spotted black groper on soft
coral, Kermadec Islands, New Zealand; Coral
reef with tropical fish, Vatu-i-Ra Passage, Fiji;
School of juvenile demoiselles; Banded coral
shrimp, Kermadec Islands, New Zealand; Striped
boarfish, Kermadec Islands, New Zealand;
Lionfish; Short-tail stingray, Poor Knights Islands,
New Zealand; Large school of blue maomao and
demoiselles, Poor Knights Islands, New Zealand.

Further Reading and Notes

Where to begin listing the sources? There was a book of short stories with a magnificent tale of a giant octopus, name now forgotten. I read it many times when I was about eight. Then three years later there was Arthur Ransome's *We Didn't Mean to Go to Sea*. This children's novel about a sailing adventure of 1937 was devoured, and built the false belief "I could do that." It may have inspired, shortly after reading, a careless row into the path of an ocean liner traveling up a major shipping lane. The coast guards remedied the situation and I lived to tell the tale, unlike the drowned surfer I saw around the same time, lying on the beach like a model in a life-saving exercise. The sea is a magnificent and dangerous place: We must be careful visitors to this world of another element, for its sake and ours. The sea is seductively accessible and yet elusive. It may be free to encounter, but few venture below the surface, and even less to a significant depth. So it is to the words and images of intrepid explorers, sailors, scientists, and other adventurous minds that we must turn, and it can be delight and inspiration that inspires us to return to the waters with new objectives. I remain humbly in great debt to the learning and experience of others. Their works were vital to the writing and curating of this book. Here are just some of the documents or references I drew on.

Fiction

Coleridge, Samuel Taylor. *The Rime of the Ancient Mariner*. In *Lyrical Ballads* by Coleridge and William Wordsworth. London, J. & A. Arch, 1798.

Conrad, James. *Lord Jim: A Sketch. Blackwood's Edinburgh Magazine*, London, October 1899–1900. Conrad wrote many other stories in which the life of the sea is key.

Hemingway, Ernest. *The Old Man and the Sea*. New York, Scribner's, 1952.

Melville, Herman. *Moby-Dick or, The Whale*. New York, Harper, 1851.

Nonfiction

Auden, W. H. *The Enchafèd Flood: Or The Romantic Iconography of the Sea*. New York, Random House, 1950.

Carson, Rachel. *The Sea Around Us*. New York, Oxford University Press, 1951.

Daley, Ben. The Great Barrier Reef: *An Environmental History*. London, Routledge, 2014.

Hamilton-Paterson, James. *Seven-tenths: The Sea and its Thresholds*. London, Faber & Faber, 2007.

Kunzig, Robert. Mapping the Deep: *The Extraordinary Story of Ocean Science*. London, Sort of Books, 2000.

Kurlansky, Mark. *Cod: A Biography of the Fish That Changed the World*. New York, Walker, 1997.

Lucas, Joseph, and Pamela Critch. *Life in the Oceans*. London, Thames & Hudson, 1974.

Monbiot, George. *Feral: Searching For Enchantment on the Frontiers of Rewilding*. London, Allen Lane, 2013.

Raban, Jonathan. *Passage to Juneau: A Sea and Its Meaning*. London, Picador, 1999.

Roberts, Callum. *The Unnatural History of the Sea: The Past and Future of Humanity and Fishing*. London, Gaia Books, 2007.

Winchester, Simon. *Atlantic: A Vast Ocean of a Million Stories*. London, HarperCollins, 2010.

Online

HMS Challenger Report. Access to parts of the twelve-volume report and information around the pioneering research trip can be found at the Natural History Museum site: http://www.nhm.ac.uk/nature-online/science-of-natural-history/expeditions-collecting/hms-challenger-expedition/index.html

MarineBio Conservation Society: http://marinebio.orgmarinebio.org

National Oceanic and Atmospheric Administration: http://www.noaa.gov

Oceans at MIT: http://oceans.mit.edu

Scripps Institution of Oceanography: https://scripps.ucsd.eduscripps.ucsd.edu

Smithsonian National Museum of Natural History Ocean Portal: ocean.si.edu

Thomson, C. Wyville. *The Depths of the Sea: An Account of the General Results of the Dredging Cruises of HMS "Porcupine" and "Lightning" During the Summers of 1868, 1879, and 1870 Under the Scientific Directions of Dr. Carpenter, F.R.S., J. Gwyn. Jeffreys, F.R.S., and Dr. Wyville Thomson, F.R.S.* London, Macmillan, 1873. This can be downloaded at http://docs.lib.noaa.gov/rescue/oceanheritage/Gc75t481873.pdfhttp://docs.lib.noaa.gov/rescue/oceanheritage/Gc75t481873.pdf

References within this book

Beginnings

Ghose, Tia. "Oldest Animal-built Reef Discovered in Namibia." *LiveScience* (June 2014). http://www.livescience.com/46556-oldest-animal-reef-discovered.html

Gorder, Pam Frost. "Researchers Find Origin of 'Breathable' Atmosphere Half A Billion Years Ago.' *Ohio State University Research News* (October 2007). http://researchnews.osu.edu/archive/oxypulse.htm

Heim, Noel A., Matthew L. Knope, Ellen K. Schaal, Steve C. Wang, and Jonathan L. Payne. "Cope's Rule in the Evolution of Marine Animals." *Science*, February 20, 2015. http://www.sciencemag.org/content/347/6224/867

Mulkidjanian, Armen Y., Andrew Yu Bychkov, Daria V. Dibrova, Michael Y. Galperin, and Eugene V. Koonin. "Origin of First Cells at Terrestrial, Anoxic Geothermal Fields." *PNAS*, vol. 109, no. 14 (April 3, 2012). http://www.pnas.org/content/109/14/E821.full

Smith, Paul M. and David A. T. Harper. "Causes of the Cambrian Explosion," *Science*, vol. 341, no. 6152 (September 20, 2013). http://www.sciencemag.org/content/341/6152/1355.short

Connections

Lindsey, Rebecca and Michon Scott. "What are Phytoplankton?" Earth Observatory (July 13, 2010). http://earthobservatory.nasa.gov/Features/Phytoplankton/

Rincon, Paul. "Oldest Evidence of Photosynthesis." BBC News (December 17, 2003). http://news.bbc.co.uk/1/hi/sci/tech/3321819.stm

Thomas, Gregory. "Surfonomics Quantifies the Worth of Waves." *Washington Post* (August 24, 2012). http://www.washingtonpost.com/surfonomics-quantifies-the-worth-of-waves/2012/08/23/86e335ca-ea2c-11e1-a80b-9f898562d010_story.html

Wild

Coelho, Sara. "Sponge Competition May Damage Corals."
Planet Earth Online (May 3, 2011).
http://planetearth.nerc.ac.uk/news/story.aspx?id=972

de Goeij, Jasper M., Dick van Oevelen, Mark J. A. Vermeij,
Ronald Osinga, Jack J. Middelburg, Anton F. P. M. de
Goeij, and Wim Admiraal. "Surviving in a Marine Desert:
The Sponge Loop Retains Resources Within Coral
Reefs." Science 342, no. 6154 (October 4, 2013).
http://www.sciencemag.org/content/342/6154/108

Deep

For more information about the Census of Marine Life, visit
http://www.coml.org/about-census

Edmond, John M. and Karen Von Damm. "Hot Springs
on the Ocean Floor." *Scientific American* 248, no. 4
(April 1983).

Morelle, Rebecca. "'Supergiant' Crustacean Found in
Deepest Ocean." BBC News (February 2, 2012).
http://www.bbc.co.uk/news/science-environment-16834913

Murakami, Haruki. *The Wind-Up Bird Chronicles*. Translated
by Jay Rubin. New York, Vintage, 1998.

Piccard, J. and R. S. Dietz. *Seven Miles Down: The Story
of the Bathyscaph Trieste*. New York, Putnam, 1961.

"Oceanlab Film Deepest Fish and 'Super-giants' in the
Mariana Trench." Oceanlab, University of Aberdeen.
http://www.abdn.ac.uk/oceanlab/research/mariana-
trench.php

You can watch Oceanlab's series of films showing life in
the Mariana Trench on Youtube:
https://www.youtube.com/watch?v=6N4xmNGeCVU

Rehbock, Philip F., ed. *At Sea With the Scientifics: The
Challenger Letters of Joseph Matkin*. Honolulu, Hawaii,
University of Hawaii Press, 1992.

Tame

Black, Richard. "'Only 50 Years Left' for Sea Fish."
BBC News (November 2, 2006).
http://news.bbc.co.uk/1/hi/sci/tech/6108414.stm

Carter, Robert. "Boat Remains and Maritime Trade in the
Persian Gulf During the Sixth and Fifth Millennia BC."
Antiquity 80, no. 307 (March 2006).

Morwood, M. J., R. P. Soejono, R. G. Roberts, T. Sutikna,
C. S. N. Turney, K. E. Westaway, et al. "Archaeology and
Age of a New Hominid from Flores in Eastern Indonesia."
Nature 431 (August 18, 2004).

United Nations Division for Ocean Affairs and the Law
of the Sea. *Oceans: The Source of Life*. United
Nations Convention on the Law of the Sea, Twentieth
Anniversary, 1982–2002. http://www.un.org/depts/
los/convention_agreements/convention_20years/
oceanssourceoflife.pdf

"Wartime Population Faced 'Eating Plankton to Avert Food
Shortages'." *Telegraph* (February 23, 2012).
http://www.telegraph.co.uk/history/world-war-
two/9099881/Wartime-population-faced-eating-plankton-
to-avert-food-shortages.html

Beauty and the Beast

Hemingway, Ernest, *The Old Man and the Sea*. New York,
Scribner's, 1952.

Homer, *The Iliad of Homer*. Translated by Ennis Rees,
Oxford, Oxford University Press, 1991.

Homer. *The Odyssey*. Translated by Allen Mandelbaum,
New York, Bantam, 1990.

Melville, Herman. *Moby-Dick or, The Whale*. New York,
Harper, 1851.

Beyond the Horizon

Shakespeare, William. *The Tempest*. Editor Peter Holland.
New York, Penguin, 1999.

For more information about The Seasteading Institute, visit
http://www.seasteading.org

Special thanks

The Life and Love of the Sea would be nothing without
the ceaseless exploration of our subject by talented
photographers past and present. The challenging journeys
and discoveries of many artists have been drawn on to
make the work you have in your hands. This book would
also be impossible without the great team at my creative
partners, PQ Blackwell. Finally, my personal thanks to
Jan and Caledonia, who are always keen to go to sea
(subject to conditions).

For Abrams:
Editor: Laura Dozier

For Blackwell & Ruth:
Publisher: Geoff Blackwell
Editor in Chief: Ruth Hobday
Editor: Rachel Clare
Additional editorial: Abby Aitcheson
Book designer: Helene Dehmer

Original concept design: Cameron Gibb
Library of Congress Control Number: 2014959561
ISBN 978-1-4197-1862-5

First edition published in 2015 by Abrams
Concept and design copyright © 2015 Blackwell and Ruth Limited
 (formerly PQ Blackwell Limited)
Text copyright © 2015 Lewis Blackwell

Video produced by Steve Hathaway
www.stevehathaway.com
www.youngoceanexplorers.com

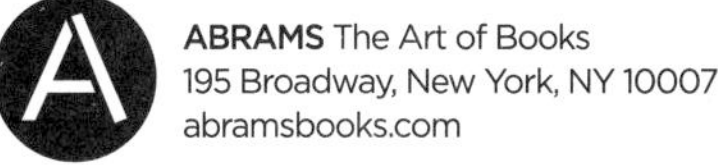
ABRAMS The Art of Books
195 Broadway, New York, NY 10007
abramsbooks.com

Image Credits

The work of many photographers inspired, informed, and influenced our work to arrive at this selection. The final selection has been captioned and credited and here it remains to thank the sources from which we were able to license the imagery for use in this book

Nature Picture Library for the images that appear on inside front and back cover, pp. 26–27, 29, 32, 34, 36–37, 44, 48, 58–59, 60–61, 62–63, 64–65, 86–87, 88, 99, 102–103, 108–109, 110, 111, 112–113, 114 (all images), 116–117, 131, 135, 138, 141, 142, 143, 146–147, 152–153, 154, 166–167, 169, 182–183, 186, and 204–205

Art Wolfe for the images that appear on pp. 30–31, 40–41, 47, 132–133,180–181, 184, 198–199, 200, and 208–209

We acknowledge the use of Rapid Response imagery from the Land, Atmosphere Near Real-time Capability for EOS (LANCE) system operated by the NASA/GSFC/Earth Science Data and Information System (ESDIS) with funding provided by NASA/HQ for the image that appears on p. 57.

Front cover, back cover, case, and all other images that appear in this book have been licensed through Getty Images.

Footage credits
Underwater footage: Steve Hathaway; NatureFootage; Getty Images/Artbeats. Music: MUSICBED. Still image: Peter Chadwick/Getty Images.

This book is made with FSC-certified paper products and is printed with soy vegetable inks. The Forest Stewardship Council (FSC) is a global, not-for-profit organization dedicated to the promotion of responsible forest management worldwide to meet the social, ecological, and economic rights and needs of the present generation without compromising those of future generations.

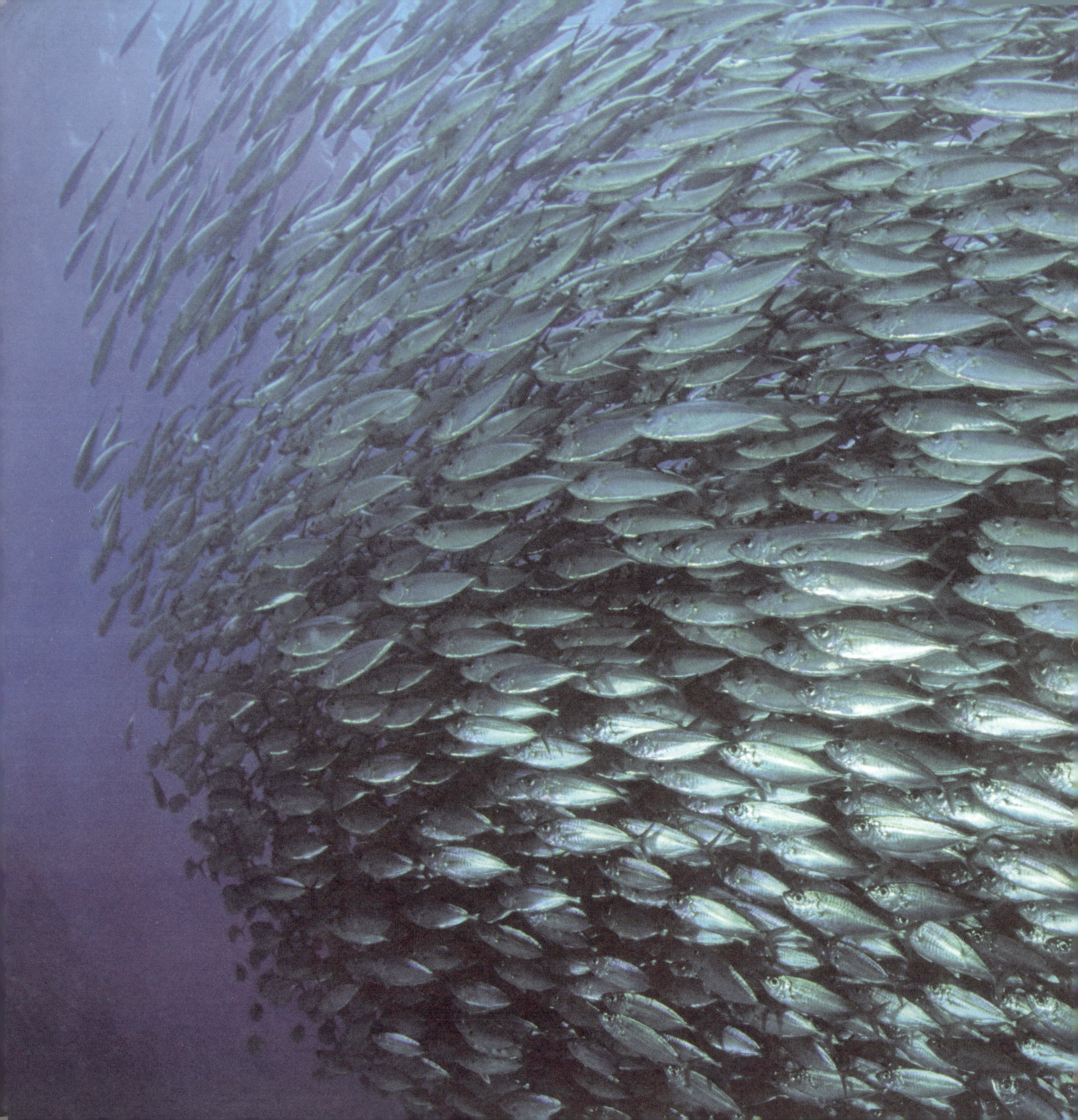